John Grigor

Arboriculture

A Practical Treatise on Raising and Managing Forest Trees and on the Profitable

Extension of the Woods and Forests of Great Britain

John Grigor

Arboriculture
A Practical Treatise on Raising and Managing Forest Trees and on the Profitable Extension of the Woods and Forests of Great Britain

ISBN/EAN: 9783337812515

Printed in Europe, USA, Canada, Australia, Japan

Cover: Foto ©berggeist007 / pixelio.de

More available books at **www.hansebooks.com**

ARBORICULTURE

OR A PRACTICAL TREATISE ON RAISING AND MANAGING FOREST TREES

AND ON

THE PROFITABLE EXTENSION OF THE WOODS AND FORESTS OF GREAT BRITAIN

BY JOHN GRIGOR

THE NURSERIES, FORRES, N.B.

AUTHOR OF THE HIGHLAND AND AGRICULTURAL SOCIETY'S PRIZE ESSAYS 'ON RAISING FOREST
PLANTS,' 'ON FOREST PLANTING, AND ON TREES ADAPTED TO VARIOUS SOILS AND SITUATIONS,'
'ON RAISING AND MANAGING HEDGES,' 'ON FOREST PRUNING,' 'ON THE NATIVE PINE
FORESTS OF SCOTLAND,' 'ON PLANTING WITHIN THE INFLUENCE OF THE SEA,' 'ON
THE DEODAR,' 'ON THE VARIETIES OF THE LARCH CULTIVATED IN GREAT
BRITAIN,' 'ON THE LARCH PLANTATIONS OF SCOTLAND,' AND ON
VARIOUS OTHER SUBJECTS CONNECTED WITH ARBORICULTURE.

EDINBURGH
EDMONSTON AND DOUGLAS
1868.

DEDICATED

To the Highland and Agricultural Society of Scotland,

WHOSE OPERATIONS

HAVE GREATLY EXTENDED AND IMPROVED

THE WOODS AND FORESTS THROUGHOUT THE KINGDOM.

JOHN GRIGOR.

PREFACE.

In writing the preface to a book it is usual for its author to state what gave rise to the work, or to indicate its supposed superiority to the works in circulation on the same or on similar subjects. In the present case, it may suffice to state that, during the last forty years, while the author successfully established and conducted the Forres Nurseries, he has written numerous papers on arboricultural subjects, for which the Highland and Agricultural Society of Scotland has awarded prizes.

The articles on the timber trees adapted to the climate of Great Britain published in that excellent work *The Agricultural Cyclopædia* (by Morton), were also furnished by the author, as well as papers from time to time on the same subject to the leading periodicals of the day; and as far back as 1838, Loudon, in his preface to the greatest work that has yet appeared on the subject, " Arboretum et Fruticetum Britannicum," acknowledges his obligations to the author for important communications respecting that useful tree, " the Scotch Pine in Scotland."

These publications have, for a long period, given rise to numerous applications by friends and customers of the author for copies of his writings on Arboriculture, which, from their publication in detached portions at various

times, could not easily be supplied. To meet this desideratum, it was at one time intended to collect the papers as
they stood in the various publications, and embody them ;
this, however, was found unsuitable, as it would have
occasioned numerous repetitions, and swelled the volume
to too great a size ; besides, the experience of upwards of
a quarter of a century had, to some extent, changed the
author's views on subjects of importance. Many of the
articles therefore have been re-written to adapt them to
the present time, and some others have been added to
make the work complete on the various subjects, from the
harvesting of the seed to the felling of the full-grown
timber.

During the present century, not only have changes taken
place, to some extent, in the opinions of writers on the
best mode of cultivation and subsequent management of
our timber trees, but in several instances the trees themselves, with respect to some of the most common and
useful species in cultivation, are changed, and are found to
be less hardy and ill adapted to the climate of North
Britain. This is more especially the case with the Scotch
Pine and the Larch, and it arises from the importation, for
many years past, of large supplies of foreign seed grown
in warm districts on the Continent, which produce plants
so tender that in this country they form a very precarious
crop.

It is a remarkable fact, that the same care is not manifested in the formation of timber plantations in general,
which may last for a century, that is usually shown in the
laying down of any of the common crops in agriculture.
In the latter, though the crop is only to last for a season,
the purity and the productiveness of the variety are carefully attended to ; while in timber cultivation, Scotch Pine

and the Larch for instance, are looked on as possessed of no varieties or difference in hardiness.

So often does the formation of extensive plantations fall into incompetent hands, that it may be stated, that in the specifications for two of the largest plantations that the author ever knew to be formed by contract, and belonging to different noblemen, the American Spruce, a dwarf tree which only attains to about one-third of the size of the common or Norway species, was the only spruce mentioned, and the proportion required involved the planting of it in each plantation to the extent of several hundred thousands.

It seldom falls to the lot of any one to raise plants and form them into plantations, and to be able to report on these plantations after they have become valuable ; but the author of the following pages having begun business early in life, and formed several large plantations nearly forty years ago, can refer to them, standing in great vigour, an ornament to the scenery of the Highlands, mollifying the blast, and after paying every expense, yielding a revenue equal to that of the finest arable land in the country, where the ground previously to these formations was not worth a shilling an acre.

At a time when timber throughout the country is everywhere enhanced in price ; when woods and forests are fast disappearing, chiefly by the construction and maintenance of railways, and by the facility they afford for its transit from place to place, it is believed that this work will be appreciated. The favoured and distinguished approval already vouchsafed from time to time on the appearance of some detached portions of it, encourages the hope that it will be esteemed both by landowners and practical men, as pointing out the best mode of raising and extending the

profitable cultivation of the timber trees that are tried and
known to be adapted to our climate, and of averting those
casualties which at various stages of their growth are apt
to assail them.

With respect to numerous trees of recent introduction
from various quarters of the world, many of them are no
doubt interesting and ornamental, and it is hoped many
will prove valuable to the country; but until their hardiness and vigour have been ascertained, and until they
become far more plentiful and cheap than they now are,
they cannot be recommended for profitable timber.

THE NURSERIES, FORRES,
October 1868.

CONTENTS.

xii CONTENTS.

VIII.

Pit Planting, and Planting on Prepared Ground, chiefly with Broad-leaved Trees, ... 65

IX.

On Planting in a Grassy Vegetation—On Planting where Timber has recently been removed, ... 71

X.

On Plantations on Bog or Peat Soil—On Ground overspread with Furze, or Whin, or other Rough Herbage—On Planting exposed or barren Ground at a great Altitude, 77

XI.

On Thinning Plantations, ... 81

XII.

On Sea-side Planting in the county of Norfolk—On the Sands of Culbin and Kincorth in Morayshire—Cost of Formation, and the Value of some of these woods after Twenty-four Years—Mode of growing Plantations on the Coast of France—Plantations on the West Coast of Scotland—Preparation of Ground—Time of Planting, etc., . . 93

XIII.

Hedgerow Timber, ... 117

XIV.

Coppice, ... 123

XV.

On Harvesting Bark, ... 129

XVI.

On Pruning Forest Trees, and Grafting in Plantations for Embellishment, ... 134

XVII.

On Raising and Managing Hedges, ... 143

XVIII.

The Pine Tree—Scotch Pine, and its Varieties, and other hardy Pines, ... 156

I.

CALENDAR OF OPERATIONS.

A CALENDAR OF OPERATIONS will, it is hoped, be a useful guide in recording the work that is usually performed in nursery and forest management throughout the year; but as much depends on the state of the weather, and on the climate of the district in which the work is going on, the precise dates cannot in all cases be adhered to.

The opening up of one season may, during spring, be early, and accompanied with drought; the next may be the very reverse; and that not unfrequently makes the difference of a month in the period for the safe transplantation of forest plants.

With respect to all plants which come early into leaf—such as hawthorn, larch, etc.,—if the nursery ground possesses a climate warmer than that where the plants are to be inserted, when the planting cannot be performed sufficiently early, the plants should be removed to the later district before the buds are expanded, and should be carefully secured in the ground against drought till they are planted out.

Both in the nursery and in the forest, the busiest period by far in the twelve months occurs in spring, during open weather in March and April. It should therefore be the aim of the manager to leave no description of work to be performed at that period, which can well be done at a time when the work is less pressing.

Though much may be accomplished by practical information of the time and the circumstances most suitable for the operations with respect to forest trees, yet directions in writing generally fall far short of the assistance and success derived from the exercise of common sense and experience.

A

JANUARY.

Continue to fell and cut up timber. Thin plantations, prune hedges, clear out all ditches and conduits which may not have undergone a clearance since the fall of the leaf. In frosty weather at this season, drive to the depot all accumulations of decayed leaves for forming vegetable mould, the heaps of which should be completed by this time.

Open weather at this season is suitable for planting all sorts of forest trees, particularly in dry situations.

As the plots in the nursery become clear of plants, dig the ground over roughly, or dig it up into ridges, so that it may be exposed to the influence of frost.

In removing from the nursery one-year transplanted larches or pines, which generally stand in lines nine to twelve inches apart, if a portion of them is required to be kept for another year, remove only every second or alternate line ; and if the plants in the remaining lines stand closer than about three inches apart, they should be loosened with a spade or fork, and thinned to about that distance, thus leaving the lines about eighteen to twenty-four inches apart, and the plants about three inches asunder—the proper distance for their becoming two-year transplanted plants. When thus thinned out, dig the ground between the lines. Similar treatment and thinning out should be practised on all one-year transplanted plants where they have not sufficient space for growing, and are intended to become two years transplanted.

If the weather is open and dry, the sowing of all seeds which were kept in pits during summer should now be completed ; such as hawthorn, holly, mountain-ash, yew-tree, etc.

If acorns or chestnuts are not yet sown, this work should be completed as soon as possible.

Insert cuttings of deciduous plants that are propagated by that means, such as poplars, willows, elders, etc.

Scotch and other pine cones are commonly ripe at this season; severe frost changes them from a deep green to a grey colour, giving them a hard surface, which indicates their maturity, after which the seeds are extracted with a very moderate degree of heat compared to what is necessary for such as have been gathered before undergoing the influence of severe frost ; all sorts of pine, fir, and larch cones may now be collected and the seeds extracted. If gathered in dry weather they may be stored in a dry place to any depth, but larch cones suffer if stored in a wet state. The extracting of the seeds from cones can be accomplished during weather unsuitable for most out-door operations. The modes of procedure are detailed under the names of the trees—Scotch pines, larch, etc.

If the texture of the soil in any part of the nursery ground requires to be improved, frosty weather will be found suitable for carting and applying heavy loam to such parts as are of a loose and sandy description, and light sandy or peat soil, etc., to such parts as are tenacious and apt to harden in dry weather.

Nursery ground that has been frequently cropped with various plants, is greatly renovated by a coating of fresh clean soil ; that from a field after a potato or turnip crop is very suitable, and where it is applied early in winter it has the best effect.

As hawthorn springs earlier than almost any other nursery plant, its transplantation both into hedges and nursery lines should be completed now, or before the buds expand.

Trench ground, and forward all operations preparatory for the hurry of spring.

Turn over manure-heaps, compost and leaf heaps, that they may become pulverized.

FEBRUARY.

The operations recommended during last month are suitable for a like state of weather during this month. If the weather is suitable, continue to transplant all sorts of forest trees into the forest, and also into nursery lines ; plant hedges, etc. Complete plantations on dry ground.

Oaks, when two years transplanted in nursery lines, are generally removed to the forest; and although they would improve very much in appearance during the third year in nursery lines, yet at this period they run so much to tap-roots, that on being afterwards transplanted they are apt to suffer from the operation, become stunted, and frequently require to be cut down. It will therefore be found of great benefit to the plants, when they are of a sufficient size for forest planting, to loose and lift them from the nursery lines, and prune off with a strong knife all straggling root fibres, without shaking the earth from the roots, replacing them immediately into the same ground, to remain for another season before inserting them in the forest. This operation of disturbing oaks in the nursery is speedily performed by casting out an opening a stamp deep alongside the outside line, so that after lifting and pruning the roots, the plants are immediately inserted into the adjacent opening and a stamp of earth turned over on their roots, thus forming an opening for the next line, and so on, until the whole lot of plants is replaced. It is advisable during this operation to throw out all plants that are much under the ordinary size, as such would continue inferior, and are of no value. In thus removing the plants it is not necessary, as in the ordinary mode of transplanting, to tramp or firm the plants with the foot, as they are removed with a considerable quantity of earth adhering to their fibres. It is found that the increase of small fibres becomes more abundant in loose than in firm ground. It is of no consequence for the plants to occupy a broader line, or not to stand up with such exactness as formerly. Plants thus disturbed will not grow much during the ensuing summer; but on removing such into the forest in the following year, their roots will be found compact and bushy, and will require no pruning. Their success compared with that of plants which have not undergone this operation will be very apparent.

The foregoing method of preparing oaks for going out is well adapted for two-years transplanted beech, whether they are intended for hedges or forest trees; and as all sorts of plants are apt to suffer most from removal after having grown vigorously, the operation will be found very useful on every kind of plant that is apt to produce bare roots destitute of that fibrous bushiness so necessary to successful transplantation. It is also necessary in all cases where plants stand so close in lines as to be destitute of the space required for the growth of another season; and in such circumstances ample space should be afforded beyond that which the plants formerly occupied. It is also useful in cases where the forest ground is not ready to receive the plants, as it retards their exuberance, improves their roots, and keeps them in a fit state for being transplanted a year or two afterwards. Open weather in any of the winter months is suitable for this work, which should now be completed.

If the weather is open and the ground sufficiently dry, the end of this month is a good time for sowing beech, ash, sycamore, elm, etc., as thus they are generally less liable to be injured by late spring frosts than if sown at an earlier period.

With equal quantities of clay and of horse-droppings, thoroughly mixed with water, prepare for grafting. Thus prepared, the mixture suits the purpose far better than if only made up at the time of grafting, which is generally in the end of March.

MARCH.

The thinning of plantations should be completed by this time, for after the buds begin to expand, trees are much more sensitive of cold, and the sudden admission of cool air has an injurious effect. This is often apparent, particularly on larches when thinned late in spring, compared with those that have undergone that operation at an earlier period of the season.

The formation of plantations of all deciduous trees should be completed in this month, or as soon as possible, and all vacancies or failures in former plantations filled up.

There are some sorts of ground, such as bare moorland, partaking of moss to too great an extent, in which it is difficult to plant so successfully in any other month, on account of the soil retaining a great deal of moisture, and swelling during severe frosts, thereby raising and ejecting the plants, notwithstanding the most careful drainage. All plantations on soils of this kind are most successfully made during the opening of the season.

In the nursery, during favourable weather, this is one of the busiest months in the year, when everything fit for the forest should be cleared out, and the plots filled up by transplanting seedlings, or by digging over the ground in preparation for seed-beds. All plots that are exhausted should receive a dressing of well-rotted manure or vegetable mould. In such parts as have become wild, or stand in need of renewal, plant potatoes or turn it over preparatory to a turnip crop.

At this season nothing is more necessary to be guarded against, both in the formation of forests and in the nursery, than the exposure of the roots of plants to sunshine, or to the drying influence of the weather. In lifting nursery plants in bright weather, it is of great advantage to form them speedily into bundles, and puddle the roots with earth and water made to the consistency of thin paint, and to expose them as little as possible until they are again planted.

Elm seeds which have been kept dry since their ripening, if not yet sown, should be sown early this month. Birch also, and all other crops, except those of the pine tribe, should be put into the ground before the end of this month.

Graft the different kinds of forest and ornamental trees propagated by that method, such as beech, elm, oak, laburnum, etc.

Dig between nursery lines; this will encourage the growth of fibrous roots, and will be found the best method of keeping down weeds.

APRIL.

All plantations of hardwood trees should have been completed before, or in the early part of this month ; but it is still suitable for transplanting all sorts of the pine tribe, particularly by the notching mode. Larches are also successfully transplanted this month in a late season, or in late districts, provided the plants have been removed carefully, and protected before their buds have been expanded ; and it sometimes happens that such plants escape injury from late frosts better than those inserted at an earlier period, particularly in plantations having a southern slope.

Complete the grafting of forest trees ; look over and repair the clay where any defects appear in those grafted last month, when the clay has dried and become sufficiently firm. Earth up the lines of dwarf grafts, leaving the scion only above ground.

In the nursery the transplanting should be finished as soon as possible ; and in South Britain the sowing of all sorts of coniferæ seeds should commence about the middle of this month. In the North of England, and in Scotland, the most approved time is about the last week of this month, or early in the next, when the weather is in a settled state ; for in some soils excess of rain immediately after the sowing, and before the soil gets dry, is fatal to the crop of larch, Scotch fir, etc.

Immediately after sowing, all sorts of coniferous crops are liable to be picked by birds ; the beds should therefore be securely netted over or overlaid with even drawn straw, fern, or the spray of silver or spruce fir. As protection from birds by the latter methods forms a shade on the surface, the cover of earth on the seed should be made very slight. When none of the foregoing methods of protection is adopted it is necessary to set a watch on the crop. The destruction by birds is always greatest early in the morning and late in the afternoon.

Weed all seed-beds of last year, and dig the alleys ; dig between the lines of transplanted plants, raking and smoothing down the ground ten days or a fortnight thereafter, which will effect the destruction of seedling weeds. The surface of the beds of oak, chestnut, beech, etc., which were not finished off at the time of sowing, should now be raked smooth, which will have the effect of removing clods, destroying the weeds, and adapting the soil for the springing of the crop.

Prune laurels and all sorts of evergreen shrubs.

The casualties which forest trees suffer by frost are generally much greater during this and the two subsequent months, than at any other season of the year, and they are most apt to happen after warm weather in March and April; it is therefore advisable to protect the beds of one-year-old seedling larch, silver-fir, also those of hawthorn, which were sown early in spring, and towards the end of this month the recently sown beech, sycamore, ash, chestnut, etc.

I have found that the easiest way of protecting these is to collect the spray or small twigs of broom, spruce, silver-fir, Portugal or common laurel, and to stick the twigs into the beds in an upright position, which forms a shade and shelter for the young plants. The cuttings of beech hedges are also suitable when scattered on the surface of the beds, as such admit the influence of the atmosphere, and do not retain much moisture.

It is found that although the twigs do not form a complete cover, yet if they form a shade from the warm sunshine succeeding a night of frost, the plants are generally safe. In the absence of this precaution I have found that in many situations, larch and silver-fir of one and two years old are on an average headed down or deprived of their leaders once every three or four years, which reduces them more than fifty per cent. in value. Although nothing makes a more complete or handy protection than the spray of the silver-fir, yet it should not be adopted where the slightest fear exists of the trees being infected with disease.

Fork out perennial rooted weeds where they appear throughout the nursery ground.

MAY.

See that the fences around plantations are in a secured state ; and that no herbage or undergrowth interferes with the progress of the plants. Clean hedges, clear out open drains, and turn over decayed leaves and compost heaps.

Drain land for the plantations of next season, and erect sufficient fences.

In planting land which has formerly produced timber, the soil requires to be much exposed to the influence of the weather; capacious pits should therefore be formed at this season ; and all old roots cleared out which appear at the edges of the pits. Form similar pits throughout all woods where the standing trees are not close enough to develop the resources of the soil. Prepare for barking oak. Towards the end of this month in some situations the sap will circulate sufficiently to render coppice and standard oaks fit for that operation ; the earlier in the season the timber is removed, and the stools dressed, the more vigorous will be the succeeding growth.

In the nursery, complete, if possible, the sowing of all sorts of pine and fir seeds during the first or second week of this month (see directions for last month). During dry weather employ the hoes and rakes vigorously in the killing of weeds. Weed seed-beds. Crop vacant lots with turnips or other green crops, using byre or stable manure well decomposed ; this is least apt to create disease in the ground.

JUNE.

Continue the forest work recommended for last month ; urge forward the barking operations with all possible speed, taking every precaution to prevent the infusion of bark by wet weather.

The seeds of the wych or Scotch elm are usually perfected and fit for sowing during the first or second weeks of this month ; collect them fresh from the tree and sow them into beds; there is always a larger or smaller proportion of the seeds empty.   The thickness in the bed should be regulated by the quality of the seeds, which is always easily ascertained by pressing them between the finger and thumb, which will indicate whether they are full or empty.   They should, how-ever, be spread thick on the ground, so as to insure a crop, as the seeds are not valuable ; and the plants when too close are easily thinned out at the first weeding.   If the weather is dry, the seed-beds should be watered and kept moist for a week, when the seeds will begin to germinate.   If required, a further supply of seeds should be collected towards the end of the month ; these should be thoroughly dried by the in-fluence of the weather, so as to prevent fermentation, when they may be packed up and kept from mice, and sown in February or March.   If sown now, after being dried, they will not vegetate till spring.

The great work in the nursery at this season is the cleaning of the ground, and keeping down weeds, so as to prevent them from shedding their seeds.   By the end of this month, if all appearance of frost is gone, the twigs protecting the seedling beds may be removed.   Vacant lots of exhausted ground should be well manured, and cropped with turnips.

JULY.

Continue to drain and 'fence land for plantations.   Where ground is to be planted which formerly yielded timber, dig out the pits of a large size, soften the ground along their edges, and expose the whole to the fertilizing influences of the atmosphere.   Perform the same operation where forest trees are to be planted into pits in all soils containing iron, or having a substratum of moorband or ferruginous crust ; so that these soils may be rectified before the season of planting. This precaution should be taken with all soils which do not appear perfectly sweet and congenial.   In all such land where plants are to be inserted by the notch system, this is a good season for going over the surface and disturbing the soil in each spot where a plant is to be inserted, by the operation of a tramp-pick, or common shoulder-pick ; this will bring into play the purifying and fertilizing influences so necessary for the successful growth of plants, and if the operation is well performed, the spots broken up will appear quite distinct throughout the ensuing planting season.

Look over young plantations, and clear away whins or rank vegetation of any kind that is likely to injure the progress of the plants ; where tall ferns are apt to prevail over plants recently inserted, it is found that breaking down, or jointing the ferns, at this season, placing their tops flat on the surface, is more effectual than cutting, as in the latter case they generally spring again, and in the former they seldom do so, but lie flat on the surface of the ground during the season.

All barking operations should now be finished for the season.

Prune forest trees, look over those which have been a few years planted, and those more advanced, and with the pruning-knife shorten all competing shoots in order that the leader may be clearly in the ascendant ; and shorten also, or remove all other growths throughout the tree that are likely to run away with the sap, and that bear too great a proportion to the main stem.

Prune hedges of all sorts of vigorous growth ; the switcher is the most speedy and efficient implement. Pruning at this season is particularly useful to hedges which are apt to become too bare at the bottom ; in such cases the top of the hedge, consisting of its upper half, should be pruned once or twice during summer to induce a growth lower down, which latter part needs only be pruned once a year, but should be carefully kept clear of weeds and vegetation of every kind.

In the nursery, the great business of this month is to keep down weeds ; a little neglect in this respect generally entails a great amount of labour, as weeds so readily mature and shed their seeds at this season, and unless they are very small and young, this is sure to happen when the ground is hoed and the weeds not immediately removed.

Look over, untie, and secure grafts, rubbing off all under-growths. Turn over pits of hawthorn, mountain-ash, holly, and other tree seeds, to decompose them regularly, and keep them in a wholesome state. The end of this month is in general the best season for budding ornamental trees (for the operation of budding, see Lindley's *Theory and Practice of Horticulture*, p. 303). The bandages which secure buds should be slackened and retied about ten days or a fortnight after the operation, as they are apt, if the stocks are very vigorous, to become too tight, and about the same period thereafter it is generally safe to remove them altogether.

Turn over compost heaps ; and heaps of weeds which have matured their seeds should be turned over and built up in a compact form so as to induce fermentation, that their seeds may be completely destroyed.

AUGUST.

Continue the operations of draining and fencing land for plantations. As formerly recommended for all inhospitable soils, or where former plantations existed, cast out the pits of a good size, to expose the ground to the influences of the weather. Break up with the pick the spots in ground where notch-planting is intended. Prune forest trees ; although large branches should never in ordinary practice be removed in well-managed plantations, yet, in case of necessity, this is the best season for doing so, as the wound more speedily heals at this than at any other time. See that no wild vegeta-tion interferes with the growth of young plantations. Con-tinue the pruning and cleaning of hedges.

Towards the end of this month transplant evergreens of all sorts, preferring wet or cloudy weather. In the nursery, be careful that all weeds are subdued and kept from seeding, which is particularly ruinous to ground which may next season be laid down in seed-beds.

The end of this month is the proper time for inserting cuttings of laurel, box privet, yew, and all sorts of evergreens propagated by that means. Towards the end of this month the temperature of the earth and of the air is generally nearly the same, and then plants root more readily than under any other condition. Sharp sandy soil, partially shaded or having a northern exposure, is most suitable.

The end of this, and the next month are best adapted for transplanting evergreens. Immediately after the insertion of the plant at this season, the heat of the ground excites its roots so that they at once produce numerous young fibres or spongioles, which process of nature insures its safety, and fixes it in the ground before the approach of winter.

Complete the budding of ornamental trees as formerly recommended.

SEPTEMBER.

Continue the operations preparatory to forest planting as formerly recommended ; prune forest trees, thin plantations, thin out the supernumerary shoots of young copse-wood, leaving the most promising. This is the time most advisable for ornamental or evergreen planting ; where specimen plants are to be removed, remove them without delay, giving a copious watering if the weather is dry.

Plant hedges and screen fences and underwood of holly, yew, privet, laurel, etc., and all sorts of evergreens, preferring moist and cloudy weather.

In the nursery the cleaning of the ground should be urged forward, so long as the weather is dry and suitable for that purpose ; dig between the rows of evergreens and other transplanted plants, finish the transplanting into nursery rows of seedling holly, evergreen oak, and other evergreens.

In nursery management it is found a good method to remove at this season all sorts of evergreens that are fit for

going out.   This is most speedily accomplished by casting out
a stamp or trench along one side of the lot, after which one
man precedes with a spade and loosens the plants thoroughly,
another follows lifting them up with one hand, and with the
other pruning or switching off all very extended or straggling
roots of unnecessary length, and placing them into the prepared trench or opening in a slanting direction, with as much
earth as adheres to the roots; the digging up of the earth and
the covering of the one line prepares an opening for the next,
and so on until the whole is finished.

The growth of the season being nearly over at this time,
the plants receive no harm by being placed in a slanting
position; and in removing them it is not necessary to fix
them in the ground by treading them with the feet; and as
the tops of the plants of one row extend over the roots of the
former line, their position keeps the ground warm if they
should remain throughout the ensuing winter, and if the
situation should be exposed, the foliage of the plants will
retain its greenness, and from their position will be exempt
from injury by snow or severe frost.

Evergreens which are thus disturbed in September, when
the temperature of the ground is greater than that of the
atmosphere, immediately throw out fresh spongioles, and
acquire a bushiness in root, which retains the soil and adapts
them for being successfully transplanted thereafter *during
any month of winter or spring.*

This treatment is particularly advisable in sale nurseries,
as such plants are lifted with great facility at a time when the
work of a public nursery is sometimes difficult to overtake,
and such as are not disposed of require to be again replanted
in rows before or early in summer; but the fresh appearance
of the plants, and the valuable state of their fibrous roots, render them well worth the labour bestowed on them. If the
cuttings of evergreens were not all inserted last month, this
work should be now completed as formerly recommended.

Birch seeds usually ripen at this time, and should be collected; as they are generally mixed up with leaves and some-

what moist, care should be taken that they are not confined in sacks, as they are apt to ferment, which destroys their vitality. Spread them in an airy loft until dry, when the leaves should be sifted out, and the seeds kept until the time of sowing in March or April.

Collect sycamore seeds, which are ripe about this time, and the other varieties of maple, which in an early season are matured in this month; these, when dry, may be kept in a heap within doors until the end of March; if sown earlier, they are apt to come up too soon, and perish with spring frost.

Cherry-stones should be collected and sown immediately in dry ground; they do not vegetate in the first spring, but the most successful crops are generally those that are early committed to the ground.

OCTOBER.

The draining and fencing of ground, and the making of pits for plantations, should be continued. Where the surface of the ground is bare, and readily admits frost, these operations should be completed first while the weather is still open, as those parts possessed of surface vegetation can be worked during frost.

As already stated, it is of advantage in some cases to pit and expose the soil to the influences of the weather for some time before the plants are inserted. In other cases, where the soil is congenial, the planting may follow with perfect safety immediately after the pit making. Trench ground for spring planting.

By the beginning, the middle, or the end of this month, according to the season, all sorts of forest plants are ripe and fit for planting. The present is the proper time for planting dry ground of every description, whatever the mode of planting may be. Plantations in such places should therefore be proceeded with as speedily as possible, both by notch and pit planting. Such grounds, however, as are composed to a considerable extent of organic matter—for instance, mossy or boggy parts, which retain moisture,—should be left for spring planting, as on account of this description of ground swelling during frost the plants are apt to be ejected, particularly those that are inserted by notch-planting.

Fell timber; thin out plantations; grub out furze on plantation ground. Many of our most important hardwood trees ripen their seeds at this time.

Underneath the oak, and particularly after frost with a breeze of wind, the ground will be covered with acorns ; these should be collected and spread out on a barn floor, so as to prevent fermentation ; or if the nursery ground is ready, they may be sown immediately.   Prefer the largest ; these are most readily got by using a wire riddle of a size suitable for allowing the small to pass through.

Beech-mast is also shed about this time, when it should be collected and passed through a barn fanners, or exposed to a steady breeze of wind in an open situation to effect a separation between the sound and the empty seed.   When sown immediately, it comes up, if the weather is fine, in the end of March, or early in April, when the crop is endangered by frost; it is therefore better to spread the seeds on the floor of a loft, turning them occasionally till they are dry, when they may be kept in a heap, and sown in the end of February or in March.   The crop will then appear at a time when it will generally be exempt from injury.

The seeds of maple of sorts, if not formerly gathered, should now be collected ; also chestnuts, walnuts, and all other sorts as they become ripe.

As the ground becomes empty in the nursery it should be dug deep and turned over roughly, or cast into ridges, to admit the influence of frost.   Previously to this being done, such parts as are exhausted should receive a coating of well-decomposed manure, particularly after crops of ash, elm, beech, spruce, or hawthorn, which generally leave the ground in a very poor state.

Of all sorts of deciduous plants none require to be attended to in early transplanting more than hawthorn or *quick*.

Transplanting this plant into hedges and nursery lines this month or next affords the plant an early start in spring-time. Turn over dung-hills and compost-heaps to induce decomposition.

NOVEMBER.

The operations in the forest which should be carried forward at this time are very diversified. Fell timber, thin out plantations, grub up furze and all sorts of wild vegetation on land for plantations, and to relieve those recently formed. Repair drains and fences. Forward plantations of all sorts, and all the operations recommended for last month as may be found suited to the state of the weather.

Collect and cart leaves to the depôt. Frosty weather will be found suitable to cart to the nursery grounds such accumulations of vegetable mould as have been prepared in pits or recesses in the forest, and these pits or recesses should now be filled up with fresh leaves as speedily as possible.

Plant hedges of thorn, beech, etc.

Frosty weather will be found most suitable for carting manure, compost, timber, and all heavy materials, over soft ground. Ridge up all plots of ground in the nursery as soon as they become vacant, forwarding the nursery operations recommended for last month as the weather becomes suitable.

Remove layers from stools, and plant them into nursery lines. Dig around and dress the stools, giving them a coating of vegetable mould mixed with fine sand and well rotted manure, and lay down the young shoots, the produce of last summer.

Collect tree seeds.

DECEMBER.

Continue to form pits and trench ground for plantations, and forward planting, and the operations recorded for last month which may be found adapted to the state of the weather.

The berries of hawthorn, mountain ash, holly, and common ash, should be collected by this time. They should either be mixed up with sandy soil and pitted where they may be turned over every two or three months; or, as these do not vegetate the first year, they may be now sown into rich prepared nursery ground; and on the surface of the same ground an annual crop may be raised during the first summer of a light nature, as onion, radish, cauliflower, or cabbage plants. These being carefully removed, the tree crop will appear in the second, and to some extent, with respect to hollies, in the third summer.

Turn over roughly or ridge up nursery ground, that it may become pulverized and enriched by the influence of frost.

Prune and dress hedges, forking out and removing every perennial weed in the ground on each side, within a yard of the fence.

II.

THE RISE AND PROGRESS OF BRITISH
PLANTATIONS.

In this country the name Plantation is chiefly applied to grounds planted with forest plants for the purpose of producing useful timber; and it also distinguishes artificial woods or forests from those that are of spontaneous growth. The uses of plantations are very various. Not only is timber absolutely necessary in every description of architecture and rural occupation, but trees in a living state improve the climate, in yielding shelter to lands exposed to rough winds, and shade to those under a burning sunshine. Of the native trees of Britain, there are only about twelve genera and thirty species which attain the size of timber trees, or trees of above thirty feet in height, and only three of these are evergreens— namely, the Scotch fir, the holly, and the yew, if we suppose the last to be a native.

It was during the sixteenth century that plantations began to be extensively formed for timber and embellishment; but long before that period many of the timber trees had been introduced, and, in the absence of any distinct account, it is generally believed that they were brought to Britain by the Romans, or by the monks of the middle ages, with many of our cultivated fruits and vegetables. Of our earliest introduced trees we have the chestnut, the lime, the English elm, and the beech, with the apple, the pear, the peach, etc. In England previously to the reign of Henry VIII., the timber required for the purposes of construction and fuel was supplied by the native forests of the kingdom. The first accounts we have of the introduction of many of the timber trees are given by botanists and apothecaries in London, who gathered together every description of foreign herbage, and formed the most extensive collections of medicinal plants extant at that time. In Turner's *Herbal*, published at different times about the middle of the sixteenth century, he notices the introduction of the common spruce fir, the stone pine, the evergreen cypress, the sweet bay, and the walnut. Towards the end of that century, Gerrard, who had a physic-garden in Holborn, London, published the first edition of his Catalogue, which gives an account of the introduction of the pineaster, the laburnum, and a considerable number of smaller trees and shrubs. The evergreen oak and the arbor-vitæ were also introduced during that century.

In the seventeenth century, it appears that Dr. Compton, who was Bishop of London from 1675 to 1713, introduced a considerable number of exotic trees, and advanced this branch of rural improvement more than any other individual of his time, having imported many of our best trees, chiefly from America. Botanic gardens began to be established throughout England about the middle of this century, which greatly facilitated the introduction of hardy trees.

· In Scotland, the Botanic Garden of Edinburgh was formed in 1680, and in 1683 the cedar of Lebanon was introduced into it; and in the same year it was planted by Bishop Compton at Fulham, and also in the Chelsea Botanic Garden. According to the *Hortus Kewensis*, the most important foreign trees introduced in this century, besides the cedar, were the silver fir, the larch, the horse-chestnut, the acacia (locust-tree), the scarlet maple, the Norway maple, the American plane, the scarlet oak, the weeping willow, balsam poplar, balm of Gilead, fir, the cork-tree, and the black and the white American spruce firs, besides a great many smaller trees and shrubs. During the early part of this century, the British arboretum appears to have been greatly indebted to Parkinson, a physician in London, who possessed a large collection of plants, and was appointed apothecary to James I. Parkinson is the first to record, in 1629, the introduction of that valuable tree, the larch, and at the same time the horse-chestnut, but the introducers are not known.

The number of species of foreign plants introduced into Britain during the eighteenth century was very great, amounting to nearly 500, but three-fourths of these were shrubs. More than half the number were natives of North America. The timber trees consisted chiefly of oaks, pines, poplars, maples, and thorns,—species or varieties of trees formerly introduced. Botanic gardens by this time were established in different countries throughout the world, and the interchange of plants became general. Nurseries for every plant in demand were established, and the taste for planting foreign trees rapidly spread among the landowners of England. This taste was greatly influenced by the Princess-Dowager of Wales, who established the arboretum at Kew, and from the celebrity of the plantations formed by the Duke of Argyll at Whitton. Large plantations were formed at Croome, Syon, Claremont, and at Goodwood. At the last-mentioned place the Duke of Richmond planted 1000 cedars of Lebanon, five years old, in 1761, which form part of the second generation of the tree grown in England, having been produced from one of the first trees known in the country. Among the other English plantations of valuable timber trees may be mentioned that made at Pains' Hill, Woburn, Strathfieldsaye, and Purser's Cross. Syon had long had an established fame for trees before this period, having been greatly enriched by Henry Earl of Northumberland in the beginning of the seventeenth century; and in 1750, an accession to the arboretum was made of every kind of tree to be found in the kingdom.

Notwithstanding our numerous importations of foreign trees during the eighteenth century, it is remarkable that that of the larch to Scotland from England, where it had existed for a century, and yielded seeds for generations, should be the greatest acquisition, and distinguish the period beyond any other circumstance connected with British arboriculture. In

Scotland, early in the eighteenth century, a spirit of plant-ing on a large scale, for profit, began to awaken. Among the chief promoters of this description of planting was Thomas Earl of Haddington, who formed large plantations, and wrote a treatise on the subject. It is recorded that the Countess of Haddington, at the same time, became so enthusiastic in im-provements by plantation, that she sold her jewels to enable her to plant Binning Wood, which comprehended 1000 acres, and was formed in 1705. The plantations of Archibald Duke of Argyll were also at the period very extensive in Scotland as well as in England. By these extensive improvements the taste for plantations is said to have been imbibed by the Duke of Athole, the Earl of Panmure, Sir James Naysmith, Sir Archibald Grant, and others; and by the example of these landowners, planting became very general throughout Scot-land. The great success of the larch forests of the Duke of Athole occasioned that tree to be planted to some extent on almost every property in Scotland.

The introduction of new timber trees during the present century has been extensive, and in some instances very valu-able, particularly in the Coniferæ. Some of the best of these have been introduced by Douglas, a native of Scotland, who went to North-west America as a botanical collector, and introduced from the banks of the Columbia the *Abies* (spruce fir) *Douglasii*, the *Picea* (silver fir) *nobilis*, and many other firs and pines of great beauty, though perhaps less valuable. From other sources there have been introduced the *Welling-tonia gigantea, Cupressus Nutkaensis* (known as the *Thuiopsis borealis*), and the *C. Lawsoniana*, all of which are hardy, of rapid growth, and of American origin. Of Indian importa-tions, the *Cedrus deodara* is the most celebrated. A silver fir introduced from the mountains of the Crimea, *Picea Nord-manniana*, is said to be hardier than the common species.

From the great size that these trees generally attain in their native countries, it is reasonable to expect that many of the recent introductions will become valuable additions to the British forests, particularly after a few generations of the various species have been grown from seed and thereby acclimatized.

During the present century, the influence of the Highland and Agricultural Society of Scotland has greatly facilitated the spread of plantations throughout the country, by awarding various premiums for the introduction of new timber trees, for the cultivation of the native Highland Scotch pine, and for the formation of extensive plantations throughout Scotland.

From the published reports of this Society, it appears that the late Earl of Seafield was by far the most extensive planter in Scotland during the first half of the present century, he having, at the date of the last report on the estates, planted 30,000,000 of plants, in a space exceeding 8000 acres. This extent of plantation, perhaps, has not been approached by any British landowner since the formation of the extensive plantations by the Duke of Athole during last century. But extensive as these plantations are, they are not greater than those formed recently, and now in course of formation, by the present Earl of Seafield, in proportion to the time that his Lordship has been in possession. Since the Honourable T. Bruce became commissioner on these estates, plantations have been formed, chiefly in Strathspey, in a congenial soil, and to an extent rarely equalled in this country.

The other extensive planters in the north of Scotland during the early part of the present century have been the Duke of Richmond, the Duke of Sutherland, the Earl of Fife, Lord Lovat, Sir William G. Cumming of Altyre, Bart.; and more recently may be added Sir George Macpherson Grant of Ballindalloch, Bart., Mr. Mathison of Ardross, and Mr. Fletcher of Rosehaugh, etc., all of whom have planted large forests on their properties.

Such is a glance at the progress of the formation of plantations up to the present time.

III.

ACCLIMATATION.

This is a subject on which both practical and scientific men of the past and the present time have entertained opposite opinions. This may arise from different kinds of plants having been experimented on by different individuals. Although all plants require a certain peculiar range of temperature, moisture, light, and atmospheric pressure, which in some kinds cannot be greatly interfered with without proving fatal to their existence, yet the limits are very different in different tribes of plants; they are widest, as might be expected, in those whose native habitat embraces a wide geographical range.

It has for long been supposed that the sensibility of plants may be diminished by habit, by gradation of climate, and by succession of generation. This theory will be found quite correct, generally, respecting ligneous plants, but it does not hold true with regard to agricultural crops, such as the cereals, which do not exist during the winter. Unfortunately, almost all the recorded experiments, that I am aware of, have been in relation to annual crops, or crops which are naturally of short duration, or greenhouse plants, the hardiness or tender nature of which was not previously well known.

Among the greatest authorities for acclimatation we have Sir Joseph Banks and Dr. Macculloch. In advocating the hypothesis Sir Joseph relied upon the following case :—" In the year 1791, some seeds of the Canada rice-plant, *Zizania aquatica*, were procured from Canada, and sown in a pond at Spring Grove, near Hounslow. They grew and produced strong plants, which ripened their seeds. Those seeds vegetated in the succeeding spring, but the plants which they pro-

duced were weak, slender, not half so tall as those of the
first generation, and grew in the shallowest water only. The
seeds of these plants produced others the next year, sensibly
stronger than their parents of the second year. In this
manner the plants proceeded, springing up every year from
the seeds of the preceding one, every year becoming visibly
stronger and larger, and rising from the deeper parts of the
pond, till the last year, 1801, when several of the plants were
six feet in height, and the whole pond was in every part covered
with them as thick as wheat grows in a well-managed field."

Respecting this experiment Dr. Lindley, who was no believer
in the doctrine of acclimatation, says, " It is to be remarked
that in the very first year this Canada rice grew as vigorously
as afterwards ; that its first progeny was feeble, that it only
recovered its vigour after it had been reproduced often
enough to establish itself in the deeper parts of the pond,
and that at last, after many generations, it was only as vigorous as at first. The case was not one of naturalization, but
of deterioration succeeded by restoration, not improvement.
Many reasons might be assigned for the temporary deterioration ; but that the plant was not naturalized is sufficiently
proved by its having disappeared long since." And he adds :
" Dr. Macculloch, who ably advocated the doctrine of acclimatation, rested his case upon two grounds ; the one, that many
sickly greenhouse plants acquired great vigour in Guernsey
when turned into the open air, and that seedling guavas were
productive there, although their parent was sterile ; and the
other, that hardy varieties of the vine, the pear, and other
fruits, are well known to all cultivators. But this reasoning
was unsound; Dr. Macculloch did not show that the greenhouse
plants in question had become more hardy than they were before ; it only happened that they became more healthy ; and
in regard to the so-called hardy fruits thus alluded to, there is
nothing to show that their constitutions are at all hardier than
those of their parents ; what is called hardiness in these
species consisting in an alteration in their time of flowering or
fruiting.

" The constitution of such plants does not appear to be more capable of resisting an unfavourable climate than that of their forefathers. It is perfectly true that many so-called greenhouse plants are now known to be hardy ; but such species have not increased their power of resisting our climate; they never were tender;" and it is added, " If no good evidence can be produced of plants having become acclimated by repeated sowings of their seed, the facts on the other side are numerous and conclusive. The Peruvian annual called Marvel of Peru, or *Mirabilis*, the common Indian cress or *Tropæolum*, the scarlet running kidney bean, the tomato, the mignonette, an African plant, the Palma Christi, or *Ricinus*, all natives of hot climates, have been annually raised from seeds ripened in this country, some of them for two hundred generations ; yet have in no appreciable degree acquired hardiness, but the earliest frost destroys them now as formerly.

" Potatoes, long as they have been cultivated from seeds, are in no degree more hardy than those which are now brought to us from Peru and Mexico ; indeed, some garden potatoes, imported in 1846 from Lima, and planted in November, stood the severity of the succeeding winter, when the thermometer fell to 3° Fahr., rather better than the English varieties which had been obtained from repeated seed-sowing during a century."

It is well known to market-gardeners that the earliest varieties of the potato grown in the best climate of South Britain when imported into the North become degenerate, and later in the open ground, and this most readily when grown at a high altitude. The same is observable respecting potatoes grown in hot-beds ; the first crops they produce are earlier than if the parent roots had been grown in the open ground, and a climatic change is effected, although it does not enable the potato to resist frost.

A similar change of lateness takes place in the growth of the early varieties of garden pease. The first generation of the early frame pea from seeds grown in Essex, when raised in the north of Scotland will yield a variety that cannot be distinguished from the charlton, a later kind, or second early, and repeated generations add to the length of the straw.

In the case of early cabbage, imported from the South to the North, and grown for seed, the early quality disappears to some extent in the first generation, and the cabbage becomes larger in size, so that to raise the early York variety, from seed, it is necessary to plant the early May or dwarf. Lateness is accompanied with hardiness in this plant.

In the case of turnip imported to the North they are apt to lose their early ripening qualities, and to grow larger, and later into the end of the season. This is a valuable quality in the brassica tribe. It has the singular and valuable property of giving hardiness to the esculent, enabling it to withstand the severity of winter better than those of earlier maturity, which are more subject to frost.

Thus the Aberdeen yellow bullock turnip takes the first place in the list of the British agriculturist,—Aberdeen being the depôt for the seeds saved in the cool and elevated districts throughout the north of Scotland. Now though we look in vain for any acclimatizing influence imparted to resist frost in annual crops, yet, in these biennials, the cabbage and turnip, we think we can perceive distinctly enough, in a generation or two, the power of the plants to adapt themselves to the climate they occupy. The bringing forward of our remarks on these plants may appear a digression from the object of this work, but they have been referred to for the purpose of showing that little can be expected to be done in acclimatizing a plant that is an annual, and has no existence during the months of winter, and that biennials are the shortest-lived plants that can be expected to be inured to a low temperature.

It is pretty well understood that the purple laburnum (*C. Adami*) is a hybrid between the common Scotch laburnum and the dwarf, shrubby, purple cytisus (*C. purpureus*). The original was produced in France. It is propagated by being worked on the stock of any laburnum; as being a mule, it (the purple laburnum) does not produce seed; but the tree

often throws out blossoms of the two parent species, along
with the hybrid, so that the three sorts are found all in flower
on the tree at the same time.

About twenty years ago I planted on the margin of a
stream that runs past my house a plant of the purple laburnum.
It occupies a cool soil, only about three feet above the rise of
water ; for the first ten years it grew luxuriantly, and generally flowered very well, but it did not ripen its wood well, and
being tender, it required to be frequently divested of dead
wood, particularly after a severe winter. After being about ten
years planted, it broke out here and there through the tree
' with tufts of the purple dwarf cytisus; these tufts or bushes generally died down to the branches from which they emerged, and
were pruned off along with the other dead wood throughout the
tree. After this process of discharging, or throwing off the tender element for a few years, twigs began to appear of the common laburnum, which yielded blossoms of unusual size ; the
racemes produced seed which grew very vigorous plants of the
common laburnum, although the three sorts continued for a
few years very visible on the same tree, all from the one
scion or graft of purple laburnum, yet the cold summers
succeeding 1860 caused the hardy common tree variety to
shoot ahead, and it has now completely extinguished the
tender grafted kind, and ripens its shoots in the coldest
season, and continues to grow vigorously. What renders this
tree so interesting is the facility with which it adapts itself
to the character of the, seasons. It was only in the warm
months of 1865 that the tufts of the cytisus purpureus again
became perceptible, the tree having during the previous four
or five cold seasons almost wholly returned to the hardy common Alpine or Scotch laburnum.

There is no tribe of plants with which I am acquainted that
is so susceptible of climatic influence as the Coniferæ. In the
celebrated native pine forests in the highlands of Morayshire
any variety among the trees can hardly be distinguished. But
I have taken seeds from these, and after raising them, have
planted them on the warm sands only a little above the level

of the sea, where a variety of foliage and habit became perceptible ; when these had yielded cones, and another generation of plants had been reared near the sea level, I have
found many of them so far removed from the ordinary type,
that some individual plants could scarcely be recognised as belonging to the species. This tree is found to accommodate
itself to circumstances, producing long or short yearly growths
in proportion to the ripening influence of the climate which it
inhabits. A few generations of the tree existing in a high
temperature would no doubt render its progeny nearly as
tender as our greenhouse plants. It is many years since I have
experienced the worthlessness of the plants of Pinus sylvestris,
or Scotch fir grown from imported seeds, on account of their
being too tender to withstand the severity of winter in the
north of Scotland, even in a nursery with some shelter.
Nevertheless the quantity of native Scotch fir seed sown in
Britain of late years, while a scarcity of seed in this country
prevailed, has likely been less than a tenth part of that imported from the Continent, and sown throughout the country.
Hamburg is the principal depôt for Continental seed of this
tree, and the chief forests are Hagenow and Hagueneau.
These forests of similar name are far apart, and although the
difference in their altitude must be very considerable, the
seeds of both produce plants too tender for Scottish moorland.
Even those English nurserymen, whose grounds are considerably
elevated, find that the Pinus sylvestris from Continental seed
does not produce plants sufficiently hardy to endure the severity
of an ordinary winter. Compared with plants produced from
Scotch seed, even from a good climate, the difference is great.
It is perceptible in the one-year-old seedlings, but much more
so after the second year's growth, when the plants from foreign
seed become quite brown, with a faded or scorched appearance
from the effects of the winter, and so damaged by the month
of March that they are often unsaleable ; while the plants produced from the seeds grown in the native Highlands of Scotland, under the same treatment, and standing alongside, have
a fresh green appearance ; and such is the contrast, that if

placed on elevated ground a plot of the one can be known from the other at the distance of a mile. In some warm situations in England, however, the plants from Continental seed are found to succeed, and with ample shelter they grow rapidly; while as timber trees in Scotland they are utterly worthless in exposed situations.

The larch is another tree of great importance whose antecedents stand much in need of investigation. We often see it stricken down when young and in the vigour of growth, in the absence of any visible disease, assuming all the appearance of an exotic of too tender a constitution to endure the climate of this country. This sometimes occurs in the vicinity of plantations of the same species, which luxuriate in the same description of soil. The difference is occasioned by acclimatation. No doubt the tree is to be found in some parts of its native Alpine regions inured for generations to all the exposure and cold that the species is capable of enduring, and we might reasonably expect that the seeds of such trees would produce plants quite suitable for the climate of this country. On the Continent, however, the larch has now been cultivated for generations as useful timber—in France, Germany, Prussia, etc. etc., in a climate adapted to the vine; so that seeds grown in such districts and imported have of late failed to such an extent as to cause the species, in some parts of our country, to be altogether abandoned as a timber tree. The former generation of trees having been acclimatized or inured to great heat, their offspring are unfit for at once enduring the extreme change of Scottish moorland. Hence the sad spectacle to be met with in many extensive plantations, of hundreds of acres of larch of no value, where the cost of plants and planting, in some cases, is required to clear the ground of trees in which life is only apparent.

Many years have elapsed since I first observed the advantage of acclimatization on the larch. In late seasons the tenderness of plants grown from imported seed is readily observed at every age, in nursery treatment; not only with respect to the white larch of the Tyrol, which is the tenderest variety, but in plants of the red, or mixed variety, from different parts of the Continent. The demand, however, for Tyrolese and Continental larch plants, and the frequent failure of the larch seed-crop in Scotland, has often given rise to large importations, especially during the cold wet seasons succeeding 1859.

In this part of the country—Morayshire—we have had the most ripening autumn (I now write, 15th November 1865), and the driest ever experienced. In these circumstances it is interesting to compare the appearance of seedling larch from imported seed, with those produced from home-grown seed. Of my one-year-olds all are from home seed ; but of two-years-old seedlings I have a few thousands from imported seed, standing in the same lot with plants raised from home-grown larch seed. Both were sown in the same hour ; the soil and the treatment were in every respect similar. Both sorts have now for a season ceased to grow, and are of the full size for their age, and are very equal, but the colour of the foliage is at present very dissimilar. Those from Continental seed appear quite green and succulent, having the terminal bud hid among the leaves on the top of the shoot, while the foliage of those grown from Scotch seed has a rich yellow and ripened appearance up to the top, disclosing the terminal bud full and plump, and prepared to withstand any degree of frost that may occur throughout the winter, or that might have occurred for weeks past. The difference between the two sorts is very marked, and from the colour of the foliage can be at once distinguished as far off as the plants are visible. This circumstance is by no means new to me. I have always found it so, but in a season so dry and ripening I would have expected that the contrast between the two sorts would have been greatly diminished.

I have occasionally, after severe frost in October and November, seen plants grown from imported seed that had been transplanted in lines for two years, standing with their tops drooping, with unripened foliage adhering to the plants for many months, while plants from Scotch seed in the same lot, and of the same age and treatment, stood scathless.

The influence of soil and climate on many species of Coniferæ alters their character, even in one generation, and sometimes produces as important a difference as that which exists between one species and another. I have sometimes observed a very marked difference in the hardiness of seedling araucarias, which I believe they inherit from the situation or climate in which the seed was produced.

I do not think that the influence of a hot climate is so readily impressed on our hardy deciduous trees as it is on the various species of coniferæ, yet I have observed it to some extent, on plants grown from Continental seeds, of two of our 'hardiest deciduous native trees, the alder and birch, which are decidedly more tender than those from home-grown seed.

I observe a paper presented to the Botanical Congress, on the raising of peaches, nectarines, and other fruits, from seed, by Mr. Thomas Rivers, Sawbridgeworth. He states that by repeated generations from seed they are produced " of a more hardy nature than the old sort," and that he has more than one proof of the fact. He adds, "I may be accused of enthusiasm, but I look to the future for new races of fruits with qualities far superior to the old, and the tree of so hardy a nature as to resist some of the unfavourable tendencies of our climate. I have formed this opinion on the solid basis of observation during a lifetime devoted to the cultivation of fruit-trees in all stages of their growth."

It is also worthy of notice that his late Majesty Leopold, King of the Belgians, entertained a sound opinion of the importance of acclimatation, and practically acted thereon.[1] In

[1] "Leopold of Belgium, whose loss we lament in this country almost as much as do his own subjects, had a character of shrewdness, which the subjoined extract from a pamphlet of our friend, Professor E. Morren, of Liége, will go towards justifying. Speaking of the progress of horticulture and botany, the good old King remarked on the benefits conferred on the world at large from the alliance of the two branches of science, and expressed his opinion that we need not pay so much attention to the discovery of plants likely to be useful as food for man, as to those capable of being employed as forage plants. The human race, spread throughout the world, must be in possession of nearly all the plants profitable as sources of food for man, but with reference to those indirectly useful there is more scope. Moreover,

the formation of his most extensive larch plantations in Belgium, he specially ordered the plants from Forres nurseries, the produce of the north of Scotland, as detailed under the article LARCH.

The influence of acclimatation is very perceptible on the Portugal laurel. In favourable seasons this tree ripens its berries in the north of Scotland, and plants grown from these are far more hardy than those produced from the seeds of a warmer climate. The severity of winter and spring 1867 gave very clear evidence of this being the case. I have trees of the Portugal laurel ranging from four to upwards of twenty feet high, raised from seed which ripened in Morayshire. Some of these trees occupy unfavourable situations, the soil being too moist, and only a little above the rise of water. Notwithstanding the wet weather of autumn 1866, so unsuitable for ripening the wood, the young shoots are not only safe, but the severity of the weather did not even discolour the foliage, although the thermometer stood about zero ; while other trees of Portugal laurel raised from imported seed, like many of the exotic evergreens, are cut down to some extent, or rendered very unsightly.

I had lately a consultation with the owner of one of the largest and longest established nurseries in the west of Scotland, respecting the tender nature of the larch of late years, both in the nursery and in the forest. He said that so far as the nursery was concerned he was well aware of the fact, from dear-bought experience ; that for several years his crops, grown altogether from imported seed, had been so severely damaged by frost, that on an average of years only about a fourth part of the seedlings had escaped with their tops.

continued the sagacious monarch, it is not necessary to ransack the whole world : China and Japan are the most important countries for us in this particular. In them is to be found a very ancient civilisation and skilful culture, carried on in a climate like our own. It is more advantageous to seek what we want under such circumstances than to begin anew with wild nature. We shall find in those countries plants adapted for cultivation and for our requirements, offering less resistance to our proceedings than those that we procure direct from their native wilds. Centuries are required to acclimatize plants."—*Gardener's Chronicle*, 1865, p. 1178.

In future he was fully resolved to sow none but Scotch-grown seed, which produced plants far hardier than those from foreign seed, and in the event of a failure of Scotch seed, he would be obliged to purchase young seedling plants raised in a better climate than that which his grounds possessed.

Most people acquainted with the commonest operations in gardening, have experienced the great difference in cauliflower plants subjected to the influence of cold during winter, compared to those protected in a higher temperature. Even our hardiest weeds that spring up under glass, in a higher temperature, suffer greatly when exposed to the severity of the weather in the open ground. This influence, which is so perceptible in the succulent plants of a season, is, with respect to trees, assuredly transmissible by seed to their future generations; and it is reasonable to suppose that that hardiness, or tenderness, will be more or less fixed, according to the length of time, or the number of former generations during which the tree had been subjected to such temperature.

Great are the advantages of international commerce in many commodities; yet it is to be feared that the importation of the seeds of plants that are required to stand in exposed situations is not destined to benefit either our forests or our fields. Whatever may be the fate of annuals or the crops of a season, that law in nature which I have experienced to stamp its influence so deeply on the trees of the forest, may be expected to be impressed at least on the perennial and biennial plants of the field.[1] Of course it cannot be expected that plants under the most skilful precautions will be exempt from frost or the casualties of seasons. The hardiest indigenous plants sometimes suffer, but generally their recovery is speedily effected by seasonable weather.

[1] "The unfortunate circumstance which attends clover is its being extremely apt to fail in districts where it has been long a common article of cultivation. The land, to use the farmers' term, ' becomes *sick* of it.' After harvest he has a fine plant, but by March or April half or perhaps more of it is dead."—*Farmer's Calendar*, p. 155.

To prepare plants of any description for the formation of plantations in bleak and exposed situations, it is necessary that they should be raised in an open and airy situation, and that, by standing well apart, they may be stout in proportion to their height, and by frequent transplantation be well furnished with an ample supply of young roots.

IV.

ON NURSERY GROUND, MANURES, etc.

Exposure.—Nursery ground should be moderately exposed and elevated, with a free circulation of the atmosphere, that it may be exempt as far as possible from the severity of frost, which in spring and in summer is often found to prevail in low, damp, and hollow situations.

The aspect or direction of the slope of the surface is not of very much importance. Yet it is desirable to have various exposures when such can be easily obtained ; but level land of good quality is found to suit all purposes. Although a northern aspect in some situations is apt to retain frost till late in spring, which is sometimes inconvenient for the insertion, or removal of plants at that season, yet it is generally the best for raising evergreens from cuttings, and the safest for the growth of all plants like the larch and silver fir, which are apt to spring early, and which suffer from late frost, in the opening up of the season. (Directions, however, shall be given under the names of the respective trees, by which such casualties may be averted, or greatly mitigated on ground of any aspect.)

Soil.—The best soil for the ordinary purposes of a nursery for forest trees is a sandy loam two feet deep, friable and easily worked, free from stones, with a dry subsoil or thorough drainage ; all sorts of moorland, ferruginous, or retentive subsoils, have an injurious influence, by retarding the ripening of the summer's growth of plants. It is, however, an advantage to have a variety of soil in a nursery ; a heavier or stronger soil than that described, though not suitable for seedlings, is well adapted for larger plants. A sandy peat soil, and all soft open soils that do not get bound up with alternate moisture and drought, yield the most fibrous roots, and from such, seedling plants are transplanted more successfully than from ground of a stiff clayey tendency.

Drains.—If drainage is necessary, the drains should be sunk at least four feet deep, and kept at some distance from the line of hedges.   When this cannot be avoided, as in crossing underneath a hedge for instance, to prevent the roots of the hedge from choking up the drain, the tile or built eye of the drain should be encased with broken stones, extending at least a foot on each side, and above the conduit.

Nursery Fences.—Unless the ground is naturally very well sheltered, hedges make the best nursery fence.   If the shelter is ample, wire fence is to be recommended, and in all cases where hares and rabbits abound, wire netting is the most convenient security, unless where there is on the spot a supply of timber adapted for close upright paling.

In the formation of hedges the hawthorn generally makes the most satisfactory outside fence.   If much shelter is needed beech is preferable to any others ; and for the interior divisions of a nursery, yew-tree and holly are found most suitable.   These two evergreens are of compact growth, very ornamental, and not subject to disease, and though of rather slow growth, yet, when properly nursed for hedges, they can be transplanted three or four feet high, when they at once form a shelter and screen.   Evergreen privet is also suitable for the same purpose.   Excess of shelter is injurious to the hardiness of the stock, and should be avoided.   The shelter of woods often renders nursery ground too confined, and their vicinity is frequently infested with disease and the prevalence of insects injurious to young plants.

In ground of any considerable extent a cart-road should give access to the interior, for the convenience of carting manure, and the easy removal of plants.   The ground is most conveniently cropped when the walks are laid out at right angles.

The best mode of adapting new ground for nursery plants is to have it well trenched, eighteen inches or two feet deep, the surface carefully placed in the bottom, in autumn, and laid down the first season with a well-manured green crop—turnips or potatoes. This renders the ground clean and soft, to a sufficient depth for any crop.

In manuring for nursery crops without a previous green crop, the manure should be well decomposed, and whether the crop intended be plants or seeds, it suits best for the manure to be dug roughly into the ground in autumn or early in winter; or to have the ground ridged up is preferable, as this admits more readily the pulverizing influence of frost, which acts favourably on ground for any crop.

The evil effects of fresh manure often appear very conspicuously if applied to a crop of any species of coniferæ. Hardwood plants are not so easily injured in this way; but it is not unusual to see nursery plots of pines and larches sadly damaged by manure too fresh and full of urine being applied. The mistake sometimes occurs of laying down the manure in heaps on the ground for some time before it is spread and dug in. Where this is the case, the spots in the seedling Scotch pines and larches, where the manure-heaps stood, are generally marked by the absence of plants, as distinctly as in agricultural crops, where the vegetation is generally the strongest.

Even when manure is well decomposed and comparatively dry, if it is allowed to remain in heaps on the ground, and get washed into it by rain or snow, the effect is injurious. If, therefore, manure is carted out in frost, it should be placed in some spot adjacent till it can be dug down, or on ground where it may not injure the ensuing crop.

Farm-yard manure is preferable to town manure. If town manure is employed, it is the better for being mixed up with a proportion of rank farm-yard manure sufficient to produce fermentation in a heap, which destroys the seeds of weeds before it is used for any crop. Wood or peat ashes in a dung-hill form an excellent manure; coal ashes, to a great extent, are bad for seedling crops; and manure composed of sawdust is still worse, as it tends to engender fungi, which are the prevailing enemies of the pine tribe.

A very good manure is formed by collecting the weeds of the nursery. After they are dissolved in a heap, give one bushel of lime-shell hot from the kiln to every four or five bushels of rotted weeds; these should be formed into a heap of alternate layers of lime and weeds; in a few days, when the heap has swelled out, it should be carefully turned over and intimately mixed while it is in a hot state, so that all the seeds of weeds may be completely destroyed. This mixture may be applied at the time of transplanting in the nursery, or at the time of digging for seed-beds. In the latter case it forms a good top-dressing, by applying it thinly on the surface of the ground immediately after the ground is dug, and before it is raked for seed-beds; about ten tons of mixture per acre will be sufficient, and at that rate it may safely be repeated every three or four years. If a much greater quantity is applied, the intervals in its application should be longer, unless the ground is rather strong and stiff.

Leaf mould is a valuable manure for seedling crops, if it is well dissolved, and has been exposed to the atmosphere by repeated turning. It is sometimes to be obtained in pits and hollows, in hardwood plantations, into which the leaves drift in rough weather; and it may be carted out, nearly made, from time to time, as it accumulates.

It should be turned over twice or thrice, with an interval of two or three months, before being applied, that it may be thoroughly dissolved; it is best adapted for seed-beds, and the mode of application is that recommended for the lime mixture, as a top manuring laid on between the digging and raking of the ground. It forms a fine mixture of soil for seeds, and is not apt to cake or harden on the surface after heavy rain, which hardening is much against the growth of small seeds.

Peruvian guano cannot be recommended as a nursery manure. In some soils and seasons it excites a rapid but rather feeble growth; and in late seasons plants stimulated by it are apt to fail in maturing their young wood, and are consequently more liable to the influence of frost. Plants appear to require nourishment of a more steady, bulky, and permanent nature, though not so stimulating, in order to produce well ripened wood, full and sound, with their terminal buds well developed. Pines and firs grown on very poor nursery ground, particularly during the second year of their life, assume a yellow colour in their leaves; and although nothing the worse, yet the appearance is often objected to. In this case a dressing of guano, followed by rain in August, or its application in a liquid state, will change the foliage to a darker green in the course of a few weeks.

It is desirable that nursery ground should be large enough to admit of the plots being restored to good condition every fourth or fifth year by a well-fertilized green crop of turnips, or potatoes, etc. All sorts of two or three year lined plants exhaust the ground considerably; among hardwood, ash, elm, and sycamore are the most exhaustive, oak the least so ; and among the fir tribe, the Norway spruce reduces the condition of the ground more than any of the other species.

Some of the most valuable crops of seedling Scotch pine and larch that I have ever seen, were produced on plots of land very much exhausted by nursery crops, but renewed or enriched a few months before being sown, by receiving a good coating of fresh mould from a field in the vicinity, which during the previous summer had yielded a crop of turnips, which had been well manured, and kept clean and free from seeding weeds during their growth ; this soil was carted and spread on the nursery plots, about two inches deep, and roughly dug over during the months of January and February.   A mixture of various soils is very invigorating for all tree crops ; and it is found profitable to give an extra quantity of manure to the field in exchange for soil·well adapted for crops so important, particularly where the carriage is not very distant.

In the private nurseries throughout the country, seedling plants are seldom raised, as it is generally found more profitable to purchase young plants one or two years old.   These we have been in the practice of raising for many years to numerous customers both in Scotland and England, who order them *a year before they are required*, whereby we avoid to some extent the risk of growing an overstock, and are thus enabled to sell them generally about 20 per cent. under the usual rate.

V.

ON DRAINING GROUND FOR PLANTATIONS.

In Scottish moorland more plants perhaps have been lost by being inserted into ground too wet than by any other cause, and it is seldom that any considerable extent of ground is found adapted for plantation, without some parts requiring to be drained. Open ditches are generally best adapted for this purpose; every other kind is apt to be closed up by the growth of the fibrous roots of the trees. In hilly ground it sometimes happens that an open ditch cannot easily be formed, owing to the depth required in some parts to allow the water to escape, and the great width required at the surface to admit a sufficient slope on the sides. Such places are generally small in extent, and the best method is to build an eye of stone, or to introduce a tile on a scale sufficient to form a discharge, surrounding the conduit, in either case, with an accumulation of loose stones, or rough harped gravel, which tends to intercept the roots, and prevent a stoppage. Although land may not appear very wet, yet if the surface or the subsoil be impervious, and particularly if the surface is inclined to a great space, catch-water drains should be excavated, and the materials placed on the lower side, that the flow of water into the drain may not be impeded. Inclined moorland, or peat, which retains moisture, may, in many cases, be relieved of surface-water in the same way, by the operation of the plough. I have seen this sometimes very expeditiously and effectually performed by a trench plough, particularly where oxen can be employed that are accustomed to the reclaiming of waste land.

In laying out the course of drains in hilly moorland, it is

often necessary to exercise caution, and not bring too much into any common discharge, along a great declivity, before reaching a well-swarded channel or stream, as an accumulation of water to one point of exit, particularly in ground not adhesive, has the effect during floods of cutting ravines, overlaying lower and more fertile grounds with débris, and shutting up and diverting the course of streams, and the like. To prevent such inroads it is sometimes necessary to causeway, in a concave form, the course of a drain along a steep declivity. Stones for such a purpose are generally found abundantly on the ground, and even when carted loosely into ‚ the bottom of a ditch they have the effect of preventing the water from deepening the trench, although they are not built; but, if the descent is rapid, building is generally required.

Where stones are scarce, well-swarded turf, closely built, by being set on edge, is found to suit the purpose, where the discharge of water is not constant; in such the herbage grows, and becomes quite fixed, and sufficient for the discharge of a flood, if only of a few days at a time. For such places a shodding of stone and turf alternately make the strongest fence of any against the inroads of water. The stones, if flat, should be set on edge, if boulders, on end ; and although the stones are generally round, yet by introducing a fresh turf set on edge between each course of stones, the turf grows, and the roots, belonging commonly to a mixture of grasses, fill up all crevices underneath, and the wild native vegetation spreads above, and thus encases the stones securely. I have seen this description of work resist the inroads of streams in flood, where more costly mason-work was of less use. The more gentle the cavity in the bottom, or the slope on the sides, the greater the strength of the work.

In forest planting, deep-rooted plants require a soil more elevated above the rise of water than the pines and other surface-rooted trees. Young plantations recently formed are more apt to suffer in the absence of drainage than plantations that are more advanced ; because, when trees (particularly the pine tribe) become so far advanced as to form a cover on the ground, they have a tendency to dry it; they intercept the slighter showers of rain and the humidity of the atmosphere, and expose a large surface to evaporation. On a large scale, however, forests have a tendency to increase the humidity of the atmosphere and strengthen the springs of water in their vicinity. In soil surcharged with water, where it cannot be properly drained, the mode of ridging it is sometimes resorted to, and successfully adopted, for the purpose of forming a cover of plantation, which improves the appearance of the place, and under coppice is frequently attended with profit. Ridging is performed by excavating every alternate space of five or six feet in breadth; the space which forms the ridge is thus raised in proportion to the depth of the excavation. The work should be performed in summer; and the operation of planting in this description of ground should be in spring.

Fences.—It is hardly necessary that anything should be said on the necessity of fences around plantations, or of the best mode of forming them. The materials to be employed in their formation will of course depend on the resources of the district where they are required. In some situations the expense of fencing a plantation is equal to the cost of plants and the operation of planting; on a small extent it is sometimes much more. Perhaps the least expensive mode generally applicable for waste land, and to protect from the inroads of cattle and sheep, is to form a turf-dike on the brink of a ditch. The sward from the surface is built to form the outside of the dike, and the earth from the ditch forms what is called a *backing*, being a bank or slope of earth falling from the top of the dike to the surface of the enclosed ground. This of itself is generally a sufficient fence against cattle; but a formation of this sort, to prevent the inroads of sheep, is usually surmounted with a single line of wire, fixed with wooden supports, and placed a foot above the top of the turf. In districts where thinnings of young plantations abound, these are used, instead of wire, to form the top bar. The expense of construction must no doubt depend on the adapta-

tion of the soil for a formation of this description. A usual
price, apart from the upper fixture or rail, is from 4d. to 5d.
per lineal yard of dike five feet high, measuring from the
bottom of the ditch to the coping-turf. The cheapest dike of
turf I have known to be erected, suitable as a fence or protection against cattle, was formed five feet high, at the cost
of 3d. per lineal yard. A common method is to form the
fence with earth, and to face up the outside with stones,
where such material is conveniently obtained.

I have lately built by contract in moorland about two miles
of dikes, of two sorts, where stones are found conveniently,
partly from newly improved land, and partly from cairns,
which abound along many mountainous districts in the north
of Scotland. These cairns form heath-covered mounds throughout the moors, and require little or no labour to excavate. A
dry stone dike, five feet high, was built with a scarcement of
three inches on each side near the surface of the ground;
above that it started at the thickness of two feet, and rising
to four and a half feet in height, where it tapered to sixteen
inches; here it was surmounted by a projecting cope, making
its entire height five feet. The cost of building was 10d. per
lineal yard, carting stones, 3d., making the price of the dike
1s. 1d. per lineal yard. This was the fence adopted where
the ground was level, or had only a gentle slope.

Where the ground was steep, the fence was formed of a
different construction; there the surface of the moor was
excavated to the average depth of twelve inches, and to the
breadth of five or six feet. The surface turf was built up
to the height of about four feet, the earth excavated from
beneath the surface turf was cast over in the progress of
building to form the *backing*. This turf wall was then faced
up with stone from the bottom of the excavation to the
height of four and a half feet, and a stone coping made the
faced-up side of the dike stand five feet high, with a scarcement at bottom of a few inches, and with a slope or batter of
four or five inches. The surface of the embankment on the
opposite side was formed with a slope of not less than two

to one. Where this fence was formed at a right angle to the natural slope of the ground, catch-water drains were formed, which furnished additional material for the slope or back of the dike, where required. These drains form a safeguard for the dike during floods. The cost of this fence, including the catch-water drain, amounted to 1s. 2d. per lineal yard; but of course the cost of all such fences depends on the supply of materials and of workmen in their vicinity. It sometimes happens during winter that dikes of every description become drifted with snow, over which sheep find an easy access into plantations; for this reason wire fences, as they are less apt to accumulate snow, have become common for the protection of plantations throughout the Highlands of Scotland.

Roads in Plantations.—It is usual, before planting land to any great extent, to line out the roads which may afterwards be required; this is most easily accomplished while the ground is bare, and the inclination of the surface can be brought under the eye. It is seldom necessary at this stage to do anything further than mark off the roads, remove obstructions, and form side drains, where the soil is wet, to admit a carriage drive throughout the ground. These are commonly formed from fourteen to eighteen feet wide, and it is seldom that the use of such roads justifies any great outlay in their formation, until they are required for the removal of timber, which seldom occurs, to a great extent, before eighteen or twenty years after the formation of the plantation.

The carriage of timber, on account of its vibrating motion, is more severe on roads than that of almost any other commodity; this accompanied with the circumstance, that roads in woods have a tendency, in all sorts of soil, to retain moisture from want of air, and from their seldom having been formed with any substantial body of materials to support heavy loads, accounts for the bad state in which they soon appear after having been subjected to any considerable traffic. Few operations are attended with greater outlay than the formation of good roads, and when such are only occasionally required in the removal of timber from the forest, the cost forbids their formation. I shall therefore only indicate what I have found the easiest mode of forming a temporary road sufficient under such circumstances.

In some soils the surface vegetation forms a tough sward adapted for sustaining temporary traffic, particularly when it is not confined to the same tracks, but shifted by being spread over the surface of the roadway. In the absence of a firm or gravelly soil, which continues some time passable, advantage should be taken of frosty weather when the traffic is heavy. I have found that the best method of making a road with a soft surface passable, is to overspread it with a close thatching of the branches of trees, each branch six, eight, or ten feet long; larch, spruce, silver fir, beech, birch, elm, etc., all lie flat, and can be built into a compact cladding across the surface. Scotch fir is of use for this purpose, but inferior to many other sorts. Having placed the branches in this position, the subsoil from ditches on each side of the road should then be cast up on the cladding of brushwood, giving the road a convexity of a few inches in the centre, and of a depth sufficient to cover over above the branches about two or three inches. If the subsoil is good gravel, this will make a strong and lasting road; or if gravel is to be had in the close vicinity, it is generally worth while to cart it; but whatever be the nature of the subsoil, the branches prevent the loaded carts from sinking to a great extent, and form a very passable road for the purpose, and at the smallest outlay, as the materials are always at hand.

In plantations formed for the growth of hop-poles, propwood, or the like, that are to be cleared off at an early period, —say ten, twelve, twenty, or twenty-five years,—roads may be made narrower, or they may not unfrequently be omitted altogether; but where extensive plantations are intended for heavy timber, roads are indispensable, and their advantages are well worth the space they occupy, as they impart a healthy influence by the admission of air, and open up the interior of the plantation, and allow the state of it to be ascertained.

VI.

ON SELECTING NURSERY PLANTS.

TABLE SHOWING THE NUMBER OF PLANTS REQUIRED TO PLANT
AN IMPERIAL ACRE FROM ONE FOOT TO THIRTY FEET, PLANT
FROM PLANT.

IMPERIAL ACRE.		IMPERIAL ACRE.	
Distance.	No.	Distance.	No.
1	43,560	12	302
1½	19,360	12¼	270
2	10,890	13	257
2½	6,970	13½	239
3	4,840	14	222
3½	3,556	14½	207
4	2,722	15	193
4½	2,151	15½	181
5	1,742	16	170
5½	1,440	16½	164
6	1,210	17	150
6½	1,031	17½	142
7	889	18	134
7½	774	18⅜	127
8	680	19	120
8½	603	19½	114
9	537	20	108
9½	482	22	90
10	435	24	75
10½	395	26	64
11	360	28	55
11½	329	30	48

On Selecting Nursery Plants.—It is but reasonable to suppose that with the view of laying down a crop destined to
stand for generations—it may be for upwards of a century,
—every precaution would be taken to secure its vigour and

success, by selecting plants of the most approved varieties of the species ; in many instances, however, this is not done. Indifference, in this respect, with the trade, or with plant merchants, who pass the commodity from hand to hand in course of a few weeks, is not so surprising ; but with those who are to own the plants in their final destination, the selection is surely worthy of the exercise of thoughtfulness and care. Seldom, however, is this care taken; seldom is the same vigilance exercised here, which the agriculturist displays in laying down a crop destined to last only a few months. In arboriculture the result stands far away in the future, whereas with the farmer 'it is close at hand—the character and quality of his crops are readily ascertained, and the difference between good and bad is realized in a few months in a tangible form.

In the formation of plantations, great or small, the work is generally proceeded with as if every tree or plant of its name were equally good, without regard to variety, pedigree, or climatic influence. All Scotch pines, for instance, are often treated as one sort, and larch and spruce in the same manner, yet in each of these there exists a diversity greater than that found in wheat, barley, or any other grain. Although the varieties and qualities of the trees will be found pretty fully detailed under their respective names, yet as those named are important leading kinds, as a guide to the selection of the most suitable plants, I may here make a few remarks on each of these trees.

The Scotch pine is a tree very susceptible of climatic influence ; when removed from its native mountains to a warm country, and grown from seed, it changes its appearance to some extent, showing many seedling varieties. Every succeeding generation produces softer timber ; and away from its native habitat the tree thus degenerates and becomes tender. Since 1860, many tons of Scotch pine seeds have been imported from the Continent into Britain, as already noticed, and many millions of plants have been thus produced, far too tender for the exposed moorland of this country. In the best protected nursery ground they often perish in a severe winter. These plants abound in British nurseries. The seeds are obtained at a cheap rate, labour throughout the Continent, the principal expense attending them, being far cheaper than in this country. Plants from Continental seeds succeed only in well-sheltered ground at a low altitude. On elevated and exposed moorland they are worthless. Unless frequently transplanted in the nursery, they are barer in the roots than the native plant, and consequently more apt to die in being transplanted; and where they have taken root and lived for a few years, I have seen them in moors standing about knee height in the first of summer, brown, of a scorched appearance, and twiggy, being topped by frost, and worse than nothing in the ground.

Pinus sylvestris montana is another variety of Scotch pine which is imported from the Continent, and grown in this country under that name. The plants of this tree have a close resemblance to those of the native pine while a plant, which is apt to lead to a serious mistake, as the *Montana* of the Continent, which is introduced and propagated to a very great extent, is a dwarf—worthless as a timber tree, as at most it only becomes a spreading bush. With respect to this plant the confusion is the more perplexing, as Don, in his writings on the native varieties found in Scotland, has given the name *Montana* to a valuable variety of the Scotch pine, which differs widely from that which has of late years to an unusual extent found its way to the nurseries.

The *Montana* of commerce, though imported from the Continent, is a very hardy plant, owing to its seeds being gathered from the tree in its native habitat, which is always at a great elevation. Unlike the loftier sorts of Pinus sylvestris, it is unsuited for profitable cultivation in the warm plains of foreign countries. The plants should be avoided for all plantations in the course of being formed for the sake of timber. Yet I have known this dwarf inserted with great care, extensively, ten to twelve feet apart, with the view of ultimately suppressing all the other plants associated with it in the forest.—(For further particulars see Pine-Tree.)

D

The selection of larch plants is a matter of great importance; hardiness is the great desideratum, and the produce of
Scotch seed should have a decided preference to those grown
from foreign seeds, of which so many tons are yearly imported, that during the last few years the seeds of this tree
saved and sown in Britain did not form a tithe of the quantity
imported from Hamburg and other Continental ports,—seeds
which are readily obtained in the planted woods in the
valleys and warm slopes of Germany, Prussia, and throughout the Netherlands.   Plants raised from these seeds are very
subject to frost; this is very apparent in the nursery treatment, compared to plants from home-grown seed.   The leaves
of both sorts sometimes unfortunately perish on very severe
occasions of summer frost, but seldom to the same extent.
The blighted leaves of early summer places the plant in the
most susceptible condition for the attack of the *coccus laricis*
or *larch aphis*, which lives on the juice.   In the absence of
the leaves in winter the insect may be detected by the plants
appearing here and there dwarfed, and darkened in the colour
of their bark, as if subjected to the influence of smoke; such
plants ought to be avoided.—(See LARCH.)

In selecting *Spruce Fir* for forest planting, it is necessary
to know that the Norway species is the only sort fit for
becoming a large tree and yielding valuable timber, except
the *Douglasii*, which is easily known.   But the Norway
species and the White American bear a close resemblance to
one another; and as the seeds of the latter are often imported from America, and grown by the hundred thousand,
care should be taken to avoid it.   The Norway and the
White American, grown together under the same circumstances, are quite distinct.   The Norway is darker in the
foliage, more vigorous and robust in growth, particularly in
the leading shoot.   The White American is paler in the
foliage and bark, shorter and more slender in its growth, its
foliage is closer on the branches, and the leading shoot is
seldom very vigorous; it is a dwarf tree, and its timber is of
little value.   Where the Norway species is grown on poor

hard soil, the plants have very much the appearance of the White American species.—(For further details on this subject, see the article on the various species of SPRUCE FIR.)

I have now said enough to show the mistakes that are most likely to arise in selecting plants of a few of the most common kinds of the Coniferæ. Respecting many of the other kinds of this tribe, we depend almost entirely on foreign countries for the supply of seed; and although there is no doubt that a great difference exists in the hardiness of the produce of the seeds of the various kinds grown in their native heights, compared with those from seeds produced in the lower and warmer districts, yet these circumstances, notwithstanding their importance, cannot be easily controlled, and afford no choice in selecting newly introduced kinds in a nursery.

It is of great importance to obtain plants grown from a good stock, or from the most approved trees of their species. For several years I cultivated the variety of oak known as *Q. robur pedunculata.* This tree yields the best description of oak timber. It also attains to as great a size as any other variety of the British oak; although when in the nursery the sessile variety generally gives the stronger plants, as of the two its acorns are the larger. I could not, however, obtain a better price for the sort selected than for those grown promiscuously; and as there are intermediate sorts between the two, the difference is not very important, and the kinds are now seldom grown apart.

In selecting oak plants from the seed-bed, there is commonly a considerable difference in their size, unless the very small acorns have been sifted out and not sown.

In transplanting young oaks from the seed-bed, those very inferior in size should be rejected, as, although they are transplanted, they generally continue comparatively dwarfish, even with the best treatment, in good soil.

In making choice of hardwood plants of any sort, good roots are as conducive to their success as good tops, and much more so with oaks, beeches, and birches, and with all sorts that are apt to get bare and destitute of fibres. When these sorts have acquired the proper size for transplanting into the forest, they are much improved, particularly the oak, by being removed or disturbed in the nursery lines a year previous to their being finally planted. This removal, if it should only be from the site of one nursery line to the next, insures a mass of fibrous roots, requiring no pruning on being planted out, which gives the plant a great advantage over those not so treated. And although, in consequence of being thus disturbed, they are not so full in the display of young wood, yet they are far more valuable than plants that have not been recently removed, though the latter appear the more vigorous. When an excess of vigour appears in hardwood plants, they seldom take readily to the ground on being removed, particularly if they have been nursed in a sheltered situation. Plants of every description, and especially those of the pine tribe that are to become the inhabitants of bleak exposures, should, in the nursery, have the benefit of open and airy ground, to admit of the play of the wind on all sides; and such as have been transplanted the preceding year will have the best roots and the hardiest tops, and are best adapted to survive under adverse circumstances.

VII.

MODES OF FOREST PLANTING.

In no work in general practice is there a greater diversity of methods than in that of forming plantations.

This arises, to some extent, from the difference in the soil, and in the herbage which overspreads it, in different parts of the country. On the soil and exposure should depend the kinds of plants, and on the herbage the size or description of plants that should be employed; and by far the most important step in arboriculture is to adapt the plant to the soil and climate congenial to its growth. Like every other operation that does not yield a speedy return, the profit of a plantation must always depend very much on its original cost. It is not unusual to see plantations formed at an outlay of £2, £3, or £4 per acre (for plants and planting), while another equally valuable is formed on ground of the same description, and with plants of the same kind, at less than half the cost. This sometimes arises from the size of the plants employed, and also from the mode of inserting them. Sometimes the most costly mode will altogether fail, while the least so is a complete success; but I have also observed the reverse of all this—where the modes of planting reckoned least expensive failed again and again, and the costly method had to be adopted, which, under the circumstances, should at first have been resorted to. The success of plantations, therefore, greatly depends on the skill and experience of the forester. I shall now detail as minutely as I can the different modes of inserting the plants, and the sizes of the plants, which I have found most successful under various circumstances. The mode of operation, the time or season of performing it, the kind and the size of plants, etc., are all dependent on circumstances, such as the surface soil, the subsoil, the state of the native vegetation, the climate, and exposure of the district, etc.

There are three methods of planting generally practised throughout the country, namely, *notch* or *slit* planting; pit planting; and planting on trenched or prepared ground.

Notch planting in moorland.—The implements by which this operation is performed vary much in different parts of the country. Two sorts, however, are sufficient for notch-planting under every circumstance, namely the small planting spade or hand-iron, and the common garden spade. The latter is improved for the purpose by being well worn down.

The hand-iron here figured has a blade about seven inches long; its entire length is eighteen inches, and its weight is about two and a half pounds. The person using it should have a small bag, for carrying plants, tied round his waist. This spade should be struck into the ground with one hand in a slanting direction, which will make it penetrate more easily than when it is kept perpendicularly; the plant is inserted by the other hand, and is placed on the farther side of the hand-iron; and, by turning the turf a little to one side with the hand-iron, an opening is made for the roots. When the plant is put in, the ground forced up should receive a stamp with the foot, to make it firm. Moor or peat ground is naturally apt to contract and shrink during the heat of summer, and when plants are inserted by making a larger opening, the incision in some soils opens, and exposes the roots at a time when they are most apt to suffer. Persons planting with the hand-iron advance regularly, each keeping the exact distance from his neighbour that is required between the plants. An overseer follows after every ten or twelve planters, to see that none perform the work carelessly or slightly, in order to keep pace with those who are most expert at the work. In a dry, sandy, or gravelly moorland, with a cover of open heath, not exceeding five or six inches in height, this mode of planting is not only the most speedy, but also the most successful and economical. It is not practised with plants of a great size ; but it is adapted for Scotch pines and spruce two-year seedlings, or for such one-year transplanted, and for larches, either one or two, year seedlings, or for one-year seedlings, one year transplanted, and for other plants not exceeding the size of these, at the ages stated. In England, the use of the small planting spade or hand-iron is almost altogether unknown; and there the system of notch or slit-planting has been termed *a coarse operation*, because the plant is inserted among the herbage without the soil being prepared or pulverized ; but the cause of its being spoken of thus disparagingly arises from ignorance of the advantages connected with the system. It is necessary here to explain that the herbage with which the plants are associated by this mode of planting is that of the heath, which, when of a moderate size, is far more favourable to the young plants than if they were inserted on a bare prepared surface. A cover of heath affords great protection, while its open stems do not retain moisture to rot the plants, nor do its roots injure them like those of a grassy vegetation. Moorland of the usual description, situated at a considerable altitude, when much disturbed and pulverized, is generally less adapted to the growth of young plants than soil less expensively prepared. This arises not only from the prepared soil and pulverized spots being deprived of a shelter of heath, but also from the circumstance of these places, thus prepared, absorbing an excess of moisture, which causes the ground to swell during frost, and subside in open weather, whereby the plants are ejected. For this reason, the planter of experience, in the progress of notch planting, will diverge from the regular distance if the spot happens to be bare, with a broken surface, to take advantage of the nearest point possessed of a heathy sward, into which he will insert the plant. In the native-

grown forests, it is not in the bare pulverized spots that the
young plants are found; in such they are ejected, and die ;
but, associated with the brown heath, they spring up and
prevail without any artificial aid ; and it is into this description of soil, which nature has adapted for the seed-bed of
pines, that plants are usually inserted by notching.

The first over-stock of plants that I had in the nurseries
at Forres occurred in the year 1830. During that year I
planted, by contract, about two millions of plants, which were
all inserted into moorland by the hand-iron. The largest ot
these plantations was made on the estate of Ballindalloch ; it
measured about 400 Scotch acres, and contained 1,400,000
native Scotch pine and larch plants. They were notched
into the. ground at the average distance of about four feet
asunder. The plants were distributed in kind and size to
suit the quality of the ground and the state of the herbage ;
and on an average each Scotch acre contained, as under, at
the following prices, for plants, carriage, and planting :—

	£		
500 one-year transplanted larches, . .	£0	1	9
1500 two-year seedling do., . .	0	3	0
500 one-year transplanted Scotch firs,	0	0	9
1000 two-year seedling do. .	0	1	0
Carriage of plants to the moor, . .	0	1	2
Expense of planting 3500,	0	2	4
Total expense per Scotch acre,	£0	10	0

This plantation was well fenced, partly with stone, and partly
with turf dikes, and as no drains had been formed, I had
liberty in planting to avoid all spots too wet for the growth
of timber; these, however, were of no great extent. The
soil has generally a good slope or declivity, and ranges from
200 to 550 feet above the river Spey, which runs near to it.
The ground is probably from 350 to 750 feet above the
sea level, and about 25 miles distant therefrom. This plantation was begun during the last week in February, and the
weather being altogether favourable it was completed in

thirty-four days. Its formation was reported in the Prize
Essays and Transactions of the Highland Society of Scotland
for 1832, and being the first plantation, and consequently the
oldest formed by me, I inspected it lately, with the view of
ascertaining its progress and value as timber in thirty-five
years. I was accompanied by two practical men, the one a
forester in the constant practice of selling timber, and the
other a timber merchant; both well acquainted with the
purposes for which such timber is applicable, and with the
present state of the timber market.

I found the plantation generally in a very vigorous state,
and containing many remarkable specimens of rapid growth.
Near the base of the plantation, and where judicious thinning
had been practised, the best of the larches stood from fifty to
upwards of sixty feet high, and from four to five feet in girth
at the ground, and from three feet six inches to four feet two
inches at the height of six feet. These contained from twenty
to twenty-five cubic feet of timber, and were worth not less
than 20s. each tree. Some of the best parts in the wood were
worth £80 per acre; and altogether the value of the timber
covering 400 Scotch acres, equal to about 500 imperial acres,
at a low estimate amounted to £31,600.

The native Highland pine presented some very fine specimens of thirty-five years' growth, containing from fourteen to
sixteen cubic feet of timber. Every year after this age a
plantation, when carefully thinned, makes a great accession
to the bulk and value of the timber.

As there has been a succession of several proprietors,
factors, and foresters on the estate since the date of the formation of the plantation, I am unable to give any correct detail
of the thinnings obtained from it up to the present time,
further than that the free revenue must have been quite sufficient to have paid the cost of its formation, with interest, and
a rent for the ground many times greater than that obtained
for the best description of hill pasture. It had formerly been
only common heath or moorland.

During the next twenty or thirty years the repeated thin-

nings of this plantation cannot fail to yield many thousands
of pounds, with every prospect of the standing timber at the '
end of either of these periods being worth more than double
its present estimated value.

At the time of the formation of this plantation the ordinary
transit for timber to market in the quarter was by floating in
rafts on the river Spey and shipping at Speymouth. Mountain streams everywhere abound, and afford convenient power
for the manufacture of timber by saw-mills.

Other means of transit have lately been established. In
this remote district the Strathspey railway runs close to
'the south side of the plantation, and cries aloud for traffic,
which no doubt enhances the value of timber in the
district.

The other plantations which I formed in the autumn or
winter of 1830 are equally successful. They stand on ground
of a lower altitude ; but as it was very bare and exposed, and
as a considerable portion of the heath had been burned off two
or three years before being planted, and as the soil consists
generally of a sandy peat on a dry subsoil of sand and gravel,
the smallest description of plants was employed, and the circumstances rendered close planting indispensable. The following was generally the number of plants inserted at about
three feet asunder :—

2000 one-year seedling larch,	.	.	£0	2	6
1000 two-year do. do.,	.	.	0	2	0
3000 two-year native Scotch firs,	.	.	0	3	0
Expense of planting 6000,	.	.	0	3	0
Total expense per Scotch acre,	.		£0	10	6

The above was the contracted price per acre ; all plants for
replanting, if such should be required, were to be supplied ;
but the expense of work in filling up was to be defrayed by
the proprietor. The ground being well adapted for the
plants, no failure was perceptible, except along a small space
of grassy vegetation which required additional drainage, and

for which a few thousand plants were furnished. A few years after the formation of these plantations, the one-year seedling larch plants which had been employed could not be distinguished from those inserted at the age of two years.

Although one-year-old seedling larches only range from four to seven inches in height, yet when they are transplanted into favourable moorland, they often double their height during the first year of their transplanted growth. No seedling takes more readily to the ground or overcomes the check of transplantation with greater facility.

Before finishing my remarks on notch planting with the small spade, or hand-iron, it is necessary to state that this mode is the most suitable only for moorland where the surface is bare, and the heath not much above the height of the plants. If the herbage is grassy, or composed of other vegetation than heath, stronger plants are necessary than those adapted for being inserted by this implement; or if the surface soil is composed of moss, or pure peat earth, of a greater depth than the small spade can easily strike through, it should not be employed. Pure peat is commonly found the prevailing surface soil of moorland at a high altitude, and being destitute of a sufficient mixture of sand, the plants do not grow freely until they reach the subsoil. For this reason, the quality of the surface soil, as well as that of the vegetation, should be taken into account in deciding on the mode of planting, and on the plants to be employed.

On rocky ground, where it is difficult to obtain a cover of wood, the hand-iron is often found most suitable; it enters the crevices and makes way for the insertion of young plants, where larger tools would not penetrate.

In such places the seeds of pines are sometimes sown, but one-year seedling plants notched are generally most successful. This implement is also to be preferred in the formation of plantations on bare sandy links.—(See SEA-SIDE PLANTING.)

Notch planting by the common spade.—As already noticed, a well-worn common garden spade is far better adapted for the work than any other; being lighter, more wieldy, and being worn to a circular form in the mouth, it cuts the herbage and penetrates easily into the ground. If spades of this description are not at hand, new ones of a small size should be formed into the same figure and sharpened for the work. This spade is suitable in all rough ground where the heath is rank and forms a close cover ; or where heath is mixed with the mosses or other herbage, where the hand-iron is found too light for the operation. This mode of notching plants with the common spade is generally known as the "cross cut," the shape of the incision being made thus— The surface soil being raised, the plant is inserted into the corner and the turf pressed down with the foot. The work is most speedily performed if a boy accompanies the spadesman, carrying the plants and inserting them. Notch planting with the common spade is sometimes practised by making only a single notch or incision, into which the plant is often inserted by the person who makes it ; in this case the planter carries the plants in a bag or apron tied round his waist.

The plants must be large in proportion to the herbage they have to contend with. Transplanted plants should always be used. Two-year-old seedlings which have been one year transplanted of larch, and Scotch pine, or two-year transplanted spruce, are all fit for being notched with the common spade where herbage of the sort referred to overspreads a soft soil ; but the preparation necessary for each spot where a plant is to be inserted depends on the nature of the ground and on the vigour of the herbage, as will be afterwards explained.

Moorland Plantations, cost, etc.—In forming plantations on this description of ground, in elevated situations, Scotch pine, larch, and spruce are the kinds most profitably grown. If the exposure is rough and bare, 5000 plants are sometimes judiciously employed per imperial acre, although that number places the plants rather less than three feet apart. If, on the other hand, the ground is such as is reckoned sure for the plants to take root in, and the altitude low, or, if at a high altitude, the surrounding hills afford a good shelter, 3000 plants are sufficient ; but unless the shelter is ample, the outsides should be made closer than the interior of the woods. The proportions of the kinds depend on the quality of the soil, but that most common is two-thirds of Scotch pine and one-third of larch. With these regularly mixed the plantation may be thinned out, so that ultimately it may consist either of a mixture, or of any one of these kinds, as may, from the appearance of the trees when somewhat advanced, be considered most profitable. Where the larch thrives, it is the most profitable of any tree ; but it is subject to many casualties, and forms too precarious a crop to consist purely of the genus, except on the surest ground,—on the slopes of ravines, along the alluvial banks of rivers, and in all situations where the soil, though somewhat moist, is open and free from stagnant water. The same description of soil is also adapted for the spruce fir, but the spruce fir is not suitable for so high an altitude as the larch. Spruce grows well on a level surface, and being less subject to disease, the tree may be formed into masses by itself on favourable soil, where it may be allowed to stand at a closeness which would be ruinous to larch. Although the spruce is often planted on dry soil as shelter to hardwood, etc., yet it never becomes valuable timber in ground where its roots are undermined by drought. Respecting the Scotch pine, there is no situation, however dry, where it will not sustain itself, provided it has taken root. It accommodates itself to a greater variety of soil than any other tree, and is therefore employed in forming a mixture in all plantations on rough exposures, and where there is a doubt of other sorts growing, in consequence of the soil being of inferior quality. The cost of forming plantations varies according to the quality of the plants required for the soil, and their price at the time. Plantations of the cheapest description are formed by notching on good moorland, with a herbage of heath ; a common proportion and price per acre are as follows :—

3000 two-year seedling native Scotch pine, at 2s. per 1000, ...	£0	6	0
1500 two-year seedling Larch, at 4s. per 1000, ...	0	6	0
Planting with the hand-iron, . . .	0	4	0
	£0	16	0

In sheltered situations, where close planting is not necessary, the expense is 4s. or 5s. less per acre; 3000 plants being commonly used.

If the soil is a grassy heath, with a close vegetation, stouter plants are generally more suitable; such are also more suitable where the surface vegetation is heath, but where the soil is a pure moss to the depth of a few inches, or where the heath cover is rank or deep, in any of these circumstances the plants and cost per acre generally stand thus :—

2000 two-year seedlings, one year transplanted, native Scotch pine, at 4s. 6d. per 1000, .	£0	9	0
1000 two-year seedlings, one year transplanted, Larch, ...	0	6	0
Planting with the hand-iron, . . .	0	4	0
	£0	19	0

It is customary in the Highlands to burn off very strong heath two or three years before the ground is planted; because, *when the surface soil is favourable*, a short heath admits of seedlings being used, instead of transplanted plants, which is generally a saving of 50 per cent. on plants and planting; but where the surface soil is a pure peat, or mossy, the transplanted plants are the surest, and ultimately the cheapest. The advantage of transplanted plants on a mossy surface is, that being stronger and possessed of more rigidity of fibre, they are not so apt to rot by the constant moisture retained by this description of soil; and when inserted they come in contact with and more speedily reach the sandy subsoil, so necessary to their vigorous growth.

It is only in well-sheltered ground that burning is to be

recommended, and under any circumstance it is unfavourable for plants to be inserted immediately after burning. The play of the wind on a bare surface disturbs the plants, and the absence of all other vegetation exposes them to the depredations of vermin and casualties which do not assail them when surrounded by young vegetation.

Throughout the north of Scotland two-year-old seedling Scotch pines and larches are planted at once into moorland, to nearly as great an extent as transplanted plants; and the expense of forming plantations of this sort is generally considerably under £1 sterling per acre, including the cost of draining, fencing, etc. Where the cost is above that amount, it is commonly incurred in consequence of the plantation being of a small extent; the expense of enclosing in that case often exceeds that of plants and planting.

Of the last 1000 acres of moorland which I planted, in four or five different lots, about a fifth part consisted of one-year transplanted plants. The average cost of this extent did not exceed 15s. per acre for plants and planting, with all necessary upmaking. About one-third of the whole was larch, and the others were native Highland pine, with a small proportion of Norway spruce; and the average number per acre was 3500, which placed the plants about three and a half feet asunder. The closest planting was formed on the outsides, on hill-tops, and on bleak exposures, also along moist grassy patches, overspread with rank herbage; the progress of firs on such being always marked by a slowness of growth, until they form a cover and suppress the surface vegetation, when their effect tends to dry the ground, and convert its energies purely to the growth of wood.

In very exposed and bare ground, the description of plant most apt to succeed is Scotch pine, one year old, and one year transplanted—a plant only two years old, but transplanted into nursery lines at the age of one. This description of plant is most tenacious of life, and seldom fails; yet it is not in ordinary demand; it is sold commonly about 3s. to 4s. per 1000.

The largest description of plants that are inserted by notch planting, and fit for rough ground, with rank herbage, where the soil is soft and adapted for being *notched with the common garden spade*, generally costs, per acre, including planting, as follows :—

2000 two-year transplanted native Scotch pine,	£0	16	0
1000 two-year seedling, one year transplanted, larch, ...	0	7	0
Notch planting with the common spade, .	0	7	0
	£1	10	0

The expense of notch planting varies much, and depends on the quality of the ground, and the roughness or smoothness of the herbage. In soft smooth heath soil, a person practised to the hand-iron will insert from 3000 to 4000 plants in a day of ten hours; and the expense of labour only, in notching plants with this implement, seldom exceeds 3s. or 4s. per imperial acre. Planting with the common spade, either by th single notch or cross-cut system, is generally double the price of planting by the hand-iron. In other cases, where the herbage has to be cleared off, or where the ground is disturbed by a pick, or the spots dug over by the spade before the plant is inserted, the expense is still greater, in proportion to the state of the soil, the labour bestowed, and the rate of wages obtained by workmen in the locality.

It is usual to look over plantations of this description after the second or third year of their growth, and to make up all marked deficiencies. It is not necessary to fill up every spot where a dead plant is found; but all such places as are not possessed of a sufficient number to furnish the ground, so that they will readily form a cover on the surface, should be replanted.

VIII.

ON PIT PLANTING AND PLANTING ON PREPARED GROUND.

Pit planting.—This is the mode generally resorted to in planting the various kinds of two-year transplanted hardwood trees, where the soil is generally deeper, richer, and better sheltered than that usually planted with pines. The more capacious the pits are formed, the roots of the plants can be the better spread, and have the greater extent of loose soil, —circumstances necessary for their early growth. Pits are usually made eighteen inches wide and fifteen inches deep; and that size is usually contracted for at 1s. to 1s. 6d. per 100, according to the nature of the ground, etc. In hard soil it is usual to disturb the bottom of the pit with a pick beyond the above depth. When plants beyond the age of two-year transplanted are employed, the pits must be made large in proportion to the size of the plants. Pits should be made a few months before the plants are inserted; and, in forming them, after the bottom is made quite soft, the surface sward should be filled in first, chopped, and covered up to the depth of five or six inches with the soil, which will be considerably decomposed by the time the plants are to be planted. In case, however, the surface sward is very thick and matted, and the ground dry, and more especially in the case of larch planting, the best method is to lay aside the surface turf, and, at the time of planting, to divide it into two halves; one half, with the surface side under, should then be placed on each side of the plant, after it is planted. It is usual to plant ground moderately dry and sheltered in the beginning of winter; and when the soil is of an opposite

E

description, early in spring, during open weather : these are
the seasons most approved of. The operation of pit planting
is generally performed by a man and a boy; the man takes
out a spadeful or two from the half-filled pits formed in
summer, and the boy inserts the plant, with the roots well
spread ; the earth, which should be pulverized, is then spread
on the fibres, while the boy moves or shakes the plant as the
pit is being filled ; so that the roots are fully spread, and each
individual fibre is surrounded with the soil; after which, the
earth should be pressed down with the foot all around the
plant, which should stand perpendicularly, and about an inch
deeper in the ground than it had formerly stood in the nursery, which point is readily known by the ground mark between the root and the stem.   It is of great use, particularly
in dry soil, to observe, in finishing the upfilling, to leave a
regular concavity around the plant, suitable for retaining
moisture, which is valuable in establishing newly inserted
plants in pits, and is the more necessary on the slopes on hill-
sides, where the outer or under edge of the pit should be
carefully formed sufficiently high to intercept the rain, for the
benefit of the plant.

Prepared ground.—There is no method of preparing ground
which adds more to the vigour of a young plantation of hardwood trees than that of trenching, and when the subsoil is of
inferior quality, it does least mischief to the trees by being
placed to some extent on the surface.   In all plantations for
ornament or for shelter, or wherever large plants are to be
inserted, and rapidity of growth is desired, there is nothing
that will effect the purpose so soon as deep trenching; digging
or ploughing is of much less benefit to trees, as their roots
penetrate to a considerable depth ; it is therefore in the
looseness of the ground underneath that the great advantage
lies.   It is only the expense of trenching for plantations that
prevents it being more generally adopted ; for every plantation (except that of an open heath or moorland, where in
many cases it would be unsuitable for young plants) is benefited by the operation in a very marked degree.   The expense

of trenching is frequently compensated for by taking a crop
of potatoes, carrots, or other roots, during the first season of
the plantation; and even when this method is not resorted
to, trenching in many cases is less costly than it at first appears.   In all prepared ground where ornament and rapidity
of growth are required, it is necessary to keep the surface
clear and free of weeds during the first two or three years
after planting, until the trees form a cover sufficient to prevent surface vegetation; and the labour in doing so is always
much less on trenched ground than on that which has only
been dug or ploughed.   Another advantage arising from
ground that has been well trenched is, that in it trees of a
greater size will more readily take root and become established than in ground prepared in any other way.   But when
plants are of the ordinary size, their growth in six years in
trenched ground is generally equal to that of ten years in any
other description of ground.   Dry, sandy soil is frequently overspread with a closely matted herbage, chiefly composed of the
Fescue and other wild grasses, which, without being trenched,
render any mode of planting very uncertain.   The closeness
of the surface sward intercepts the influence of showers, and
the depth of its fibrous roots is a great hindrance to the
growth of young trees.   Furze not unfrequently forms a close
cover on good soil, well adapted for the growth of timber.
The preparation of such places is most speedily accomplished
by trenching the ground.   By any other method the furze is
apt soon to reappear and interfere with the growth of the
trees.   By trenching, the ground is softened and rendered so
clean that subsequent clearings are seldom required, as the
trees generally advance rapidly, and soon subdue all native
vegetation.   In planting ground from which timber has recently been removed, and in all hard and inhospitable soils,
trenching is to be recommended.   It should be performed in
autumn, and the planting is generally most successful after
the severity of the winter is over.

Plantations consisting chiefly of the broad-leaved trees are generally formed by *pit planting*, and the plants used are commonly

those that are two years transplanted.   In this, as in all other
modes of forest-planting, much depends on the quality of the
plants.   If they have stood longer in the lines than two years
without being removed, and grown well, their roots will be
large and bare, and destitute of that fibrous bushiness so
essential to the success of a newly transplanted tree.   Three
or four years' transplanted trees appear much stouter and
more vigorous than those which remained only two years in
nursery lines ; but the former are apt to die, and rarely grow
freely for a year or two after being transplanted.   Plants for
pits should always be of the best description, and picked in
' the nursery lines ; a considerable number being often produced
more feeble than others, such should be rejected, as every
pitted plant is expected to be more permanent than many of
the plants closely inserted by notch planting.   We have else-
where stated the advantages resulting from hardwood plants
being moved or transplanted in the nursery the season before
being finally planted out, which will generally make them
twice transplanted.   This insures fibrous roots ; consequently
they rarely fail, but readily take to the ground, and contrast
favourably with plants bare in their roots from having been
too long in the lines without being moved.

In forming plantations for profit there are several circumstances that require to be kept in view.   Almost every description of soil will grow several kinds of trees to a
considerable size ; it is therefore important to know the description of timber that.will most readily find a market in the
locality, in order that that sort may be cultivated, whether for
hoops, crates, staves, agricultural or other purposes.   In districts far from a large town, and from water, or other cheap
conveyance, some kinds of timber are of comparatively little
value.   It is therefore necessary to study the cultivation of
the kinds that will best afford a long carriage.   In situations
of this sort it is always necessary to manufacture the timber
into stave-wood, or deal, or such articles as contain the
greatest value in proportion to their weight.   For instance,
wooden bobbins are required in great numbers at all the large

manufacturing towns; in their manufacture there is perhaps a greater waste of timber than in that of any other article in ordinary demand; they are readily manufactured by machinery from any closely grained hardwood, and only weigh about one fourth or one fifth part of the timber from which they are formed; consequently, timber adapted for such purposes is valuable, even though very remotely situated.

In a piece of ground of great extent its quality is often found to vary. A loose deep earth will grow trees of any description; a dry, poor, gravelly, or chalky formation, will suit best for the beech, the birch, and the pines. A clayey soil, or a deep clayey gravel, is generally best adapted for the oak, and the most profitable tree to intersperse with it is the larch. The oak feeds chiefly on the subsoil, and the larch on the surface; the latter, being of upright growth, is not apt to injure its associate. It is usual, in the formation of plantations intended to be chiefly of oak, to begin by planting pines and trees for shelter, and to insert the oak plants after the nurses are a few years advanced; this is a good protection for the safety of the oak plants; but it is only necessary on bare and exposed ground, as the larch and pines advance more rapidly than the oak, and furnish shelter in a very short time, when the whole are inserted at once. Beech is more profitably grown alone than with a mixture. It is apt to become branchy and broad-headed; its timber is only valuable when it yields clean tall trunks, and these are most easily produced when the species stands by itself; interspersed, it is apt to prevail over more valuable sorts, and to become of a branchy and worthless figure. The ash and Scotch elm are trees which yield valuable timber, particularly for the purposes of agriculture; they grow well together, and are almost equally hardy; they require a good, deep, loose soil, and the ash prefers that which has a tendency to moisture. In low situated alluvial soil, with moisture, the *silver fir* acquires a great size; where the soil is very suitable, it is often very profitably grown. When young, it should be interspersed with faster-growing trees, such as larch, willows, etc., as nurses, closely planted; the silver fir

requires shade and confinement for eight or ten years ;
alone, the trees admit of being grown very closely to one another ; as a mixture, they are very ornamental in the forest,
and tend to break the monotonous appearance of deciduous
trees. In planting rich swampy ground too moist for trees
in general, the large growing species of poplar and of the
willow, alder, and ash, should be employed, and for evergreens, Norway spruce, Scotch and silver firs. I have else-
where named the trees best adapted for exposure at a great
altitude, and for a description of the various kinds of soils
adapted to their growth, their quality of timber, etc., and the
reader is referred to the articles on the various genera under
their respective names.

The distance at which broad-leaved trees are planted varies
very much in different districts, and should depend on the
exposure of the ground and the value of young timber in the
neighbourhood.   1500 plants per imperial acre is usual, with
nearly an equal number of larches, firs, etc., to be first thinned
out—placing the plants about four feet apart.   Sometimes only
about 500 are pit planted, being placed nearly ten feet apart,
and the intermediate space should be filled up with sorts less
valuable than the kinds that are intended ultimately to occupy
the soil.   The kinds that are interspersed are generally of
the fastest growth when young, when they serve as nurses to
promote the growth of the more valuable sorts, and prevent
their getting into a bushy form ; and though pruning is not so
necessary in close plantations as in narrow belts and in hedge-
rows, yet many plants are found to require it ; and such as
are pruned early and judiciously will be improved in quality,
and increased in their useful dimensions and ultimate value.

ON PLANTING IN A GRASSY VEGETATION.

LAND overspread with a grassy sward, whatever be the quality of the soil, is more hostile to the growth of young plants than that overspread with a cover of heath. This arises from the roots of a grassy vegetation depriving the plants of nourishment, and pressing more unfavourably on them than when associated with heath. In such places notched plants have a very poor chance of success, unless the ground is previously prepared; therefore, in the absence of trenching or pitting, each spot should be prepared for the plant by having a turf, or surface sward, about a foot square removed. If the soil is hard, it should be opened up by digging over a spadeful on each spot laid bare, or it may be loosened by a tramp pick. This preparation of the ground should take place in the autumn. If the surface herbage is not very close and matted, the ground may be at once prepared by digging over a spadeful where each plant is required, burying the surface; but if the sward is *bulky*, or the soil dry and sandy, such as will not readily decompose the surface vegetation, a turf should be pared off and laid aside, divided and placed on the top, with the grassy side downwards, after the plants are inserted. When ground is of good quality and not hard, the removal of the grassy surface is sufficient preparation. If the soil is dry, the planting may be performed early in winter; if strong or moist, the prepared ground should be exposed to the frosts of winter, and planted in spring, by notching with the common spade.

A similar preparation is required where the surface soil is composed of pure peat too deep for the plants to reach the subsoil. In such, the Scotch pine and Norway spruce are most suitable; the subsoil in such places is often at the depth of six or eight inches, and composed of a sandy gravel or clay, a mixture of which, by being brought up to come in contact with the roots of the plants is of much advantage, particularly for larch; and if the spots are prepared in autumn, and the plants inserted in spring, their success is commonly certain.

The only method of bringing up a proper mixture of soil in some parts is by the operation of the common spade. In some ground, however, the pure peaty surface is much shallower, and a stroke or two with a mattock or tramp pick will penetrate into the subsoil, and form a mixture of soil more speedily. Where the cost of preparing the ground is so great, the plants should always be of the best description, picked from those that are transplanted, and of a size sufficient to cope with the herbage.

The following is the usual number of plants for an imperial acre, and their prices :—

1500 native Scotch pine, two years transplanted, at 10s., ...	£0 15 0
1000 larch picked two-year seedling, one year transplanted, ...	0 8 0
500 Norway spruce, two-year seedling, two years transplanted, at 12s., . . .	0 6 0
	£1 9 0

The price of the labour per acre is often equal to that of the plants; all depending on the nature of the herbage, soil, and rate of wages in the district.

On planting ground where timber has recently been felled.— The formation of healthy plantations on ground that has lately been cleared of a close crop of timber is perhaps the most difficult of any in ordinary practice, particularly if the timber recently removed has been of the pine tribe. In course of time the land improves for plants as the roots and exuviæ of the old wood decay.   I have frequently observed where timber had stood very close, that eight to twelve years sometimes elapse before the soil becomes fit for the growth of young plants by notching, without any more costly prepara-tion.

Where the trees of the former plantation, however, stood thin, which is often the case before the whole are removed, spots are found here and there which admit of the growth of young plants at once; and it would have been a great improvement on any other method had such places in ordi-nary good soil, not too dry, been planted with silver fir before the complete removal of the timber.   This tree has every recommendation as a succeeding crop to larch and Scotch pine, and shade and shelter are of advantage to it in early life.

The surface soil of old fir woods is generally overspread with a considerable depth of half decomposed vegetable sub-stance uncongenial to the growth of plants.   Burning is some-times resorted to as a means of clearing the spongy surface, and as a preventive against the ravages of the wood-beetle and other insects which infest such places.   But burning is not always successful; and although it often makes a more favourable surface for young plants if it is left for a few years till a new vegetation arises, yet it is often impracticable from the nature of the ground, and from its already con-taining some plants which it may be desirable to retain. Trenching over the ground and removing the old roots is the surest method, but the expense of the operation generally prevents that from being practised on a large scale.   Ground that has produced a crop of timber is always hard and close under the vegetable remains, and excludes the influences of the atmosphere to such a degree that, unless it is well dis-turbed, it remains long uncongenial to the profitable growth of young plants.   The birch and the willow most readily spring up on the surface remains, as their seeds are adapted for being drifted to great distances.   The mountain ash also not unfrequently appears, the seeds having been carried to the old plantation by birds, and taken root from their droppings; and the plant being tenacious of life, exists through long confinement, and appears on the removal of the old trees. All these, however, are generally unprofitable; although they often form a cover on the surface, they grow much less vigorously than if the soil had been disturbed and softened.

The mode of procedure, therefore, in such ground, is to prepare the spots for the plants a considerable time, say six or twelve months, before planting; and four, five, or six feet asunder are the usual distances, according to the exposure or 'shelter for the plants. The surface accumulation of bark, leaves, and foggage should be cleared off, not to be used, and the solid soil disturbed with a pick-axe to at least the width of fourteen inches, choosing the spots as free from roots as possible; cut and clear out all such met within the bounds to be prepared, while a second person following with a tramp pick should disturb the bottom to the depth of at least twelve inches. If the old trees have stood close, it is usual for the spots, after having been cleared of the surface, and all old roots picked out, to be rather under the ordinary level of the ground, therefore a third person should follow and dig a stamp or two of pure soil, as free from the remains of the old wood as possible, and place such on each spot where the plants are to stand. This gives the plants a great advantage in soil; it raises them rather above the level of the surface, and forms a position in which pines are found to be exempt from the attacks of the wood-beetles and of field-mice. The hole also left open, which furnishes the soil, in the vicinity of the plant, has an ameliorating influence, by admitting air to the ground, and thus preparing it for the spread of the young roots. This process has all the effect of pit planting, and is less expensive, and generally as successful for firs and larches.

When oak, elm, or other hardwood plants are to be inserted after old wood, trenching or pit planting should be practised, and the made pits should be exposed to the weather for several months. Where there is rank jungle, or such vege-

tation as would confine or suppress the strongest description of one-year transplanted plants of pines, larches, etc.,
stronger plants must be employed, and pit planting should be
adopted; and in the inserting of larch plants, in all cases
where trees have formerly stood, it is to be recommended that
no root-pruning should be practised. The plants should be
full of fibrous roots, and carefully removed from the nursery
grounds; and in all cases this is most easily accomplished by
employing plants that have stood one year only, after having
been previously transplanted.

The pits should be formed large in proportion to the size
of the plants, and their roots should be well spread. I have
observed plantations of larch that had been inserted quite
young into such situations more exempt from the disease
known as heart-rot, or pumping, than plantations formed
with older and stronger plants. The cut or mutilated root
fibre of a larch, or any other plant-root, will imbibe any
watery substance whatever that it comes in contact with;
whereas a sound root has its fibres terminated by healthy
spongioles, having the power of selecting, at least to a great
extent, those elements congenial to the vigorous development
of the tree. Every precaution, therefore, should be taken in
replanting land liable to produce fungi, particularly where the
larch, deodar, and such other plants as have an affinity for,
and are readily destroyed by mycelia, are employed.

Ground that has yielded a heavy crop of timber is considered by many planters unfit for yielding another of the
same species. Its density and cover of matted roots is much
against the growth of any kind until the roots are decomposed,
or the soil broken up and prepared; after which, Scotch pine
is found to grow after the same as well, and often much
better, than most other trees.

The soil that is congenial to the growth of a tree continues
so, provided the influences of the atmosphere are not excluded.
In the case of other species naturally springing after pine
forests, that arises in consequence of a new formation from the
exuviæ which the pines deposit.

The oak, after yielding a heavy crop of timber, springs up from well-dressed stools far stronger than in the best plantations newly formed on virgin soil.

After a crop of any sort of pine timber is cleared off few aspects are more bare and unsightly. This suggests the propriety of interspersing such with a few oak plants, although at the distance of thirty feet apart, for which forty-eight plants only are required per acre; and although these are scarcely discernible during the vigour of the pines, yet, being tenacious of life, they keep the ground, become deep rooted, and are ready to spring up on the removal of the crop of timber. In such cases, or in the case of fire destroying a plantation, the oak acts a conspicuous part, in speedily renewing the appearance, and ultimately becoming valuable.

In all cases where the soil is ordinarily moist and good, few plants are of more profit for timber at present than the grey poplar, *P. canescens*, the Italian poplar, *P. monilifera*, and the best varieties of tree willow. The timber of all these is of rapid growth, and much sought after in the formation of railway carriages, brakes, etc.

The time of inserting plants in the prepared spots by a notch with the common spade, or by common pit planting in such ground, will depend on the nature of the soil; if dry, the planting should be performed so as that they may have the advantage of the winter moisture; if the ground is possessed of ordinary moisture, early spring should be preferred.

X.

PLANTATIONS ON BOG OR PEAT SOIL.

THERE exists a very great difference in the quality of bog or peat soil. Some sorts are much more congenial to the growth of trees than others. The prevailing cause of the sterility is generally the excess of moisture which it contains, and the quality of the soil is often greatly influenced by the nature of the springs of water with which it is submerged. In bog or moss there is also a great difference, dependent on its composition. That which is pure, free of sand, and possessed of little inorganic matter, is least adapted for the growth of wood, and it not unfrequently occurs that the barren soil is accompanied with a bare and bleak exposure. Here we have a combination of hostile elements, above and below ground, which it requires plants the most tenacious of life to withstand.

Before forming a plantation on moss land it should be completely drained. This operation is often required to be made a year or two before the plants are inserted. Open ditches are the best and cheapest for the purpose, and, if possible, they should be made down into the hard subsoil. The closer they are made to one another they will be the more effectual. I have drained moss with open ditches ten yards asunder, and on an average eight or nine feet deep, when in less than two years the soil collapsed fully three feet, and became well adapted for plantation. Where moss is much shallower, and the ditches can be easily formed into the subsoil, that material brought up forms a valuable mixture for the growth of plants in general after it has been exposed to the influence of the weather. After bog land has been drained, when it is very destitute of inorganic substance, its quality is often improved by being burned, provided there is depth of soil to spare above the rise of water. After being burned it should rest a few months at least before being planted.

Pines of every species grow better in mossy soil than most other trees. The native Scotch pine should be preferred; the Austrian and Corsican are also suitable, but the latter is least so in the absence of shelter, and notwithstanding the partiality of the pinaster to pure sand, it also grows well in boggy soil, and is found valuable for shelter in a rough exposure. *Pinus montana*, a native of a high elevation on the Alps and other mountains, is also suitable for shelter along the outskirts of plantations in rough exposures, where it forms a compact bush, very tenacious of life. In soil of this description the other evergreen trees to be recommended are the Norway spruce, Douglas's spruce, and silver fir; in such the last named often becomes valuable timber, though after planting it requires shade and shelter from other sorts for a few years till it takes root, when it often grows much faster than other trees, and admits of standing very close. Although but little is yet known respecting the growth of the Wellingtonia in this country, yet it is ascertained to grow freely when young in mossy ground, and with an excess of moisture that would injure many trees.

Among deciduous trees, all the willows grow in mossy ground, and are often turned to good account by being cultivated and cut yearly for basket-making, and at a more advanced period for crate-wood, hurdles, etc. The best timber-tree kinds are *Salix caprea, S. alba,* and *S. Russelliana.* Among poplars, *P. canescens* and *P. monilifera* are most suitable. The common birch and alder generally succeed. Where the moss is very pure the ash only attains to a small size, but in this soil it is often valuable when cut, and cultivated as coppice wood.

Ground overspread with furze or whin, or with other rough herbage.—Where whins prevail, the usual mode of procedure is either to contract for having the ground trenched or the

whins rooted, or grubbed up four or six inches under the surface, consumed, or cleared off the ground ; the cost of these
operations varies much, according to the cover and quality of
the ground.   Where the cover is close, it generally ranges
from 30s. to 50s. per acre for grubbing, etc.

Trenching is generally much more expensive, but the ground
is at once prepared and deeply softened for the insertion of
the plants, and the whins are generally longer in overtaking
and interfering with their growth ; in all such cases the planting, whether by pit-planting after grubbing, or notching after
trenching, should be with plants of the largest size that are
likely to grow freely, and they should be inserted closely, in
order to form a cover, and subdue the native vegetation, which
is sure in such places to rise with vigour from seed, if not
from roots in the ground.   The larch and the Scotch pine are
generally the first to form a close cover in such places ; and
where these sorts are not intended to form the ultimate crop,
their services are generally useful in subduing the vegetation
—particularly the larch,—although it should be cut out at an
early period to make way for other sorts.   Two years' transplanted larches from two years' seedlings are usually about
thirty-six inches high, and I have never seen larches of
a greater age (that is, four years old) and larger size grow
better when inserted into plantations ; nor Scotch pine beyond that age (four years) when used in the north of Scotland.

On planting exposed and barren ground at a great altitude.—
Failures very often occur in establishing plantations in such
situations.   The difficulty is commonly felt in getting a commencement, or a shelter established to form a screen to future
accessions.   In all such places, where shelter is required for
residences, garden, or agricultural grounds, the soil should be
well drained and trenched, so that every advantage may be
afforded to the successful growth of the trees ; indeed, in some
cases, in order to get the trees to make an early and vigorous
start, screen fences should be employed, composed of the
thinnings of Scotch fir plantations, turf dykes, or the like.

The following kinds are the first to make an appearance and a rapid growth in the prevailing winds of a high altitude :—

Hoary Poplar, Goat Willow,	Wild Cherry,
Trembling Poplar,	Mountain Ash,
Sycamore,	Service Tree,
Weeping Birch,	Alder,

among which should be interspersed the native Scotch pine, with its varieties Montana, Mugho, Pinus cembra, and Austriaca.

I have known several residences in the Highlands where it was difficult to establish shelter.   At all of them one or other of the trees named was esteemed as the hardiest; but perhaps at none of them were all the sorts experimented on or tried. For instance, I lately visited a summer residence or lodge, situated at an altitude of 1200 feet.   Here the mountain ash grew wild in great vigour; the largest trees in the neighbourhood consisted of this species, and their round heads standing thirty feet in height were composed of a thicket of branches giving no indication of the prevailing wind.   Their berries in autumn are very ornamental, and attract singing-birds, and here this tree was pronounced to be the "sheet-anchor," or chief reliance for the protection of all other vegetable productions.

In the planting of bare moorland at a great altitude, I have found no plant superior to the native Highland Scotch pine. Its success depends greatly on the quality of plants employed. In no case of rough exposures should plants be used that have stood more than one year in the nursery lines after being transplanted.   Where the heath is quite short, one-year seedling plants transplanted into nursery lines for one year is the most reliable sort.   If the heath is too rank for such, these plants should be transplanted a second time into nursery lines; and in all cases for such exposures, the plants in the nursery lines should have the advantage of plenty of room in well-exposed nursery ground.   The same treatment should be bestowed on larch and all the other pines intended for such exposures.

XI.

ON THINNING PLANTATIONS.

THERE is no department in arboriculture more misunderstood or neglected than that of thinning plantations, particularly those of the Scotch pine and larch, throughout many parts of Scotland. No doubt the native forests are left, for the most part, to thin themselves by their own efforts; by the provisions of nature, the timber is preserved in health, and exempt from suffocation or injurious confinement; but this is never the case with respect to plantations. The native and planted forests stand under very different circumstances. These will be readily understood by my extracting the following article, which I sent to Mr. Loudon at the time he was writing the *Arboretum Britannicum.*

In vol. iv. p. 2181, Mr. Loudon says :—" After perusing Mr. Grigor's report on the native pine forests of Scotland, of which an abstract is given in p. 2165, we wrote to him for information on the subjects of thinning and pruning, as actually practised in these forests, and also in artificial plantations; and as to the effects of the neglect of either or both of these operations. To our application Mr. Grigor kindly and promptly sent us the following answer :—

" 'The old trees of the native Scotch pine forests have trunks quite clean and free from old stumps, so that the side-branches must have rotted off when the trees were young and of a small size. Some of the pines, grown on exposed situations, have strong side-branches, but not very near the ground ; such branches are commonly found above large clean trunks of from fifteen feet to thirty feet in length. When the timber of these forests is cut up, loose knots are rarely

F

met with ; indeed, knots of any importance are seldom seen, except where such were attached to live branches at the time the trees were felled. The wood of the old trees appears so clean and equal when sawn up, that in many, only very slight marks of lateral branches are visible. The young trees of from twenty-five to forty years' growth present regular tiers of decayed branches near the ground, which fall away in course of time.

"'The proprietors of the native forests sometimes prune and thin the woods, but not often ; they thin when the trees are much crowded, and of nearly an equal size, especially when situated near a road or river, where timber is of most value ; but this is not attended to in the more remote parts of the forests. I have only seen the trees pruned when they stand quite thin, or from having lost their leading shoots, by sheep pasturing the ground, or other casualty, have become bushy. In this case I have seen a considerable extent gone over in January and February, and pruned to the height of from two feet to four feet with the axe ; the whole height of the trees being from five feet to ten feet. In the Highland natural forests the young plants do not often rise of equal strength and size. There is commonly a portion of them (a sufficient crop) stout enough to overtop the smaller ones ; and the latter are of much benefit in preventing the side-branches of the former from advancing to a large size. The side-branches of the true Highland pine naturally take a wide or horizontal direction, whereby they are more subject to decay by the closeness of the trees than if they inclined to a more perpendicular figure, as do our low-country pines. In planted woods, the pine-trees are commonly of the same size and age ; and then it is absolutely necessary to thin them as their tops rise equal, and form a surface parallel to that of the ground on which they stand ; therefore, without relief by thinning, the whole are to a certain extent injured ; whereas in natural forests the difference of sizes and ages is great, and the strongest prevail unhurt. I am acquainted with many artificial plantations of pines ; and the common method is, to thin the trees gradually as they get too close or too high for their girth. Planted pines are not commonly pruned, that being considered the worst mode of treatment. Many proprietors of late have given over thinning; but the woods are much hurt by being too much confined : a good tree can scarcely be seen, except near the outside, or where a road opens up and admits air. I am clearly of opinion that we shall not have good pine plantations until they are produced from the seeds of the native Highland forests, which are more healthy and permanent than the kind commonly cultivated.'" 9th September 1837, Loudon adds, " The Earl of Aberdeen ; Macpherson Grant of Ballindalloch; Mr. George Saunders, gardener 'and forester to the Duke of Richmond at Gordon Castle; Mr. Roy, nurseryman, Aberdeen ; and other proprietors and gardeners of the north, have sent us answers to all our queries on the subject of thinning and pruning, which correspond with those given above by Mr. Grigor."

The natural provisions to prevent pines from rising close to one another, and of equal strength in the indigenous state, are very marked and interesting. The seeds on being shed are provided with wings adapted to carry and spread them throughout a wide district. They are never shed but in a warm and dry day, and as soon as the seed receives the slightest moisture it separates from the wing and travels no farther. Of such as fall into the soil a small proportion immediately vegetate; those that do not reach the soil, but are retained on the surface vegetation, may be picked up by birds, or are washed into the ground by the influence of the weather, and vegetate during some succeeding spring, for, under certain circumstances, the seeds of Scotch fir retain their vitality for several years even after they are separated from the cone. This is ascertained from seedlings appearing of various ages, years after being sown on moorland where no plants or seeds had previously existed. The diversity of situation also produces a great difference in the strength and size of the plants; but even in the best prepared nursery-bed, where the seeds are all sown at the same time, and vegetate together, the plants are always unequal in size and strength. This arises from the unequal strength of the seed; not only in different cones, but in the same cone, some are far more strong and robust than others, and are produced in the centre—those at both the base and the apex are comparatively feeble. All these provisions of nature appear to be framed to prevent trees of the same size and strength from pressing on each other, as in the ruinous struggle too often witnessed in planted woods.

The pine and the larch, if once deprived of their lateral branches, have not the power of ever replacing them; hence the safeguards established by nature differ from those relating to trees in general, which have the power to furnish ample top and side branches to increase their diameter after losing their just proportions. It is therefore an absurd theory which advocates the practice of not thinning pine plantations on the plea that the native forests are not thinned, and yet arrive at maturity, yielding the finest timber. The destruction of the tree by overcrowding had come under the notice of Gilpin in *Forest Scenery*, by Lauder, vol. i. p. 173. That writer makes the following interesting remarks :—" All trees indeed, crowded together, naturally rise in perpendicular stems; but the fir has this peculiar disadvantage, that its lateral branches, once injured, never shoot again. A grove of crowded saplings, elms, beeches, or almost of any deciduous trees, when thinned, will throw out new lateral branches, and in time recover a state.of beauty ; but if the education of the fir has been neglected, he is lost for ever."

Few descriptions of timber are more profitable or more readily sold than that adapted for railway sleepers, particularly larch timber; but without sufficient thinning it cannot be readily provided.

I have just returned from inspecting a plantation for which I furnished a superintendent and plants thirty-two years ago. I recollected a portion of the ground being apparently well adapted for larch, and, at my suggestion, a space of about twenty acres, composed of a hazelly loam, with a subsoil of gravelly clay, having been planted almost entirely with this tree. This part of the ground had a slope to the channel of a mountain stream, and it had a cover of heath from six to eight inches. The plants employed were—

2000 larch, one-year seedling, one year transplanted, ...	£0	10	0
700 native Highland pine, two-year seedling, one year transplanted, ...	0	2	6
Planting with the hand-iron, . . .	0	3	0
Plants and planting per acre, .	£0	15	6

The other parts of the ground were also planted four feet apart, with two-thirds of native Highland pine and one-third of larch; and each kind was varied in such a manner that the sort reckoned most suitable prevailed here and there throughout about 180 acres, which comprehended the whole plantation; a narrow track was left in planting as a road, but being only nine or ten feet wide, it had been filled up with the lateral branches of the trees, and being of no use, was abandoned. I found the wood far too close, and although thinning to some extent had been practised throughout the whole, scarcely a well-proportioned tree was to be found except at the outsides, or where a thinning—in consequence of the failure of a few plants by some accident—admitted air. I expected to be able to record some remarkable specimens of larch, but I was disappointed to find it an almost impenetrable thicket of tall trees, displaying a grey mass of dead twiggy branches near the surface of the ground, surmounted by a green and waving foliage some thirty or forty feet overhead. There each acre contains from 1000 to 1200 trees, and the average value of each is about 1s. or 1s. 3d. If the twenty acres were sold in one lot, it would readily fetch £50 per imperial acre.

This plantation of larch has the appearance of having been only once thinned, and that about eight or ten years ago. I cannot say what revenue the thinnings yielded; but had the wood been thinned early and judiciously, and other 500 or 600 trees been removed per acre, they no doubt would have brought £10 per acre free, even if the thinning had been begun early, when the wood was only fit for sheep-flakes, and of little value. By this means only about 500 trees per acre would have now remained, which would no doubt have been worth on an average 4s. or 5s. each; thus the plantation could not have failed to be worth double its present value. This, however, does not represent half the loss of the mismanagement. The 500 trees would have been in vigour and ready for another thinning—only in the morning of their life, possessed of that justness of proportion indispensable to profitable growth, or the speedy formation of cubical contents. In a soil so congenial, trees manifesting such vigour after the age of thirty-two years require again early and repeated thinnings, and the revenue per acre up to the age of sixty might be expected to be not less than £150, leaving in each acre, at that age, about 150 trees, worth on an average £2 each, or £300 per acre. After this period the mode of procedure should of course depend on the capability of the soil as shown by the vigour of the trees. If they still continued to increase vigorously in size, they should have ample space for the development of their foliage, which would likely give occasion for another thinning. About this time, a judicious mode of procedure would be to insert into all the greater vacancies, in well-prepared pits, plants for a succession, such as silver fir, oak, deodar, Douglas fir, Wellingtonia, or such as stand in need of shade and shelter in the first stage of their growth. But I must now return to the plantation of thirty-two years old as it really stands, and to the twenty acres of larch in particular; for although in many parts of the wood the trees stand too close, and are likely to get too tall for their girth, they have in general been pretty well thinned, particularly near the outsides. Where the soil is poor and exposed, the trees are in consequence less vigorous, ruin has not so speedily overtaken them, and by immediate thinning their health may yet be preserved. Not so, however, with the twenty acres of larches; although a thinning is now going on, it is doubtful if the trees left will make much more progress. The chances are, these will fall by the first hurricane of wind; if they escape that, their roots are sure to be strained and injured, if not broken, by the trees being top heavy. Dead roots connected with mutilated live ones give rise to Mycelia, which are imbibed into the trunks, and occasion the well-known disease called "dry-rot," "pumping," etc. The Scotch firs which were interspersed have been chiefly removed, and such as remain, although too tall for their girth, are not in so hopeless a state, and in course of time, if not uprooted by winds, may regain their vigour. Gradual thinning is to be recommended generally; and although in the present case it cannot be practised with much assurance of success, yet repeated thinnings over a space of a few years is better than doing it all at once. The only question is, whether it would not be better to clear the trees off at once by rooting them out, taking advantage of their stems to facilitate that work before cutting them. Had the ground been in a conspicuous situation, I should have had no hesitation in recommending that course; for however common, few scenes more unsightly are to be met with in woodlands than the display of unshapely trees struggling for an existence, and diseased through mis-management. In the absence of proper thinning, it is far more profitable where ground is good, with moderate shelter, to plant only half the usual number per acre, more especially if the tree is the larch.

Another plantation is worthy of notice, and instructive on account of the treatment it has received during the last six or eight years. It was formed by me in the spring of 1841, and is now twenty-five years old. It extends to 119 imperial acres, of which about 100 acres consisted of a dry bare moor-land, with a surface of stunted grassy heath and a gravelly subsoil. The other part was more or less overspread with furze, which required to be grubbed out and planted with stout transplanted plants. On the whole, two-thirds of the plants employed were native Highland pine, partly two years' seedlings, and partly one and two years transplanted; the other third was of larch of various ages, and all were inserted at the distance of four feet apart, and planted by contract at 20s. per acre; no upmaking was required for years, until the furze became strong, caught fire, and occasioned a small blank. This plantation became quite vigorous, especially just after it formed a cover and subdued the natural vegetation on the surface, thus converting the whole energies of the soil to its development. At the age of fifteen or sixteen years, thinning was very much required, as the trees had become crowded. It was at the age of eighteen and twenty, however, before the first thinnings were made by the forester on the property, and then it was not the more feeble and worthless that were removed; the primary objects appeared to be the obtaining of wood for a special purpose and the raising of money—not the relief and future well-being of the plantation. It has been thinned again and again for timber for paling, prop-wood, and for sale. The value of the timber removed is unknown, as much of it was cut down for country purposes and was not valued. I have not unfrequently witnessed with regret the same sort of procedure in woods more matured and consequently more profitable; the great profit of the future being sometimes sacrificed for a small sum to meet the present necessity. In the present case, however, no such necessity existed; the wood belongs to a wealthy corporation; yet just as the trees assumed a timber size of small value, though suitable for the present purpose, they were struck down as a gardener thins his asparagus bed in spring. The trees which should have been first removed, now for the most part fill the ground, unshapely and stunted, and the future of the plantation is dwarfed for ever. I have before seen the effects of such treatment both on Scotch fir and larch.

A proprietor who inherited an estate which he had seldom or never previously seen, was averse to the planting of Scotch pine, because woods composed of the tree sixty or seventy years of age scarcely produced a tree fit for railway sleepers, and not suspecting the early treatment of the trees, he considered the soil unsuitable for them. Another proprietor who purchased an estate twenty years ago, which contained a considerable extent of young larch plantation, then about thirty years of age, where all the best trees had been thinned out previously to the sale of the property, had difficulty in accounting for the contrast at present between the dwarf stunted old trees and the vigour of those of his own planting in soil apparently similar. The effects of early mismanagement are manifested as legibly in the vegetable as in the animal creation, more especially in the Coniferæ, which have not the power of re-establishing their lateral branches to redeem their proportional girth.

It is necessary to plant pines and larches close in exposed situations, and also in some cases to suppress a surface vegetation; early thinning, therefore, is sometimes necessary to afford a sufficient space for the trees before the thinnings have reached a size in ordinary demand; but the thinnings that are of no use in one locality are frequently valuable in another. Early thinnings are generally useful for small rustic fences, and the trees with branches (brushwood) are suitable for fences and shelter to young hedges, sea-side plantations, for embankments, to prevent the encroachment of rivers and rapid streams. In the Highlands this description of brushwood is employed by the agriculturists in forming sheds for storing turnips during winter.

It is when the trees are apt to get too tall and feeble in proportion to their girth that thinning must begin, and plantations arrive at this point at ages which vary very much, owing to the soil, the situation, the closeness of the planting, and the species. No tree is more easily injured by confinement than the larch. In some districts, however, this tree is planted closely, and grown in masses, for the purpose of hoppoles and prop-wood; in such cases, however, the ground is cleared of the whole crop while the plantation is yet young, seldom exceeding twenty-five years of age.

In plantations intended to yield heavy timber, the smallest trees should be first weeded out, giving sufficient room to

those of greater vigour and promise, as it requires a great
extent of foliage to mature healthy and sound timber. There
is no given space which trees of any particular age should
occupy. The skilful forester regulates the thinning so that
the trees may possess the figure for the purposes intended,
and the prosperity and profit of plantations depend greatly on
his skill and management.

In *hardwood plantations*, the oak, the ash, and the Scotch
elm are among the most valuable of our timber trees. The
oak is in general of slower growth than the others from the
plant, and requires rather more space to allow it to grow to
maturity in its best form; but the difference is compensated
by the value of its bark; and although it requires shelter
when young, it luxuriates in soils less fertile than that
required by most other sorts. In good soils, the oak, the ash,
the elm, and the sycamore may be grown profitably, either by
themselves or mixed. In thinning a young plantation of
hardwood, it is frequently found that a Scotch pine, or any
other tree inserted for shelter, presses too closely on a more
valuable plant, and yet that the shelter cannot be altogether
dispensed with; in such cases the branches of the pine which
press too closely on the more valuable plant should be
removed in the first instance, and the tree itself at a subsequent thinning. The distances at which hardwood trees
should stand apart in a plantation, at any particular age,
depend on their luxuriance, and on the exposed nature of the
ground; in narrow beltings they admit of being much closer
than in the depth of the forest.

Considerable loss is frequently sustained by producing
through confinement tall trunks without a proportionate
diameter; and unless the soil is very congenial, and the trees
of great vigour, they are often slow to become stout or shapely
when ample space has at last been afforded to them. In
plantations formed with plants at a distance of four feet, the
thinning should commence when the plants attain to the
height of from twelve to fifteen feet, by removing the more
worthless kinds, which press too closely on the others, and

fully half the number of plants inserted per acre should be removed by the time that the most valuable portion is twenty feet high. When they attain the height of thirty feet, they should stand on an average fully seven feet asunder, or about 800 per acre. At the height of forty feet, which is generally that number of years' growth, the trunks are formed to a considerable height; and at this stage of their progress it becomes necessary to furnish considerable space for the development of the leaves of the trees which are to occupy the ground, in order that their trunks may possess a girth corresponding to their height; therefore, generally speaking, they should stand from eleven to twelve feet asunder, or at the rate of from 300 to 350 trees per acre.

An acre of ash, of elm, or of sycamore, at the age of forty years, in favourable soil, is generally found to contain from 2000 to 3000 cubical feet of timber, and at the age of sixty about double that extent of measurable timber. This is exclusive of the thinnings, which are gradually removed up to this period; which, even while plantations are young, form in many localities a source of great profit. The oak is found to be of rather slower growth from plants than the kinds last named, but its growth from stools of former trees is equal to these kinds. When oak and other trees, which spring from the root, are felled, it is of advantage to the succeeding growth to dress the surface of the stools into a convex form with an adze, so that they may not retain water. It is also necessary, in some parts, to remove a turf around the base of the old trunk, to admit the influence of the weather and promote the young growths, the strongest of which; after having advanced a few feet in length, should be selected, and the others cut off.

At all stages of a plantation, space should be gradually allowed, according to the growth of the trees, which, with some sorts, in favourable situations, extends till the plantation is eighty years of age. But before this period, the trees, in plantations of hardwood, generally became irregular in size, and in their distances asunder, and the forest assumes the appearance of a native wood. This is commonly occasioned by filling up, and often to some extent by the growth of stools which produce saplings. Unlike the larch and pine woods, the broad-leaved plantations are generally permanent ; young saplings everywhere exist, ready to take the form of timber-trees, on receiving sufficient space by the removal of the largest trees in their vicinity, and a thorough clearance is seldom made. The profits derivable from the timber of hard-wood trees are generally very great, but exceedingly variable; being dependent on the soil, on local consumpt, on proximity to a cheap conveyance, and greatly on the management.

XII.

ON SEA-SIDE PLANTING.

THE influence of the sea spray prevents the profitable growth of plants over many extensive tracts of land adjoining the sea. But while this is the case, there is also a great space of waste land in maritime situations utterly barren, which by skilful treatment could be rendered profitable by the growth of forest trees. Nature, more especially unassisted nature, does little to tempt man to plant by the sea-side ; it is a union of the wild and the tame, which, though permitting it, she will not foster. Hence we never see trees spontaneously arise in such places. Art must therefore go to the fullest length of her resources, often in the preparation of the soil, and more frequently in forming a shelter—always in preparing the plants for adverse circumstances. Without these preliminary steps the ground had better remain as it is, for a plant that cannot readily establish itself when inserted in the soil, cannot be expected to withstand the buffeting of the far-fetched and keen-edged winds of the ocean.

The formation of sea-side plantations is often more expensive than of those formed under ordinary circumstances. This arises mainly from its being necessary to erect screen fences, to a greater or less extent, and to trench the ground in almost every case, except where it is formed of sand-drift. In all maritime situations the difficulty is most formidable where the ground is near to the sea-level, with a gradual ascent, exposed to the prevailing wind, with what sailors term a " long fetch," or great extent of rough sea. However close the position of land may be to the sea, if it is elevated a few hundred feet above high-water mark, the difficulty is more readily overcome, as the salt spray exhausts itself at a lower altitude. In such cases, where the ground forms a declivity, or falls from the coast side, or where the outside shelter is established, plantations enjoy an immunity, not only from saline influences, but from the severity of the frosts of winter. The high temperature of such places during the winter months acts favourably on plantations in general, and particularly on many of our best ornamental trees and shrubs. Plantations have recently been successfully formed both in England and in Scotland in the vicinity of the sea, in soil apparently of the poorest description, which until of late was reckoned wholly 'unfit for vegetation. These plantations, however, are not only giving promise of becoming profitable as timber, but are already spreading a shelter, and consequent fertility, over the adjoining lands. Some of the most thriving sea-side plantations in England were formed at Trimingham and Runton, the property of Sir Thomas Fowell Buxton, Bart., and stand on the northern extremity of the county of Norfolk, on the cliffs adjoining that part of the coast known as the " Yarmouth Roads." They were chiefly formed in 1840 and the three succeeding years ; the ground ranges from 200 to 500 feet above the sea; generally speaking, the surface is poor, and the subsoil a hard ferruginous gravel. In a report on these plantations, for which the Highland and Agricultural Society awarded the late Mr. Grigor of Norwich their gold medal, the success of these plantations is chiefly attributed, first, to the careful preparations of the ground, by trenching eighteen inches in depth ; second, to the fences, composed of furze and brushwood, and similar materials, having been erected as screens six feet high ; third, to the plants being of the best description, two and three years old, transplanted into nursery lines the year before they were inserted into the plantation, and consequently possessed of bushy or fibrous roots, and closely planted $2\frac{1}{2}$ to 3 feet apart ; and fourth, to cleaning, by hoeing the land for the first two years after planting, during which period root crops were produced among the young plants. These plantations embrace a space of 114 acres. The trenching per acre cost £6, and the fencing, plants, and planting upwards of £4, making the cost upwards of £10 per acre, exclusive of hoeing, which amounted to less than a fourth part of the value of the carrots, parsnips, and other crops.

In these plantations the black sallow or goat-willow (*Salix caprea*), the alder, the birch, the ash, the sycamore, the Scotch elm (*Ulmus montana*), and the pinasters, two varieties (*Pinus pinaster* and *P. p. minor*), are recommended. I have found all these very well adapted for sea-side planting, but the success of the ash and the elm is more dependent on the quality of the soil than that of the other sorts.

Respecting the sallow or goat-willow, it is stated that it was the wish of the proprietor, the late Sir Thomas Fowell Buxton, that the most of the trees planted on his estates in this quarter should give place to the English oak, so that the sallow was grown here merely for the sake of creating shelter, in which capacity it is certainly without a rival amongst deciduous trees. But though here used only as a nurse to the oak, it is fortunate that this willow has claims upon the attention of the planter as an independent object. Its claims, however, appear to be entirely hid from planters; for writers on trees, I find, refer to it continually under the character of an undergrowth, affording " excellent hurdles, and good handles for hatchets," and as used in the manufacture of gunpowder, etc. Now, the fact is, that though it is almost without exception kept down as an undergrowth, and used for fences and hurdles, it is capable of becoming a great tree, most singularly and beautifully clad in spring-time with handsome silken blossoms. On Mr. Moy's farm, East Runton, about three-quarters of a mile from the sea, is a specimen with a trunk which, at four feet from the ground, is nine and a half feet in circumference—thus proving that the tree not only grows to a large size, but that it does so in the neighbourhood of the sea.

Of underwood shrubs, the snowberry (*Symphoricarpos racemosus*) and the evergreen barberry (*B. aquifolium*) are strongly recommended. The plantations of Trimingham and Runton display a profusion of these shrubs growing most vigorously; and I can have no hesitation in stating that they are well adapted for growing along with other trees near to the sea-side.   Their merits may be summed up in a few words.   They are both eminently beautiful—the former when clad with its pure white berries from September to December, and the latter throughout the entire year.   This plant (the *Berberis*) is furnished with pinnate, shining, holly-like leaves, and bears beautiful racemes of yellow flowers, which are succeeded by grape-coloured or bluish-purple berries in great profusion. Though so unlike to each other in appearance, there is a remarkable affinity subsisting between those shrubs, which points them out as the fittest of companions; they fear not the sea-breeze, a fact which might have been anticipated from their being both natives of the north-west coast of America, from New Albion to Nootka Sound.   Both plants when in blossom are much sought after by bees, and the berries of both are greedily eaten by game.   The delightful uses to which such plants may be applied will suggest themselves to every one, and it is only necessary to remark that they are best suited for being planted close to walks, which should be introduced in all maritime plantations, for the sake of the view over the ocean.

Time of Planting.—In maritime places the young trees should be invariably planted in the spring, just immediately before the time when the plants begin to grow.   The next best time is the last week in October.   But though those trees planted early in autumn furnish themselves with small roots or tender spongioles previous to winter, these are not sufficient to support the trees in such situations during the most trying months; so that it is infinitely better to let them have the benefit of a full season's growth before the effects of winter are felt by them in their new situations.   The last week in March or the first week in April is a suitable time to plant in such places.   Planting in winter months has been tried repeatedly in the neighbourhood, but with no success.

Choice of Plants.—Experience proves, that for the particular situation under consideration, such plants as are two or three years old are better than any other, and such as have been transplanted in the nursery the year previous are to be preferred to those which have remained for two seasons.

Shelter.—However well the land may be prepared and the season chosen for planting the young trees, shelter is indispensable, both as an outwork, or round the outside, as well as an immediate agent in ameliorating the climate around each tree. The best external fence between the young plantations and the sea is furze bundles, or brushwood cut in summer-time, with the leaves on the branches; or failing these, a turf wall, very broad at bottom and tapering to the top. The best sheltering nurses amongst deciduous trees are the sallow, alder, osier, and birch, and amongst evergreens the Scotch pine; but as these nurses would be gladly accepted in many instances as permanent occupants, I would earnestly recommend them as particularly fitted for such situations. Oaks, and the finer kinds of pines, should be surrounded with the nurses, and particularly protected by them on the side next the sea; but in ordinary cases it is sufficient to plant them mixed with the nurses, so that the young trees may in a general way protect each other.

Cleaning.—The hoeing of the land for at least two years is all-important, and if a crop of carrots is taken from the ground the first year, as has been practised here, they will help to keep down the weeds, and pay the expenses for plants, cleaning, fencing, etc.

Plantations formed chiefly on sand-drift along the sea-side on the estate of Culbin.—Within the last twenty-eight years a considerable extent of plantation has also been formed on the sands of Culbin. These sands occupy several thousand acres of the north-west corner of the county of Moray, N.B., and are composed of small hills of sand, ranging from 20 to 140 feet high, the surface of which is ever changing by the influence of the wind. Along this district in rough dry weather the atmosphere is thickened with sand-drift, and from the prevailing westerly winds its course is generally eastward; but the ordinary driftings are intercepted by the river Findhorn, and the sand is carried out to sea and forms a bar, where the in-shore tidal current, running to the westward, carries it in that direction, again to be thrown on shore and blown east-ward; thus to some extent it forms an endless circuit. Along the southern extremities of these sand-hills, at the dis-tance of several miles from the sea, are beaches of rolled boulders from twelve to twenty feet above the tide-mark. Similar beaches exist in many parts along the south side of the Moray Firth, and these, with the accumulation of sand, are supposed to be due to the effects of the extraordinary inundations of the German Ocean in the thirteenth century recorded by Buchanan and Fordoun. But although large mountains of sand are said to have been formed at this time along the sea-shore, called the hills of Maviston, yet it must have been long after this period before they spread out on the cultivated land to anything like their present extent. In the beginning of the fifteenth century the estate of Culbin was in the possession of the Kinnaird family, and their descendants continued to prosper on it for many generations. After the middle of the seventeenth century the estate is known to have consisted of sixteen very regular and compact farms, so famous for their fertility that the district had acquired the distinguished appellation of " the Granary of Moray." The rental of the estate at that time was as follows :—Money rent, £2720 Scots (more than equal to sterling money now), 640 bolls of wheat, 640 bolls of bear, 640 bolls of oats, and 640 bolls of oatmeal. Such a rental now would be worth about £7000 sterling per annum. Besides, the family had valuable fishings in the Findhorn, which then ran westward through the estate, but its course, having been drifted up, was turned into its present channel during the general calamity. Towards the close of the seventeenth century the then proprietor applied to Parliament to be exempted from the payment of cess, because his estate, which twenty years before was one of the most considerable in Moray, with an area of about 5000 acres, was nearly all covered with sand,

and the mansion-house and orchard destroyed. Since that period the property passed by purchase through several families. The next rental, taken in 1733, after the estate was almost entirely blown up with sand, shows the yearly value in *money* and *victual* to be only £494, 4s. 4d. Scots. But small as that amount comparatively is, it has also been extinguished, and for many years the district with its broad and fertile fields, formerly designated "the Granary of Moray," has become altogether an arid waste of shifting sand, bearing only the wavy ripple of the wind. Singularly enough, these possessions have disappeared as completely as their occupants, and their names are heard of only when the search of the antiquary discovers them among the musty papers of a bygone age, or deciphers them on the mossy tombstones of the old churchyard, while he seeks for their locality in vain.

The south-east portion of this desert comprehends almost the entire property of Binsness, which measures 558 acres, and has of late diminished in value almost yearly, by the inroad of sand-drift. It stood in the cess-books of the county in 1667 at £390, 17s. 2d. Scots of yearly rent. It possesses no plantations whatever. Its arable fields have been abandoned to desolation, and the whole extent was sold in 1865 for the small sum of £660.

Kincorth Plantations.—R. Grant, Esq., of Kincorth, was the first to reclaim part of the sands in this quarter by plantation, and all the sandy space on his property is now completely covered with thriving wood. In a report to the Highland and Agricultural Society of Scotland, published in 1847, he says :—

"In the most advanced part of the plantation, which is now eight years old, the plants average a height of at least six feet, so that a man walking among them can scarcely be seen ; and the author may truly report, that the average growth of one year on the Scots firs was a foot, and on the larches nearly eighteen inches ; while all the younger parts of the plantation promise well to attain the same size at the same age. Whether or not trees may ever become valuable as *timber*, on such an arid soil, is a problem which can only be solved in future years (for he is not aware of there being any such existing under similar circumstances in this country); but he is sanguine that they will attain a size to be very useful for palings and other country purposes. But it was not in the expectation of profit that he was induced to plant, but entirely in the hope of obtaining shelter and ornament, and of giving an improved appearance to a very dreary prospect. In this object the author has every reason to believe, from present appearances, that in due time he shall succeed; and it would gratify him much, if, stimulated by the knowledge of his success, any other proprietor of similar dreary and sandy tracts, which are so frequent along the coasts of Scotland, should be induced to plant them."

These plantations, now (1865) from twenty to twenty-eight years of age, range from twenty-five to forty-five feet in height, and are very vigorous. They yield valuable thinnings, and although they stand on pure sand-drift, give promise of producing heavy timber at a period not far distant.

Observing that these plantations were treated in the most skilful manner, I applied to Mr. Grant for information regarding their pecuniary return, and he has kindly favoured me with the following statement :—

" *Forres House, 25th February* 1865.—In answer to your letter of inquiry respecting my plantation adjoining the sandhills of Kincorth, I write to say that, although I have not kept any precise notes of the entire management, you may rely on the accuracy of the following details. The extent of the plantation is about seventy acres, but having been planted during successive years, some parts of it are more advanced, and have been oftener thinned than others. I have constantly adhered to the rule of not cutting any trees in it except in thinning with the object of improving the plantation. The greater part of it was thinned twice without yielding any pecuniary return, and at an expense of from £15 to £20, each time, say in all of £35. From these thinnings, however,

I had garden and some field upright fences made, which lasted substantially for four years. During the last six years I have had all the ordinary flake-palings I required for the farm made from trees in the older part of the wood,—I should say at the least to the value of £10 or £12 annually. In addition to this, I find from my book that within the same period (six years) I have received for props £235, 9s. 6d., and have expended for thinning, etc. etc., £69, 16s. 4d." Although the cost of the formation of this plantation cannot be exactly ascertained, yet it is quite clear, from the cost of other plantations adjoining (which will be immediately shown), that the returns already realized are sufficient to pay all outlay, and interest at a high rate, up to the present time ; even supposing the outlay in its formation to have been three or four times the usual rate, which some small spots, that were covered with grassy surface, together with the fences, may have occasioned.

The present money value of the oldest part of this plantation is not less than £30 an acre ; and although other parts are as yet of less value, on the whole, during the next twenty or thirty years, the thinnings will pay the current rate of interest on that amount, with every prospect of the timber in the plantation increasing very much in value per acre during the period. This is a moderate estimate of the value of the plantation, yet it is clear that it will in future yield a revenue equal at least to that of ordinary arable land ; and beyond this the proprietor has fully realized the primary object of its formation, namely, ornament, shelter, and a sure protection against sand-drift.

The success of the Kincorth plantations while they were only a few years old induced the proprietor of Moy, who owns several thousand acres of the adjoining sands of Culbin, to form plantations thereon.

Plantations on Culbin belonging to Moy estate.—In the beginning of winter 1840 I planted on these sands 199 imperial acres with Scotch fir and larch, at the following cost per acre :—

1800 one-year transplanted native
  Highland pine,   .   .   £0   3   8
1000 two-year seedling,   do.,   0   1   3
 800 two-year   do.,   larch,   0   2   0
Carriage from Forres nurseries,   0   0   3
Planting, ...   0   3   2

           199 acres, at £0   10   4   £102 16   4
Building fences and planting about
   25 acres by the proprietor,   .   .   97 14   9
In January and February 1842, I added to the
above plantation seventy-four acres, but allowed only 3200 plants per acre, thus :—
2200 two-year seedling and transplanted native Highland pine, £0   3   3
1000 two-year-old larches,   .   0   2   6
Carriage from Forres nurseries,   0   0   2
Planting, ...   0   2   10

         74 acres, at   £0   8   9   32   7   6

Making in all 298 acres, amounting to   .   £232 18   7

It will be observed that the expense of these plantations is
unusually small.   This arises from several circumstances.
The ground being soft, with little or no surface herbage, rendered it suitable for small plants which were planted by the
hand-iron, by people, in the vicinity much practised in the
work, and who could plant an acre each daily on such ground
without difficulty.   The plantations were also made at a time
when the price of nursery plants was under the usual rate.
The expense of fencing these plantations was also very small :
only a part of the south and western boundary required protection, and that is enclosed by a turf wall five feet high.   The
east end is fenced by the Kincorth plantations ; the north is
bounded by a vast extent of pure undulated sands, with a surface destitute of vegetation, and bearing only the wavy ripple
of the wind, except where a clump of bent grass here and

there arises. No cattle or sheep frequent this quarter, and a fence is unnecessary. The north sides of these planta-tions stand nearest to the sea, and are formed on knolls of pure sand-drift. These knolls are situated from one to two miles from the south edge of the Moray Firth, at a part where the sea is from twenty to twenty-five miles in breadth. Their altitude varies from twelve to thirty feet above the tide-mark.

The plants on both plantations advanced vigorously, and, with the exception of a few small spots here and there covered with sand, or where it was drifted away soon after the plants were inserted, no vacancies existed. On lifting and examining some of the plants of both sorts, from the pure sand, it was found that they had furnished themselves with tap-roots, which strike right underneath the plants to a great depth ; but the greater portion of the roots run horizontally at a depth of about four inches under the surface of the sand, and extended over an almost incredible space. Many of the plants six years planted were possessed of roots upwards of twenty feet in length, which ramified into numerous lateral fibres. Where the surface had remained undisturbed, there was a remarkable uniformity in the depth of the fibres, both in flat and in steep situations. Nature adapts the plants for emergencies, and no instance was observable of their having perished from drought, or having been removed by wind after having grown a few years ; some of them stood where sand to the depth of more than a foot had been drifted from under-neath them, and continued to thrive with a great portion of their roots laid bare. In other cases the Scotch firs grew when drifted up with sand, and had only the growth of a year or two above the surface. In this position the shoots of Scotch fir do not strike root, but where the larch became sunk in drift its shoots readily rooted at a depth of four inches under the surface. Many stools were to be found resembling handfuls of larch plants inserted into one spot, showing that the soil or sand was well adapted for the growth of cuttings.

Plants of every kind of coniferæ grown from seed deserve a preference to those propagated by any other method ; but the readiness with which the lateral branches of the larch take root, when sunk in sand of this quality, suggests a very speedy and effectual method of increasing some of the more valuable coniferæ, the seeds of which cannot be obtained.

When the cover of trees begins to prevent the movement of the sand, the vegetation first observed on the surface was *Hypochœris radicata, Polytrichum commune, Agrostis vulgaris, Agrostis alba, Ammophila arundinacea,* and *Calluna vulgaris.* The first-mentioned plant strikes its root to a great depth, grows freely, seeds abundantly, is a perennial, and likely to be valuable in fixing sand.   As to the relative fitness of the two kinds of trees for the situation at first, there was some difficulty in deciding which was the more suitable.   The native fir affords the best shelter ; it stands in great health, and has much the advantage in appearance ; its deep green contrasts beautifully with the whiteness of the sand, which gives the foliage of the native plant something of the lustre of a Himalayan.   The larch, from the shedding of its leaves, which soon decay, forms a dark stratum of vegetable matter on the surface of the sand, which consolidates it and promotes vegetation, particularly the growth of the grasses, much sooner than the exuviæ of the Scotch fir ; but both have their advantages, and, in order to furnish a choice of kinds in the operation of thinning, a mixture is to be recommended, although there is now no doubt that the Scotch fir is the more reliable tree, and will ultimately, in most cases, yield the most valuable crop of timber.   Within the boundary of these plantations are several beaches of rolled boulders of primitive and transition rocks. Although the surface of these is stationary, and the soil intermixed amongst the stones, they are almost destitute of any vegetation except moss ; on these beaches the larches do not grow, and the firs assume a yellowish green, and are much more dwarfish than those in the pure sand.   This probably arises from the stones admitting the drought of summer to a greater extent than the pure sand does.

Along the south side of the ground, under these plantations, where the surface is a sandy peat, the cover is a short grassy heath; on such parts the plants grew freely, and are generally taller than on the pure sand. In other parts along the outskirts of these sandy regions, although not comprehended in the plantations here referred to, there is a considerable extent of flat land under a close, rank, grassy sward; the ground is a pure sand, with a slight mixture of decomposed vegetable matter on the surface. This description of ground is, of all others, the most difficult to get successfully planted; for, when plants are *notched* into such, they uniformly die of drought during the first summer. The matted herbage intercepts the ordinary showers, and keeps the ground destitute of moisture at the depth of a few inches. When a surface turf is pared off before the plant is inserted it has a better chance, and when the ground is pitted or trenched it is still more for its advantage. The preparation of the ground should be made several months before the planting, and the planting should be done in early autumn, or in moist weather in March or April. In places of this description, the risk of sand-drift on a trenched surface, where it is not convenient to thatch it with brushwood, sometimes prevents the destruction, by trenching, of any surface vegetation which may already exist.

The annual growth of both kinds of tree in these woods in the purest sand was upwards of fourteen inches long, and contrasts favourably with that of plants in apparently better soil, more solid, but overspread with a matted surface of the natural grasses; thus illustrating the advantage of planting on a loose open soil, with a clean surface, whether poor or rich. When such plantations reach from ten to twelve feet in height they are fit to yield a large supply of thinnings, which are well adapted for reclaiming the sands and limiting the encroachment of sand-drift. These thinnings or brushwood are valuable for the purpose of being spread over the newly planted sands in the roughest exposures; overlapping or spreading the brushwood in an imbricated position, causes it to stick to the surface, and thus it affords shade and shelter to the young plants in situations where they would otherwise perish.

The plantations on the Culbin sands belonging to the Laird of Moy are now twenty-two and twenty-four years old, and form a very compact wood of nearly 300 acres, affording great shelter in their vicinity, and forming a striking contrast to the barren hills in their neighbourhood. The trees continue to advance most vigorously. The tallest larches are about forty feet high, and the best Scotch firs about twenty-eight feet high; generally the wood stands from twenty to thirty feet high. Roads have been opened up through the plantations, and thinning has been practised, first to a small extent in 1855 and 1856, which yielded little or no return; but from a statement furnished by the factor in charge of the property, prop-wood yielding a return of £482, 8s. 6d. was sold in 1864, and a further sale of £400 was expected before the necessary thinning was effected. I have not ascertained the cost of the operations of thinning, road-making, etc., but it is clear that the return already received for the prop-wood thinnings is more than sufficient to pay the original formation of the forest, with interest and all expenses up to the present time. From its present healthy state, 1865, the forest cannot be valued at less than £22 per acre, or altogether, £6556. Frequent thinnings will be required from time to time, at intervals of only a few years, as it is easy to see that the vigour of the trees has a tendency to draw them up too tall in proportion to their girth. Allowing trees to become drawn up through want of thinning is the ruin of many of the woods of our country, and should always be guarded against, particularly since it unfits the timber for railway sleepers, so constantly in demand, and so profitable. Of late years a few hundred acres have been planted in addition to those already described, but these have failed to some extent, from the ravages of rabbits, which overrun the place.

The market for pit-props has of late become somewhat depressed in this country, in consequence of their abundance and cheapness on the Continent, particularly near Bordeaux, where they are produced on sandy regions similar to that I have attempted to describe, and to which I shall now refer.

Mode of growing plantations on sands along the French coast.— The greatest triumph of arboricultural skill in reclaiming sand-drift was recorded by a Commission appointed by the French Government to report on the pinaster forests formed by M. Bremontier, of the Administration of Forests. In 1789 M. Bremontier commenced his operations at the Gulf of Gascony, where the downs offered nothing to the eye but a monotonous repetition of white wavy mountains, destitute of vegetation, and agitated by the wind. In 1811 the Commission reported that 12,500 acres of downs had been covered with thriving plantations by means of sowing the seeds. The process is as remarkable for its simplicity as for its success. It consisted in sowing two pounds of the seeds of the pinaster, mixed with four to five pounds of broom-seed per acre, and immediately covering with branches of pine or other trees, with the leaves on, commencing at the side next the sea, or that from which the wind generally prevailed, and sowing in narrow zones in a direction at right angles to that of the course of the wind; the first sown zone being protected by a line of hurdles, this zone protecting the second, the second the third, and so on. After sowing, the ground was immediately thatched with branches, overlapped, to protect the seed, with a hurdle fence erected to intercept the progress of the sand. In a word, wherever the seeds were sown the surface of the sand was thatched. A thatching of rushes, reeds, or sea-weed was also used, and was quite as effectual as the branches. In six weeks or two months the broom-seeds are said to have produced plants six inches high, which attained three or four times that height during the first season. The pinaster plants do not rise above three or four inches the first season, and it is generally seven or eight years before they overtop the broom, which often in these downs attains to the height of twelve or fifteen feet. At the age of ten or twelve years the pines have in a great measure suffocated the broom;

they are then thinned, the branches cut off being used for the
purpose of thatching downs not yet recovered, and the trunks
and roots cut into pieces and burned to make tar and charcoal.   These plantations, and others in the sands of Bordeaux,
and between that city and Bayonne, constitute the principal
riches of the inhabitants, whose chief means of gaining a live-
lihood arises from the preparation of resin and tar from these
pinaster forests.   The pinaster consists of numerous varieties,
all of which grow remarkably well in sand, or in dry, poor
soil, and endure the influence of the sea better than most
trees.   In wet or fertile soil it is less hardy than in sand; in
the former its shoots become more succulent, less matured,
and are thus unable to resist the severity of the winter in
Scotland away from the influence of the sea.

Plantations on the West Coast of Scotland.—Throughout the
islands of the Hebrides the Gulf Stream has a very perceptible
influence in ameliorating the climate in winter, and adapting
it to half hardy trees and shrubs in all cases near the sea,
particularly where plants are exempt from disturbance by the
prevailing winds, accompanied by salt spray.   In such places
our hardy trees prosper in a manner similar to those in inland
situations, dependent chiefly on the quality of the soil on
which they stand.   In less favourable situations, where the
sea is broad, and where a prevailing wind blows over it, the
most suitable maritime trees only can endure the exposure,
and the sorts to be employed should be judged of from the
quality of the soil they are intended to occupy.   If it is a dry,
barren sand, or gravelly soil, the pines should be employed—
P. maritima, sylvestris, austriaca, laricio or *corsicana,* and for
underwood among pines we have *montana, pumilio,* or *mugho.*
Among deciduous trees few kinds succeed in soil of this description.   The best adapted are the birch, the goat-willow
(*Salix caprea*), and the grey poplar, *P. alba canescens,* which is
the hardiest of all the white poplars.

Where the soil is of good quality, either a good, sound,
sandy loam, or a sandy peat, which is frequently the case
along the west coast of Scotland, all the pines, and also the.

deciduous trees above named, are quite suitable, together with the sycamore, and other hardy maples, the Scotch elm, Scotch laburnum, evergreen oak of all sorts, the mountain ash, and the common ash. The last-mentioned tree is found to be very dependent on the quality of the ground. It becomes vigorous in places where the soil is rich, loamy, or partly mossy, and moderately moist. The tree is very hardy, and stands the wind very well. It forms its growth during a few of our warmest months, and is not readily injured during any other period of the year. But being bare for a long period, the ash does not furnish a good shelter where such is most required.

Many years ago I inspected a large plantation made in one of the islands in the Hebrides, and found it a great failure. It was formed on a scale of great magnitude, and was composed of a considerable variety of plants. The ground ranged apparently from 40 to 100 feet above the sea, and extended from the sea for several miles inland, and was exposed to all quarters, except here and there, where a stream had formed a channel, and afforded sheltered slopes along its bed. Seldom is a more trying situation for a plantation to be met with, as the surface generally affords an extensive and uninterrupted sweep to the wind ; nevertheless, there was a greater obstacle than the exposure to contend with in the successful growth of timber. The soil was peat moss of the purest quality, of immense depth, often eight, ten, or even twenty feet,—and so pure that sand could not readily be found in its composition. Numerous ditches had been formed to drain the ground, and deep as they were, they seldom reached the subsoil, and if they had, their influence would have been of little avail to the plantation, on account of the purity of the vegetable substance which formed the soil, and the humidity of the atmosphere continually acting on its spongy surface. The vegetation on the ground formed a rough grassy heath ; and scarcely any part of the soil was so dry that water would not be found to drop by squeezing a portion of it with the hands, even in moderately dry weather.

When I inspected this plantation, it had been formed from five to eight years. Life lingered in a plant here and there, but none gave any promise of becoming useful wood, except a few narrow beltings which covered the slopes along the sides of streams. Here the plants stood under very different and more favourable circumstances; the sandy subsoil had been penetrated and disturbed by the action of water, and pulverized and ameliorated by the influence of the weather. The plantation was thus entrenched some thirty or forty feet under the ordinary level of the ground; the plants had their roots in a congenial soil, and with their tops comparatively undisturbed, they grew vigorously, and formed a very marked contrast to those on the level surface.

Of all the plants inserted in this plantation under the adverse circumstances already detailed, only two or three sorts maintained an appearance of health. The tallest of these was the *P. pinaster maritima*, distinguished for growing in pure sand; yet here, in soil of the opposite description, in which sand forms little or no proportion, it had attained the height of from six to seven feet in that number of years, even in the most exposed situations. The other plants were varieties of the dwarf pines *pumilio* and *mugho*, which do not become more than dwarf spreading bushes anywhere; they appeared in perfect health. These plants are indigenous to the mountains of Central Europe. They are found on the Alps beyond the limits of trees, and are seldom met with higher than 7500 feet, or lower than 4000 feet of elevation, where they prefer a swampy soil. These plants are only of use as a change of surface vegetation or shelter for game. In this plantation the Scotch fir appeared unusually tender, but whether it was raised from the seeds of native forests or from seeds imported from the Continent, I was unable to ascertain; but its appearance looked like the tender plant of imported seed. Instances were to be seen of a few older trees near to this plantation, where in the same bare exposure they had advanced to the height of fifteen to twenty feet in a short *stocky* form, although standing alone. The kinds were Scotch elm, ash, birch, and alder. These, however, all stood on ordinary good sound soil.

An excess of moss or bog in the soil, although dry, has the effect of producing a softness in growing trees, which renders them less able to resist any unfavourable influence, such as frost, or continued agitation by wind in a bare exposure; and young plants, as well as timber, produced in such soil, are always short of the ordinary specific gravity.

No effort was made in this extensive plantation to form any description of screen-fence or shelter. Reliance appeared to be placed on the extent of the surface planted, which under more favourable circumstances would have gone far to produce valuable timber.

The configuration of ground along the sea-side which is most difficult to plant successfully is that which is exposed to a rough sea, and is only situated a little above high-water mark, with a regular slope towards the sea. With an undulated surface, a portion of the plantation has always some protection; but on land nearly level, or having only a gentle slope, the biting influence of the spray precludes the growth of trees to a considerable distance inland. Where the ground is sufficiently elevated, even on the brink of the sea, and on the roughest coast, the effects of the spray are but little felt, as its influence is exhausted at a lower altitude. This is illustrated by the thriving plantations on the coast of Norfolk, by those on the " Sutors," which guard the entrance to Cromarty harbour, where heavy timber has been produced, and on many other high cliffs which border the German Ocean.

The preparation of the soil for plantations has, under any circumstances, a great influence on their future prosperity. This requires to be particularly attended to where plants are exposed in bleak and inhospitable situations. Where the soil is not thoroughly loose to a considerable depth, it should be made so by trenching. At a short distance from the sea light sandy ground is often found overspread with a close matted surface, composed chiefly of the coarser sorts of native grasses; this surface herbage, with its numerous roots, which abound everywhere in the soil, has a very marked effect in impeding the growth of every kind of tree,—the native vegetation deprives the plants of nourishment; and of all soils producing herbage along the sea-side, this description of land is perhaps the most difficult to cover with healthy plantation. In this case the injury to newly inserted plants by the herbage is equal to that arising from a dense and retentive subsoil, and the remedy in both cases is that of trenching the ground previously to the formation of the plantation. It is to be recommended that in either case the trenching should take place a few months before the inserting of the plants, to allow the herbage time to decompose, and that the retentive subsoil may become purified and pulverized by the influence of the atmosphere.

Regarding the preparation of the ground I have only to add, that under the adverse position of many sea-side plantations, unless the soil is made thoroughly pervious to the roots of the plants, every other effort will fail in producing vigorous trees. Wet ground should be thoroughly drained, and in all cases where an addition of good fertile soil can be made and applied to the roots of the plants, success is thereby rendered most certain.

The time of planting depends very much on the nature of the soil. I have transplanted pines successfully in every month from the beginning of August till the end of May. If the ground is dry sand the months of August and September are very suitable for Scotch and other pines, provided they are grown close at hand, and can be lifted and planted without much exposure to the weather. If wet weather occurs in April the planting of pines in sand is then generally very successful, as they lose no time in fixing themselves in the ground, and then start with the calmest months of the year. This holds good with all the kinds of plants recommended, but the fear of a want of moisture in many places renders it necessary to plant earlier, so as to take advantage of the sap of winter or early spring. Close planting is absolutely necessary; from two to three feet plant from plant is a usual distance in severe exposure, the principal part being pines, which are generally obtained at a small price.

Screen fences for protection are of the greatest value in plantations exposed to the sea-breeze. These fences may be composed of turf, brushwood, or any material most convenient that will afford shelter, and the higher they are built their influence will extend over the greater space. Where the thinnings of a young Scotch fir plantation are at hand nothing will be found more effectual or more cheap in the erection. At the distance of every 20, 50, or 100 yards, according to the severity of the exposure and the figure of the surface, another screen fence should be raised, and so on till the more sheltered ground is reached where screens are unnecessary. Stone dykes are least efficient, and cannot readily be formed of a great height. They are sometimes, however, of great use in supporting the thinnings of young fir-trees laid against them, which raise a shelter to a considerable height; the wind acting on such material loses its force, and spreads a mild and mollifying calm around the space, like the influence of a hedge. Besides the ordinary screen fences in severe situations, an open cover of brushwood, spread on the surface of the ground, is often found of great use in a newly formed plantation, even where the soil is not apt to drift. Although the outskirts of plantations, under the most skilful treatment, on an inhospitable exposure, are often sadly disfigured, yet at every pace as you advance to their interior the trees become taller and more shapely; and on a level surface the tops of the trees form a gentle ascent until the shelter becomes perfect.

The preparation of plants for sea-side planting is a very important matter. Not only is it necessary that the plants should be grown in an open and airy situation, but they should stand individually at such distances apart as to afford free scope for the play of the atmosphere all around them. And in all cases, particularly with respect to the sorts that are apt to produce bare roots, such as the various species of pines, they should be removed in the nursery the season before they are required for the plantation; by this means

H

the plants will be well supplied with fibrous roots of one
year's growth, which most readily take to the soil, establishing the plant on its removal. The pinaster is perhaps the
most difficult to remove in safety of any species of the Coniferæ. This arises from its naturally long and bare roots. A
two-years' seedling plant seldom takes root when removed;
and in forming plantations with plants, those that have been
transplanted when one-year-old seedlings, and nursed in
lines for a year—namely, plants of two years of age, or
those that are three years old and have been twice transplanted,—are·the only plants that are worth inserting. But
that peculiarity of organization which renders it difficult to
remove the plant with safety adapts it the more for being
successfully grown *from seed in sand.* Its long roots readily
strike to a great depth, and become wide-spread, and are
thereby the more serviceable in supporting the seedling plant
against drought and the casualties of a shifting surface.

The kinds of plants must be regulated by the quality of
soil. In a pure, dry, drifting sand, the native Scotch pine,
pinaster, *P. austriaca, P. laricio,* are the most suitable. If the
situation is not very much under the influence of the sea
spray, the larch may be inserted as a mixture among the
pines, provided the sand is not pure, but mixed with vegetable
or other substances. The silver fir is seldom used as a sea-
side tree on account of its requiring, even in the best situations, a few years before the plant takes to the ground and
grows freely; besides, in early life it requires more than usual
shade and shelter. After it becomes established in suitable
soil, which is a moist or mossy loam, and attains the height of
five or eight feet, few trees will keep pace with it. Gilpin,
who differs from other writers respecting the appearance of
this tree, says—"There is a sort of harsh, stiff, unbending
formality in the stem and branches, and in the whole economy
of the tree, which makes it disagreeable." He then continues—
"I may add that the silver fir is perhaps the hardiest of its
tribe. It will outface the south-west wind; it will bear
without shrinking even the sea air; so that one advantage

.at least attends a plantation of silver firs : you may have it where you can have no other, and a plantation of silver firs may be better than no plantation at all."—(*Forest Scenery*, vol. i. p. 90.) I know many fine specimens of the silver fir not very far from the sea, but in its immediate vicinity I have never observed its powers of endurance to be superior to those of the pines ; and for the reasons stated it is seldom planted in bare and exposed situations.

The pines suitable for maritime plantations, and more particularly for dry and sandy links, I have already noticed. In a heavier description of soil a variety of trees as well as the pines have also been named. Among willows, the goatwillow or sallow (*Salix caprea*), the white Huntingdon willow (*Salix alba*), the Bedford willow (*Salix Russelliana*), grow in a saline atmosphere better than most other plants. The variety of hardy willows is very great ; they are tenacious of life, and will grow to some extent in soil of any description, dry or moist. Several species are frequently cultivated on the seaside by fishermen, who manure them richly, and raise luxuriant crops of *withs*, which are cut down every autumn for the manufacture of fish-creels, etc. The black elder (*Sambucus nigra*) and the scarlet elder (*S. racemosa*) rank among the best trees for screen fences, and for forming a thicket in nursing up more valuable sorts. Of the maples, the common sycamore and the Norway maple are good maritime trees. The common alder and the beech should also be employed, and in good soil the ash and Scotch elm. None of the poplars except the grey (*P. canescens*) can be recommended. When the soil is of ordinary quality, evergreen oak (*Quercus ilex*) and its numerous varieties endure the influence of the sea better than most trees. From the tree being naturally very bare rooted, it is apt to fail on being transplanted, unless it is carefully prepared in the nursery, by being yearly removed for a few years previously to enable it to acquire bushy root-fibres, so numerous that the earth adheres to them in transplanting. Such plants inserted in soil well softened or disturbed, soon take to the ground, and form a desirable ornament near the sea-side, retaining their green and glossy appearance throughout the severest winter. In such places the temperature is always comparatively high. In the islands of the Hebrides many half hardy plants endure the winter in the open ground. In the island of Lewis the *Auralia japonica* ripens its yearly growth, and there I have seen the common fuchsia standing from six to eight feet high, with a trunk, the growth of many years, covered with a rough bark, and more than a foot in circumference.

Among the shrubs most suitable for the sea-side are the snowberry (*Symphoricarpus racemosus*), sea buckthorn (*Hippophae rhamnoides*), the tamarisk (*Myricaria Gallica*) and *Germanica* and the evergreen barberry (*Berberis aquifolia*). In dry soil during winter the Laurustine is seldom met with anywhere in greater beauty than within the influence of the sea. And as an ornamental and efficient dwarf screen-fence for garden protection few plants can be turned to better account than the common whin or furze (*Ulex europæus*).

The kinds of trees and plants detailed, the preparation of the ground by trenching, the selection of the plants, their protection by screen fences, and surface cleaning where herbage is apt to cover the ground and impoverish them, are the best means by which sea-side plantations may be established. In such situations, particularly in loose sand, plantations of the coniferæ often become of great value in a short period, especially in undulated land, where any shelter is afforded by the surface rising to some extent between the plantations and the sea. Those of Scotch pine are often ripe for being thinned for prop-wood at the age of twenty and twenty-five years. Timber of this sort is in constant demand (*see the article on* PINE-TREE), and although there are frequent discouragements in establishing plantations in the vicinity of the sea, yet it should always be borne in mind that as timber of this description forms the return cargoes of coal vessels along our shores, its value, compared to that of many plantations, is greatly enhanced by the absence of a long carriage from the interior of the country.

XIII.

HEDGE-ROW TIMBER.

TREES are cultivated in hedge-rows for the sake of their timber, for shelter to the adjoining fields, and for embellishment; and in many situations all these valuable objects are obtained in the same locality. It is true, that where timber generally arises to the greatest size and value, the situation which produces it is that which stands least in need of shelter; but where trees fail to become specimens of excellent growth, on account of the climate and exposure, the value of the timber is often compensated for by the shelter which the trees impart to the fields in their vicinity.

The quantity of timber grown in rows along roadsides, around the extremities of estates, and in the division of fields, throughout England, is supposed to be greater than that produced in close woods and forests. Many of her sheltered plains are overcrowded, and present the appearance of one continuous forest. In Scotland the case is very different; and in numerous instances, for want of the shelter and embellishment of timber, the country assumes an aspect bare and uninteresting. With respect to agriculture, both extremes are to be avoided. An excess of shelter exhausts the soil, enfeebles the crops, and renders their safety uncertain during the humidity of autumn. On the other hand, bareness in exposed situations is attended with many disadvantages. In unfavourable weather, pasture and crops of every kind are retarded, particularly in the opening of the season; and fields are found ill adapted for the more tender kinds of animals which are now to be found everywhere throughout the country. The injury also sustained by winds, both to the stems of the growing plant during summer, and in the shedding of grain in autumn, is often very considerable. We know districts of light friable soil, where scarcely a season passes without injury being sustained by the turnip crop in consequence of exposure or want of shelter ; and it not unfrequently happens that gales occurring in the end of June or beginning of July, not only disturb and injure the young plants, but drift them away. The casualty is generally most severe when rough weather ensues immediately after the young plants are singled out. (In such places, the old method of sowing on a flat surface is that most likely to be exempt from injury.)

There are however many situations where the cultivation of hedge-row timber cannot be recommended. Many bleak and weather-beaten tracts of cultivated land throughout the country first require plantations formed, of considerable breadth, on all exposed points, in order that single trees may grow freely, and afford the shelter of hedge-rows. It is only under the wing or protection of such woods or beltings, that in many districts single lines of trees can be successfully cultivated. However congenial the soil may be to the species, unless the severity of the blast is temperate, and the cutting winds mollified, hedge-row trees will fail to attain a size valuable as timber, or useful in continuing the shelter from field to field, so as in any degree to equalize the climate throughout the year, or to combine utility with beauty—the purposes for which such plantations are generally formed.

The failure of hedge-row trees throughout Scotland is of more frequent occurrence than that of any other description of plantation. This often arises from the trees employed having been allowed to grow to a large size without having their roots adapted for removal by frequent transplanting ; sometimes from their not being sufficiently protected from cattle ; often from the exposure of the ground, and the unfitness of the plants for the situation ; and not unfrequently from a combination of these circumstances.

In all windy situations, plants should be employed stout in proportion to their height, and with lateral branches down to the surface of the ground. The figure of trees varies considerably, according to their kinds, their age, and according to the physical circumstances in which they are placed; such as soil, situation, climate, and above all, to their proximity to other trees. Their natural form and outline under different circumstances can only be known when they stand alone. The sturdy oak alone in poor soil and cold elevated situations becomes a bush; in the rich and sheltered valley plantation it rises a lofty tree with a tall trunk.

In the growth of useful hedge-row timber the English elm is the tree most generally cultivated in England. When a plant, it naturally forms a bushy root; and if properly nursed it admits of removal at a size beyond that of most trees. Its figure is erect, and the spread of its branches does not extend very far. It forms a useful pollard, admits of being frequently lopped, and yields much useful timber in a short time. The suckers which it produces around its roots certainly form one of the greatest disadvantages to its being used along cultivated fields; but that circumstance, yielding a supply of young plants, along with the advantages of the tree, no doubt accounts for its being cultivated so generally throughout England.

Next to the elm, various sorts of oak are to be recommended as valuable hedge-row trees, although generally they do not stand so erect as the English elm; yet they are less destructive to the crops in their vicinity; their roots generally strike deeper than most trees, and consequently are less dependent on the surface-soil for their support; and being late in expanding their leaves, they do not overshadow the crops in their vicinity early in the season. All the common varieties of oak are adapted for hedge-rows. The Turkey oak grows the fastest, and is of an upright figure, until very old, but it is less valuable as a timber tree than most other oaks. The ash is, in some districts, of very frequent occurrence as a hedge-row tree; but it is ruinous to grain crops within the range of its roots, and it can only be recommended along road-sides, meadow and pasture lands, and the like, in the absence of tillage. As a timber tree, particularly for agricultural purposes, it has no superior.

The larch, although seldom introduced into the hedge-rows of highly cultivated districts possessed of a superior climate, is nevertheless a very suitable tree; it forms an agreeable variety, and breaks the monotonous appearance of some districts. It is profitable, being of rapid growth, and valuable as timber, and is less subject to disease in an isolated position than in masses. No tree is less injurious to green crops; its leaves enrich the soil, and when shed are commonly deposited on the surface around its roots. In rough situations, however, it is apt to be bent by prevailing winds, and to become unsightly.

For avenues, where a depth of embowering shade and seclusion are required, the lime-tree, with its large umbrageous head, yielding sweetly-scented blossoms, has no superior. The horse-chestnut also is generally a favourite in such places. The Spanish chestnut, sycamore, Scotch elm, beech, and planes, are all of that large and spreading habit of growth which recommend them for such purposes.

For situations too rough and exposed for trees in general, the sycamore, service tree, mountain ash, beech, Scotch elm, and hoary poplar (*Populus canescens*), are most likely to succeed. The three kinds first named are remarkable for their unyielding character in cold or windy situations, and even at great elevations they grow erect and produce well-balanced heads.

Of evergreen trees, the varieties of holly and oak are the best adapted for hedgerows. Of these, the tallest growing are Turner's evergreen oak and the Fulham oak, a sub-evergreen.

Among flowering plants for ornament, the varieties of thorn, laburnum, and scarlet horse-chestnut are pre-eminent. A ready method of establishing lines of the numerous species of the first-named genus is, by selecting strong stems of the common hawthorn or quick, in a vigorous growing hedge. Such will readily train to a considerable height, when they may be grafted with the varieties and species of the tree. Those most handsome and attractive in flower are the scarlet and double-red; and several other interesting kinds are noticed in our article on THE THORN.

In planting hedge-row trees their roots should not be sunk under the surface beyond their natural depth; the upper fibres should be so situated as to be influenced by every shower. For the first few years after the tree has been inserted, its vigour of growth is much accelerated by the surface of the ground being loosened and kept clear of herbage around a space comprehending the range of its roots.

The mode of pruning trees under any circumstances is of great importance, but never more so than when they are placed in hedge-rows. In the forest, their proximity to one another generally, to a great extent, supersedes the necessity of much pruning; but when situated individually, no part of their management is more important than that this operation should be performed skilfully. It should be attended to early, so that there be no necessity for the removal of large branches. The method of pruning trees for useful purposes appears to be ill understood, or, if understood, it is seldom adopted. The common method is to clear the trunk of lateral branches to a considerable height, and allow the higher ones to take their course. This has a tendency to produce a large head, widely spread and ramified, and, where this figure of growth is desired, we know of no other method which will so speedily accomplish the purpose, because it has the effect of establishing a host of branches equal in magnitude to the leader. This retards the height and adds to the breadth of the tree. Where bulk of useful timber is aimed at, the mode of treatment should be very different. It is then necessary to direct attention chiefly to the top or leading shoot, and to the branches in its vicinity, with the view of continuing the length of the trunk, and preventing it from dividing into forks or clefts. This is accomplished by preserving one leading shoot, and in shortening the competing ones, or such as bear a considerable proportion to the leader, to about half their length; and the same method is recommended in dealing with all luxuriant side-branches throughout the tree, the progress of such branches being impeded, in a greater or less degree, in proportion to the distance from the extremities at which they are cut. By this treatment the principal flow of sap will be directed into the proper channel, which will greatly increase the height of the trunk; and the remaining portions of the side branches may be removed close from the trunk in after years, when they will occasion no blemish in the timber, they being still of small diameter. September is the best month for pruning, in general, as then the wounds 'are immediately healed by the descending sap. The timber of detached trees is generally hard and good, and if cleanly grown by being early and judiciously pruned, it commonly sells at as high a price per cubical foot of measurable timber as clean timber grown in a forest. But if it is coarse, with large knots, or mutilated by the removal of large branches, the value is often depreciated forty per cent. In some districts the value of hedge-row timber is often reduced one-half by the mischievous practice of nailing paling to trees to save the trouble of inserting stakes. It almost invariably happens that the nails become imbedded in the timber, and frequently no external mark gives evidence of the circumstance, and it is only discovered when the carpenter's tool comes in contact with the iron in cutting up the timber. Owners of timber, and those who have charge of estates, should guard against this evil wherever it is practised.

XIV

COPPICE.

COPPICE, or Copse Wood, consists of trees naturally grown or planted, which spring freely from the root, and are cropped or cut down periodically before they attain the usual size of timber trees. There are situations in which the coppice has a finer appearance, and as an embellishment is more in accordance with the accompanying scene than full-grown timber trees, such as on small islands, along the margin of lakes and rivers, in narrow belts, in broken and detached corners of land, and not unfrequently it may afford a vista in the forest, where taller trees would exclude the variety and richness of the scene. But if it is principally with the view of realizing a profitable return that a plantation of coppice is to be raised and husbanded, very few situations admit of this being practised with great success.

The change that has taken place during the last twenty or thirty years in the value of coppice-wood and bark, compared with that of ordinary plantations, is very great. The introduction of foreign bark and other substances adapted for tanning leather has reduced the British bark to about half its former value; the best oak bark now seldom exceeds £7 per ton, notwithstanding the rise of all sorts of wages, which enhances the cost of harvesting the commodity, while, during the period referred to, the timber of ordinary plantations has greatly advanced in value.

There are some soils on which coppice grows freely for ten, twenty, or perhaps thirty years, until it attains to nearly that number of feet in height, when its progress becomes almost imperceptible. This may sometimes arise from the effects of a bad climate, but more frequently from an inhospitable sub-soil; and when such soil cannot be improved by drainage, it is most profitable to devote it to the growth of coppice. Whenever trees of any kind arrive at such a period of their growth that their yearly increase does not amount in value to the interest of the money which, at the time, they would produce, if revenue is purely the object, they, of course, should be cut down. This is more particularly the case with coppice, since it not only springs again, but in some situations yields a far greater value and bulk of timber for particular purposes, when cropped two or three times in fifty years, than the trees would produce if allowed to stand during that period. The oak is a common coppice tree, and valuable chiefly on account of its bark. It is frequently found in woods, sometimes indigenous, and sometimes planted, where its growth becomes stunted even in youth, where there are no vigorous shoots on the extremities of the branches, but, instead of this, a curled and feeble termination of the spray. Dead twigs will occasionally be seen towards the top, and, above all, the bark will cease to expand, and no longer exhibit those light red and yellow perpendicular streaks in its crevices, which are sure evidence of its expansion, and of the consequent growth of the wood underneath. Stunted oak of this description is often found to grow freely for some time after being cut down, and in some soils such coppice is much improved by the insertion of larches as standards.

For the sake of its bark the birch cannot be now profitably employed as a coppice-tree. Of the kinds of birch the common tree is found superior to the weeping variety, as it springs more vigorously, and is more tenacious of life when cut down. But as all the birches are apt to shed their sap profusely on being cut down, none of them are reckoned very permanent.

Besides the oak and birch, the chestnut, the ash, the hazel, and willow are commonly cultivated as coppice-wood, and sometimes the elm, the maple, and the alder. All those spring from stools.

The oak and birch are raised chiefly on account of their bark, and are lopped at various periods, according to soil and climate, from sixteen to twenty-five, or even thirty years. The season of the year for felling trees which yield bark is confined to the end of spring and first of summer, being that at which the bark is harvested. The root-ends, or largest pieces of oak coppice-wood, are usually employed in the manufacture of wheel-spokes, and the other parts generally sell for the purposes of smoking fish, making charcoal and firewood, at from 10s. to 15s. a ton. The value of oak coppice is very variable; but at the age of twenty-five years it commonly yields from £20 to £30 per acre, after paying all expenses.

Birch is less valuable. The other kinds of coppice-wood are commonly used for poles, hoops, hurdles, handles to implements, charcoal, firewood, etc. As these timbers are bulky in proportion to their value, they do not afford to be carried to a great distance before being manufactured. The revenue of coppice-wood is much enhanced by a local demand in the vicinity of the ground where it is produced; and, next to that, it is rendered valuable by a cheap and easy mode of transmitting the timber to market. Chestnut is chiefly esteemed for hop-poles, posts, etc. ; and is durable in all purposes where the wood comes in contact with the ground. Its value for these purposes, however, in many quarters, has been diminished by the thinnings of larch plantations having recently become so plentiful; which timber not only exceeds the chestnut, but almost every other kind, in durability. Ash coppice is generally preferred for handles to implements, hurdles, and for all purposes where strength and elasticity are required. It is esteemed for hoops, in the formation and manufacture of all dairy utensils; and is cut at various periods, according to the purposes for which there is a demand.

The hazel is adapted to dry soil, and is more frequently cultivated as underwood than as coppice-wood. Although it does not attain a large size, yet it yields an early and profitable return, particularly in the vicinity of potteries, and other manufactories, where it is frequently lopped every second or third year, and used in the manufacture of crates. For barrel-hoops, it is equal to oak; for that purpose it is cropped every five or six years. The plant grows everywhere throughout the country naturally, and yields seed abundantly. For particulars respecting the genus, see HAZEL.

The willows consist of a great variety of sorts. For the manufacture of baskets it should be cut down yearly; a good crop in some districts yields a clear revenue of £20 per acre. Almost all the species of willow may be grown for the purpose of basket-making; but the *Salix viminalis*, or common osier, the *S. rubra*, *S. Forbiana*, and *S. stipularis*, are greatly preferable to many other sorts.

For hoops, poles, crate-work, hurdles, scythe and rake handles, the willow is generally cut every five, six, or seven years. The sorts most suitable for these purposes are *S. caprea*, or goat sallow, and its allied kinds. No other species of willow will produce such vigorous shoots in bad soil; and in soil of good quality, after being cut over, shoots of one year will frequently rise to the height of ten feet. The two best species, for the larger purposes of willow timber, are *S. alba*, and *S. Russelliana*. All the species of willow are grown from cuttings. Those adapted for basket-making become enfeebled when cut down yearly, and generally require to be renewed every nine or ten years. The third year they attain their greatest strength, and they commonly show symptoms of exhaustion after yielding six or seven crops—a decay which does not occur in the ordinary coppice willow when cropped at intervals of several years.

Although all the kinds of willow will grow in ground much more moist and swampy than is suitable for any other tree except alder, yet their growth is accelerated and their quality much improved by the ground being drained and thoroughly relieved of stagnant water to within a few feet of the surface.

Coppice-wood is sometimes interspersed with standard timber trees, which convert it into underwood, and are apt to diminish the quantity and deteriorate the quality, except in exposed situations, where a partial shelter may be required. A few standards of oak in favourable situations are sometimes left with advantage, as it comes into leaf at a late period, and consequently does not form a deep shadow, during the opening of the season. Although the larch is not a coppice tree, because it does not spring from the root, yet it is less objectionable than many kinds as a standard in coppice, as it rises in an upright figure, soon becomes valuable, and other trees generally thrive well in its vicinity—particularly the oak, whose roots penetrate to a great depth, while those of the larch spread over the surface ; and thus the two sorts at once bring the whole strength of the soil into operation.

We have already stated that the sorts which yield valuable bark are felled at the season most suitable for the manufacture of that article, but the other sorts of coppice wood should be removed any time between the middle of autumn and the middle of spring. In cutting the timber it is necessary to bear in mind that the stools are intended to shoot forth another crop, and therefore require to be cut clean and smooth, so that water may not lodge, and so low or close to the ground, that the shoots which form the subsequent crop may proceed close to the roots, and not at some distance over them, in which case they would be liable to be blown off. A bill or hook is best adapted for cutting small timber. In removing stout oak coppice, it is the practice to saw over all the stems which stand above four inches in diameter, and to cut the smaller ones with a hook or axe, which, in a practised hand, cutting upwards, leaves the stool unblemished. Where a large oak trunk has been removed, the top of the stool, if it be intended to spring again, should be dressed up into a convex form, sufficient to discharge water. An adze is the most suitable implement for this purpose, care being taken that no part of the bark on the stool be injured. As large stools are often more stiff to yield young growths than stools of a smaller size, it is frequently necessary to remove the surface herbage all around such, to enable them to spring freely.

The second year after the removal of the timber, the stools should be carefully gone over, and all supernumerary shoots should be cleared off. The number to be left must be regulated, in the judgment of the individual, according to the space which the stool occupies, and the purpose for which the timber is intended. In oak copse it is a common practice to leave a few more than those intended to remain until the general clearance of the timber, and these, the smallest and most crowded, are thinned out, and barked, when about eight or ten years old; with this exception, the whole of the copse-wood should be cleared off at one time, as any other method is 'injurious to the remaining crop.

In forming a copse wood of oak, chestnut, and willow, the larger growing kinds should be planted at a distance of five or six feet apart, with larch or fir interspersed as nurses. It is usual to insert oak in fir plantations, after the firs have been a few years planted, and attained the height of a few feet. As larches are of quicker growth, stout plants form a shelter when planted along with the kinds intended for copse. Hazel, and such trees as are only required to attain a small size, should be planted at four or five feet apart; and willows, for basket work, at two feet. Trenching, though generally expensive at the outset, is ultimately found to be profitable.

The inroads of cattle and sheep are not more destructive to any description of property than to copse-wood. The rubbing, the bite, and even the greasy touch of these animals, have a wonderful influence in retarding the growth of young plants. In the absence of more permanent enclosures, the coppice-wood affords materials for a cheap and effectual fence. Where it is cultivated to a considerable extent it is of much advantage to have it in perpetual rotation, so that a portion may be cleared and manufactured yearly. By this means a yearly revenue is derivable from the ground, and the labourers of a district are kept in the practice of thinning, cutting, barking, etc., which will cause these operations to advance more speedily, and in a manner superior to that usually performed by hands unaccustomed to the work.

XV

ON HARVESTING BARK.

THE oak furnishes the bark which is most esteemed by the tanner in the manufacture of leather, and the birch that which is most in repute by the fishermen in the preservation of sail-cloth and cordage. The barks of the larch, chestnut, willow, and some other kinds, are also valued on account of their tanning qualities. The operation of barking or peeling, and the mode of preserving bark, is the same in all the kinds, and the value of the commodity greatly depends on the state of the weather and the care bestowed during the time of its being harvested.

The season begins as early as the sap of the tree circulates so freely as to admit the bark to rise from the timber, which varies considerably in different trees, and is also regulated by the nature of the soil and situation, and by the earliness or lateness of the season. That first removed is found to be the strongest in the tannin principle, and consequently the most valuable. When the tree expands into full leaf and produces young shoots, the bark has deterioriated one half; nor is this the only disadvantage of late barking, for the future growths from stools, which form the following crop, rise but feebly compared to those where the timber has been removed in April or in May.

In detailing the process of barking it is necessary to re-mark, that on old trees, and particularly on the birch, a rough exterior bark or *epidermis* commonly exists, which is of no value; this is removed by an axe, or more readily by an im-plement termed a scraper, which is shaped like a common draw-hoe, but is more powerful, and much sharper. It is

I

found that this rough outside bark does not easily part with the inner bark so early in the season as the inner bark rises from the wood ; but later, when the sap flows more copiously, it is readily removed.

Before the trees are felled, a person advances with a barking-iron or bill, and forms a circular incision, cutting through the bark of the tree close to the surface of the ground, and making a similar incision at the height of two feet; between these the bark is removed. A woodsman follows and notches the tree about two inches deep all round the surface, which prepares it for being cut through by the common cross-cut saw. Immediately on the tree being felled, the smaller branches are cut with an axe or bill, into pieces about two feet long, from which, when tapped over a stone with a wooden mallet, the bark loosens, and is readily removed. The barking-iron is applied in cutting through the bark around the trunk and main branches, at places about two feet apart, and with the aid of the mallet and barking-chisel the main timbers are peeled. The tools used in the various operations no doubt vary in form in different districts ; a heavy axe and cross-cut saw for felling the timber ; a light axe and a hedger's short bill for cutting through the bark—the former also for use as a mallet ; and barking-irons of various sizes, which are blunt duck-bill-shaped chisels, flat on one side and rounded on the other, are the tools commonly used in England. Women, in some districts, and boys, are employed, six or eight being superintended by a man, who lops the branches, and assists in turning the trees as the work proceeds. As the bark is raised from the tree it is classed into two sizes—the smaller into heaps, and the larger covering them, placed with the outside uppermost.

We now come to the most important part, the process of drying, which in a great measure regulates the value of the produce, and in wet weather becomes very precarious. A bark drying-shed should occupy the most airy situation in the forest or in its vicinity. It should consist of a roof, which may be formed of deal, and supported on pillars ten feet high.

Across the house, at the distance of every eight feet, splits of wood should be erected, four tier in depth, forming shelves to dry the bark. The bark, on its removal from the timber, is immediately collected and spread three or four inches deep; the smallest should occupy the lower ranges, and the large bark the upper, with the outsides of the large bark uppermost. Around this drying-shed an open space should be reserved, capable of containing several ranges of shelves, which, when supports and rails are formed, may be set up in a few minutes, and should be taken advantage of in favourable weather. Where no drying-shed is used, the bark is harvested in the open ground, and commonly at or near to the spot where it was produced. This is indeed the more common practice. A set of straight limbs are supported on forked sticks along the surface of the land, and about three feet from it; against these, first the small pieces, then the larger are piled, and over all, forming a roof, the trunk bark is placed, sheltering the whole from the effects of the weather. If the bark be of small size, and showery weather occur while it is exposed, damage must ensue; but if a considerable proportion has been yielded by stout timber, it may, if put up thus with care, be preserved with safety even during unfavourable weather. Of course the most open and airy convenient situation should be preferred.

Another method may be described thus :—A few of the forked branches are inserted into the ground, with the prongs uppermost, to support rails or splits of wood from twelve to eighteen inches asunder, similar to the shelves described for the drying-house; with this difference, that in the open ground the rail on the one side should be placed a few inches lower than the other, so that the surface of the bark, when exposed on the rail, may form a declivity sufficient to discharge water. It is found that rain on bark during the operation of peeling, or immediately thereafter, while it possesses its own sap, does it little or no injury, though afterwards, when but partially dry, it infuses or extracts its virtues. Having erected the timbers, the small bark is laid first on the rails to the depth of about three or four inches, above which a cover

of large pieces is then placed with their outsides uppermost,
which forms a shade and protection for the small. During
the preparation of bark, the forester should bear in mind that
the influence of sunshine on its inner side causes a large
decrease of its weight, by the evaporation of its most valued
juices, which do not escape while the outside is kept uppermost in drying. After the bark has stood on the rails in the
shed, or in the open ground for a day, it is apt to get compact
and mouldy; it should therefore be shifted and disturbed in
a similar manner every twenty-four hours, for three or four
days. That in the drying-shed, when crowded, should be
removed to the outside rails every favourable morning, and
placed under the roof every night, and during rain. In unfavourable weather, two or three weeks are sometimes necessary to dry it in the open ground, but under more favourable
circumstances it becomes quite dry in eight days. It is then
removed into a house and chopped to the size of about two
inches, an operation which is commonly performed by contract,
at six shillings per ton; this fits the bark for the tanner.
The cost of preserving bark must be always regulated by
the price of labour in the district, and the size of the timber
which yields it. One person will strip from stout timber
about five or six cwt. per day; from small timber only about
one cwt.

At Darnaway forest, near Forres, where several hundred
tons of oak bark are frequently manufactured in one season,
the expense of barking from trees ranging in diameter from
six to twenty inches, including peeling, drying, and chopping,
amounts to thirty-six shillings per ton. In England, from which
the larger quantity of bark of course is derived, the process of
chopping and grinding the bark is generally done by the pur-
chaser, who buys it on the ground from the timber-merchant
or landowner. The work of barking, there, includes the felling of the trees, the stripping, and the piling of the bark;
the smaller pieces against longitudinal supports near the
ground, to be afterwards covered with the larger bits, outside
upmost. Large-sized trees may be barked merely—the pieces

being laid in heaps as we have described, for twenty shillings a ton. The value of oak bark, as of other kinds, is exceedingly various; we have known it from £4 or £5 up to £8 per ton. Birch bark is commonly about the same, but larch bark does not exceed one half that price. The sap circulates earliest in the larch, and its bark is speedily removed, but it is very light, and ill adapted for distant carriage, on account of its bulk, which is twice that of the oak bark, and it of late years has seldom been manufactured profitably.

XVI

ON PRUNING FOREST TREES, ETC.

THE utility of pruning hardwood trees is generally admitted by experienced and practical men. It is sometimes denied by those who have witnessed the bad effects of an improper system, such as carpenters and mechanics, who readily discover the evil resulting from the "lopping and boughing" of a bad system, while they are unacquainted with the advantages of early and judicious pruning, which leaves no mark on the future bole, but directs it early into the figure most valuable as timber, and in some cases its effect on the individual tree may be compared to that of the judicious thinning of a plantation, as it directs the energies of the soil to the growth of one trunk, instead of a number of smaller ones. Theorists also sometimes deny the use of pruning, overlooking the frequent necessity of directing the growth of the trunk in the way most suitable for mechanical purposes, and they contend, on physiological principles, for bulk, through the agency of leaves. Although pruning does not in ordinary cases ultimately increase the bulk or weight of wood, yet trees which are *early* and *judiciously* pruned will be improved in quality, increased in their useful dimensions and ultimate value, and will grow in greater numbers on a given space.

But although early and skilful pruning is of advantage to hardwood trees generally, it is not to be recommended for the different species of coniferous trees except under unusual circumstances ; for instance, I have seen it practised with advantage in native and planted woods in the Highlands, where trees were far asunder and bushy, arising with two or three stems, occasioned by being eaten over by cattle or sheep,

or topped by black game, or some other accident. In such a
case as this,—where on account of the thinness of the trees
there are few or none to spare in thinning,—pruning, by
reducing such to one stem, is an advantage.

The figure of coniferous trees is in general all that could
be desired, therefore close planting, and early and repeated
thinning, are all that this tribe of plants generally require to
bring their timber to maturity.

With respect to hardwood trees generally, in some situations the necessity of pruning may be in a great measure
obviated by close planting and timely thinning. These means
are generally most effectual in producing straight and well-
grown timber of every species. Where young trees stand
moderately close, their leading shoot, which is to form the
future bole of the tree, is guided upwards by its own natural
efforts, and as the lateral branches of the one press gently on
those of the others all round, they are prevented from acquiring an undue strength, and ultimately disappear, leaving
straight and clean trunks, which are always of most value,
except in the case of oak timber for shipbuilding, which
should form an exception from the ordinary mode of treatment, as will be noticed in the sequel.

All experienced foresters agree that the most beneficial
pruning is that which begins early, doing little at a time, but
repeating the operation frequently, and directing the ascendency of the leading shoot till the stem of the tree has acquired
a proper form. When trees in a young plantation have produced three, or, very thriving, two years' growth, pruning
should be commenced. The pruning-knife is the most suitable
implement, and where the work is early and frequently
attended to no other implement is required during the whole
progress of forest pruning.

The top is the principal part of the plant that requires
attention, in order that only one shoot may be allowed to
remain as a leader, the others next in size, if not very inferior,
should be headed down to about one-half their length, and all
the stoutest lateral branches shortened in the same manner.

None of these branches need be cut close to the stem, and if the plantation is moderately close this will be all that they require, as they will get enfeebled and fall away; but in more open and airy situations those lateral branches which were shortened may be in four or five years removed close to the stem, before they are beyond the size of being cut off by the pruning-knife. Young plantations should be gone over every second year, until the stems of the trees have acquired a proper form, having an eye to a sufficient girth in proportion to the height, which girth is promoted chiefly by side branches, at the same time bearing in mind that next in importance to keeping the tree in a proper figure should be the preservation of the greatest quantity of its foliage.   It is the general rule to shorten the branch likely to gain an ascendency over the leading shoot; but if the leading shoot is weak, stunted, or unhealthy it is sometimes of advantage to remove it, and prefer the more vigorous one, which through the flow of sap will readily become straight and in proper form.   A few years after hardwood plants are planted it sometimes happens that some of them are found stunted and making no progress; and in the case of oak, elm, or ash, young shoots frequently appear at the surface of the ground.   This is sometimes occasioned by the roots being too bare, or destitute of a sufficient supply of young fibres, or from their exposure to the weather in planting, or subsequent drought, etc.   In such cases the plant should be lopped over at the surface, or just above the most vigorous shoot, which·should be retained for the future tree, and the other suckers should be pruned off.   The lopping of such plants should be performed with a sharp knife by a prac- tised hand, so that the operation may be made without dis- turbing or straining the root of the plant.   It is a common error in the management of plantations to clear the stems of all side-branches to a certain height at the first pruning, and afterwards to operate only on the under branches of the tree. This tends to produce a small trunk, an irregular top, and side branches more vigorous than the leader.   When this is practised in exposed places, not one in a hundred ever becomes a large or valuable tree. Were pruning altogether abandoned, trees of fifty years' standing would generally be of more value, rough, knotty, and forked as a great part of the timber would be, than those subjected to such an injurious method.

It is in hedgerows and other open situations, where trees are apt to ramify into an unprofitable figure, that pruning is of the greatest value; but even in such situations it is not necessary to shorten all the branches previously to their being removed from the trunk, though it is to be recommended in dealing with all luxuriant branches, particularly near the top shoot, and in checking such throughout the tree; the progress of.such being impeded in a greater or less degree in proportion to the distance from their extremities at which they are cut. When trees have advanced from ten to fourteen feet, the oldest and stoutest branches (previously shortened) may then be removed from the stem. Sometimes the small pruning-saw is employed as the most efficient implement, observing that at the junction of each branch to the stem there is a swell or bulge, and the branch should be removed close to the outside of it, at which point the diameter is not so great as at the very bottom, consequently a much smaller wound is occasioned, and sooner healed. When plantations are closely attended to, however, the pruning-saw is seldom required. The knife is the safest implement, its wounds heal most readily, and where the branches are sufficiently checked by being shortened they do not acquire a diameter beyond its power. When trees are from fourteen to twenty-five feet in height, or from twelve to twenty years of age, they generally advance very rapidly, and if not standing close in a plantation, admit of more pruning than at any other period; but under any circumstances trees are much injured by being severely pruned; for, as already stated, pruning is only of much advantage when performed early in those side branches which are apt to bear too great a proportion to the leading branch, thereby modifying the tree and directing its energies gradually to the top, preserving at the same time a sufficient quantity of foliage. All young hardwood trees should have tops long in proportion to their height. A good proportion in a tree of thirty feet in height is twenty feet of top to ten feet of bare trunk; but no given rule in this respect can be exacted for all sorts, as a longer top is requisite in a rough exposure and in poor soil than where the ground is well sheltered and fertile. The skilful forester observes at a glance whether the tree is possessed of a trunk stout in proportion to its height, and, as in thinning, regulates the pruning accordingly. Where height is required he subdues the side branches; where girth of trunk is necessary, he preserves them as the speedy means of obtaining girth.

ꞌ The evil consequences of cutting off large side branches from timber trees require to be stated, as nothing can more readily deteriorate their value, particularly if the branches are cut close to the trunk. This creates a large unsightly wound several inches in diameter; the influence of the sun and weather cracks the timber, which imbibes water; during frost the fissures increase, and rottenness penetrates into the trunk, and although the wound will collapse, and repeated layers of wood in course of time will cover it over, yet the timber remains unsound and much deteriorated. The experienced timber-merchant has a quick perception of the marks of lopping or mutilation, which often reduces the value of timber one half, and where indications of the removal of large branches appear, the forester is sometimes obliged to defer fixing the price till the wood has been cut up. It is mutilation of this sort that has created a prejudice against any description of pruning whatever, particularly with wood-merchants and artisans. Where it is absolutely necessary to remove a large branch, the method most safe for the timber is to amputate it beyond its first side branches, which will generally prevent dead or unsound timber.

As to the season most suitable for pruning timber I may state that the shortening of stronger side branches and of the competing shoots near the leader may be performed at any season; the outline, and consequently the wants of the tree, are best seen during winter or early in spring, in the absence of the foliage. Spring, however, should be avoided with respect to sycamore, maple, and birch, which are apt to bleed like vines until the leaves are developed; but it is found that the best time to remove the branches from the bole is decidedly early in autumn, say August or early in September, then no sooner are they removed than the descending sap immediately cicatrizes the wounds, which close up in course of a few weeks.

Mr. Cree, an authority on pruning, says, "If we form an estimate of the comparative value of pruned and unpruned trees raised within an equal number of years, the advantage which the former possess over the latter is very great. Take twenty-five elms indiscriminately of a size suitable for making naves for wheels, it will be found, if unpruned, that the quantity of timber will not average in each above five *feet*. Twenty-five trees which I have pruned will each contain more than thirty feet of timber when arrived at the age of the trees I have described above. Pruning is equally beneficial to all sorts of deciduous trees."

On pruning oak for naval purposes it is an important object to manage it in such a way as to produce the greatest proportion of bent pieces, or knees, which are always more valuable for shipbuilding than straight timber. It is therefore indispensable that the trees have considerable space; the proximity of roads, rivers, the outside of a plantation, or any circumstance affording open space to some extent, is most suitable for this purpose, and as the close pruning of the stem near the surface of the ground has a tendency to make trees ramify, when young oaks have advanced to the height of from eight to ten feet their lateral branches should be pruned from the surface upwards close to the stem to the height of from three to four feet, and if they are inclined to grow with a straight top *the leading shoot should be cut off*, and two of the strongest lateral branches which take a horizontal direction should be left at those points where they will be least confined, and the next largest branches should be shortened at the same time. In the course of three years, or, if the plantation is very healthy, in two years, the trees should be gone over a second time. Those side branches formerly shortened should then be removed close by the stem, and the old top reduced to the point where the principal leaders take a horizontal direction.

The natural habit of the oak is favourable for the production of crooks adapted to this purpose, and where a tree of suitable description is found it should be carefully assisted in its progress by removing every obstruction in the way of its horizontal growth. When a tree or limb is found to grow in the desired figure, the removal to a considerable extent of its ' side branches having an *upward* tendency, and the leaving of such as are on the under side, tend to establish the spreading position of the limb. This method will generally occasion crooked trunks, which are always more valuable, of whatever figure, than straight trees. In many cases the shoots will incline to the perpendicular, and not grow in the desired form ; but a far greater proportion of trees thus treated will be of a superior mould for naval purposes than are found among those where straight leaders have not been removed.

The pruning of trees for picturesque effect on lawns and pleasure grounds being a matter of taste, where we have not the standard of utility to guide us, very little need be said. Some consider that trees which expose their trunks to a great height are not only most valuable but also most ornamental. But to display the peculiar outline and ramification of each species to the greatest. advantage, they should be left to their own efforts, either standing singly or in masses, without anything being done to them beyond the removal of a shoot on the young plant when it happens to be too bushy, or the amputation of a decayed branch in the more advanced state of the tree. In nature, few objects are more lovely and magnificent than a lofty tree, happily situated in a congenial soil, standing unmutilated, with foliage suspended to the surface of the earth, and unveiling only here and there a part of its glossy trunk of goodly dimensions.

For pruning, and the treatment practised on the native

Highland pine plantations, see Chapter XI., on "THINNING PLANTATIONS."

Grafting in plantations for embellishment.—In plantations adjoining pleasure grounds, along the side of drives, or where it is desirable that the ordinary species of trees should be converted into kinds more picturesque in form or more attractive in foliage, this is readily accomplished by grafting, particularly if the trees are not more than twenty or thirty years of age. The simplest and most successful method of grafting such is to saw off the top where it is only an inch or two in diameter, make a slit about an inch and a half long in the bark of the stock, raise the bark with an ivory handle, to make a space for the graft or shoot to be inserted, which may only be six or seven inches long; prepare it by a smooth slanting cut on one side, slip in the prepared scion with the cut side next to the wood to the length of the cut of one inch and a half; tie round with mat, and cover closely with grafting clay all over the wound on the stock. After the clay is dry, and all fissures filled up, the ball may be covered over with moss or meadow hay, and tied over to insure safety and exclude severe drought. When the stock at the point of grafting is older and of several inches in diameter, another, and the easiest mode, is, after sawing off the top, to tie the stock round tightly for a few inches beneath the point of amputation, and force down a peg of hard wood, or any hard substance, between the wood and the bark, in the shape of the prepared scion, then withdraw the peg and insert the scion, pressing it tightly into the incision; by this method two or three grafts or scions may be inserted around the edge of the same stock, then clay as recommended. The month of March is the ordinary season for the operation, or just as the buds are beginning to swell. When the graft has grown a few inches the clay should be removed, and the bandage retied, adding a stalk to support the scion from being broken off by wind. Among ornamental trees the oak affords a great variety of evergreen and sub-evergreen,—such as Turners, Fulham, and Lucombe; and few trees are more interesting in autumn than the brilliant foliage of the scarlet oak.   The family of maples are among the best for autumn embellishment, all of which may be inserted on the sycamore or on the Norway species.   The scarlet, double-flowering, and other varieties of thorn, scarlet horse-chestnut, purple beech, weeping ash, and weeping elm of sorts, are all worthy of cultivation.   The common trees of the species form the stocks best adapted for being engrafted with the rarer sorts, and the nearer the kinds are related, the union is rendered the more perfect and permanent.

XVII.

HEDGES.

FEW improvements enrich the general appearance of a country, or increase the value of property more, than hedges, provided they are properly managed. The preparation of the ground is the first important point. It should be trenched, if practicable, to the depth of from 20 to 24 inches, and 4 feet in breadth, with the surface placed in the bottom, and 18 inches of breadth of surface added to it from each side of the trench, care being taken that the ground from the surface be covered with at least 14 inches of clean soil, which will prevent the weeds from vegetating. The additional surface will raise the trenched part considerably higher than the former level of the ground. When the subsoil is of inferior quality, it does least injury when turned uppermost; but where gravel, sand, or any inferior subsoils are found in large quan- tities they ought to be removed, and the space filled up with soil of good quality. This preparatory process should be performed in autumn, that the frosts of winter may pulverize the soil. Some subsoils, such as stiff clay, bog earth, or whatever does not readily become pulverized, are unfavourable to the growth of plants, and should therefore be exposed to the influence of the atmosphere for a year before being planted, during which period the ground should be dug or forked over once or twice, giving at same time a liberal supply of manure. No plant advances more rapidly under the influence of manure than the common hawthorn or quick, and when manure is well incorporated with the ground before the plants are inserted, it acts the more readily.

Planting, etc.—The proper time for planting generally depends on the nature of the soil. It is to be recommended that all dry soils should be planted before winter, or early in spring. Few plants break into leaf earlier than the common hawthorn, and in all cases its early insertion gives it the most vigorous start during the ensuing summer. Whatever species of plants are employed, they should all be planted erect. When they are inserted into their places in the notch, well-'rotted manure should be applied, mixed up with the soil placed at the roots of the plants to a breadth of about two feet, unless the ground has been previously enriched.

Laying plants in a horizontal position, on the brink of a mound, always retards the growth, prevents the future application of manure, and in some measure deprives them when young of the genial influence imparted by showers; and although by this method the ditch and mounds nearly form a fence at the outset, and render the strength of a hedge less requisite, yet the plants commonly require to be protected by paling, and during my experience, I never found plants so treated keep pace for two years with those planted perpendicularly.

Where the soil is very damp it should be drained, and the site of the hedge being trenched in the manner pointed out, should be made up to a sufficient height, according to the species of the plant intended. Where dry, it should be raised on each side of the plants, so as to leave them less exposed. Having thus prepared the ground, stretch a line along the ridge where the hedge is intended, and cut out a notch close to it, of sufficient depth to contain the roots of the plants. Prune the extreme fibres of their roots, and place the plants at their proper distances against the notch. They will then stand straight, with their roots inclined to that side of the line on which they were planted, it being observed, that although no ditch is made in the meantime, yet it is contemplated to form one in four or five years, when the hedge will be of sufficient height to remain without further protection ; and therefore it is necessary in planting that the roots project to the opposite side to that on which the trench is intended. Should it be resolved not to form a ditch along the hedge in a few years after it is planted, the hedge may be continued, as it stands slightly elevated above the ordinary level of the ground. The advantage of the ditch is, that it helps much to strengthen the fence on the side on which it is formed, and by casting the principal part of the soil from the ditch to the opposite side of the hedge, it places the hedge on a ledge or slope, rendering it easily cleaned, and making it more efficient as a fence, and the ground cast out prevents the roots from injuring the adjoining crops. The ditch may be formed large or small according to the circumstances or the nature of the situation, and it is recommended to be formed when the hedge is just about to become a fence of itself, and at the time the fences for its protection are to be removed.

The mode of planting the different kinds of plants is the same, but the distances, the quality of soil suitable for each, the size of the plants, and the kinds that may be associated, are circumstances that require to be noticed in reference to each species.

Hawthorn (Cratægus oxyacantha).—This is the best of all plants for an efficient fence in any soil of ordinary quality. There are several varieties of the species, distinguished by larger leaves and berries, and of more vigorous growth, but the differences are not important. In selecting plants their strength should be attended to, and not their age or height. Plants six or eight years old, that have been twice or thrice transplanted, and are fibrous-rooted, are always more profitable than smaller ones, although they are double the price per thousand ; they admit of being planted thinner by a few inches between the plants, they spring more vigorously, are subject to fewer casualties, and require protection for a shorter period. Few plants suffer more readily by having their roots

K

much exposed to the atmosphere, particularly in the drought of
spring. The plant comes early into leaf, and should be planted
out in autumn, in winter, or early in spring. Early planted
hedges, compared with late planted, generally show a great
difference during the first summer. The practice common in
the east of England, which I have witnessed, of exposing
large lots of "quick" to the influence of the weather on
market days in spring, cannot fail at least to retard their
growth very much during the succeeding summer. Plants
two years transplanted, of the ordinary size, are generally
inserted six inches apart in a line of hedge. Those of a
superior strength, from a half inch to one inch in diameter,
may be placed from eight to ten inches, according to their
strength. Such plants should be cut over within two or three
inches of the ground-mark before being planted; after being
inserted the stumps should only be visible above ground;
thus strong plants, in ground well prepared, often produce
several shoots each, some of them not unfrequently attaining
a height the first season of upwards of two feet. After planting, nothing further than cleaning by hoeing and raking is
required for the first two years, when the hedge should be
slightly pruned or equalized with a hedge-switcher, pointing
only such extreme shoots as are higher or extend beyond
the medium growth of the plants. When the plants employed
are much smaller, it commonly requires three years before it is
necessary to prune the hedge. With small plants, a hedge
composed of a double row gets more readily into shape, and
becomes more compact at an early date, the lines being ten to
twelve inches apart. One line however of strong plants is
generally the best. After the hedge receives its first dressing
or pruning, its vigour will be greatly increased by digging in
on both sides a good layer of well-rotted manure, similar on
each side to that applied to a drill of well-manured swedes.
If the hedge is properly kept, no perennial or rooted weeds
will be found within the trenched space of four feet in breadth
allotted for it. If such, however, have been allowed to accumulate, they should be forked out before the application of

the manure, which should afterwards be dug into the clean ground; in this case a double digging is occasioned in consequence of previous neglect, by allowing weeds to get entangled with the roots—one of the common sources of failure in hedges.

Few plants can advantageously be mixed with thorn in the formation of fences; the best I have experience of is the evergreen holly. When a plant of holly is inserted here and there among thorns, although it may remain for some years comparatively obscured, yet I have found that it ultimately appears in vigour, and has a tendency to prevail and spread rather than to be suppressed. This arises from its being an excellent underwood—growing well in the shade of other trees. It has no tendency to weaken a hedge, or to produce gaps, and when regularly interspersed in a thorn hedge it adds to its shelter and embellishment throughout the months of winter. I have also found the evergreen privet thrive well when interspersed with thorn; it adds to the closeness of the hedge, particularly near the ground, and affords a freshness to the aspect in winter, but being a feeble plant it contributes nothing to the strength of the fence.

The Beech (*Fagus sylvatica*).—This is a very useful plant for hedges or screen fences. It luxuriates in dry, rich, loose soil, but is apt to become diseased when placed near the rise of stagnant water. It sometimes forms a fence in soil so dry that the hawthorn fails to thrive.

The usual size of beech hedge plants is from eighteen inches to two feet. Such plants are generally two years transplanted and four years old, and range in price from 15s. to 25s. per thousand, according to quality and the supply, etc. Plants for hedges should be grown thin in the nursery lines, and well furnished with branches near the surface of the ground, and inserted as recommended for thorns, according to their size, eight, ten, or twelve inches asunder. The plant naturally grows with a good shape for a hedge plant, it also roots well if frequently transplanted, and by affording it sufficient room it is capable of being removed in safety into soil of ordinary quality, where with little or no protection it almost forms a fence at once. The most desirable feature in those large plants is breadth or spread of branches close to the surface of the ground, which necessitates ample space in the nursery lines, and removal of the top shoots or excess of upward growth. Where fences of this description are required on a large scale, it is recommended that the nursing of the plants should be adjacent to the site of the proposed fences. The preparation of the ground for the plants should be similar to that for hawthorn, but as a greater depth of trench is required for their roots, the surface herbage of the trenched ground should be placed at a depth not to be disturbed by inserting the large plants into the line of hedge, unless it has had a sufficient time to decompose before planting. Along with well-decomposed manure lime is found to be of great advantage in the growth of beech. The hedge may be formed any time in open weather between October and April. If the situation is much exposed to the severity of the weather, planting in March or the opening up of the season is preferable to earlier planting, which would subject the fence to be disturbed by the winds for months before it would begin to take root or establish itself in the ground. Such plants should be inserted as close as their branches will allow them to stand, which is generally a foot apart; when planted it is only necessary to prune off the more straggling side branches and the more aspiring tops with the pruning-knife, and dress into shape with the switcher after the first year's growth.

The Hornbeam (*Carpinus Betulus*).—This plant is sometimes reared for fences. It grows close, and yields a great quantity of leaves. It grows rapidly with the same treatment as the beech, but is inferior in strength or resistance.

The Holly (*Ilex aquifolium*) is a very slow-growing plant when young, taking generally about eight years to attain the height of two feet, with a proportionable breadth and well furnished roots. It would be a very easy matter to produce plants of this height in half that time; this is readily effected by keeping them close together in the nursery lines, and allowing them to remain without transplanting them, but this would render them bare, and unfit for a hedge, and bad for any purpose,—for unless they are frequently removed their roots get bare, after which they are very apt to die on being transplanted.   Plants one foot high, which have been twice transplanted in the nursery, are commonly used as hedge plants; they are sold at about £2 per thousand.   When bushy plants, well rooted, about two feet high, can be obtained, they are worth £6 to £8 a thousand.   A dry, rich, loamy soil is congenial to their growth.   I find moist, cloudy weather in the end of August or in September to be very suitable for their removal; then their roots take to the soil at once, become fixed before the winter, and start in spring with all the vigour of longer established plants.   The next best time for transplanting is in spring, or in wet weather early in summer, provided the roots are not much exposed.   With respect to this the plant is very sensitive.   When healthy and of two feet in height it is afterwards by no means a slow-growing plant, and it forms a more desirable fence than any other evergreen tree.   (*See* HOLLY.)

The *Crab-apple* (*Pyrus Malus*) and the *Crab-pear* (*P. communis*) very readily make efficient fences; the only objection to them is that they are subject to the attacks of caterpillar and bug.   Strong plants are to be had in nurseries at the price of hawthorn of the same size.   Their treatment is in all respects the same as that recommended for the hawthorn, but they do not make so close and compact a hedge as that tree.   (*See* PYRUS.)

Elder (*Sambucus nigra, S. virescens, S. racemosa*).—These species of elder make a valuable screen fence in exposed situations; they retain their vigour at a great altitude and in a diversity of soils in which few other plants will exist.   In cultivated fields with a good climate the elder is seldom or never used as a fence, as it occupies a wide space, does not rise in a compact form, is too soft and yielding as a fence on level ground, and impoverishes the soil in its vicinity.   I have, however, seen the elder used as a hedge along the top of a two or three feet dike or bank of earth, and cut twice a year, in June and in October, which alters its ordinary habit of growth very much, rendering it close and compact; and in such situations it forms an efficient fence and valuable shelter.

The Common Privet (*Ligustrum vulgare*).—Of this plant there are several varieties, of which some are much better ever-greens than others. It is most readily grown from cuttings ten or twelve inches long, planted in September in shaded, sandy soil. These should be inserted two-thirds into the ground in nursery lines, and the soil firmly tramped about them, or the hedge may at once be formed with cuttings or slips. The plant as a hedge is very ornamental when well kept and often pruned, particularly in the end of August, which enables it to push out foliage afresh, which forms its winter clothing fresh and green. Of itself, the plant is too feeble where a strong fence is required. It is therefore often used on the top of a dwarf wall, and for garden and orna-mental purposes.

The Yew Tree (*Taxus baccata*).—This is the most ornamental of all evergreen hedges, but as it is poisonous it should not be placed in any situation accessible to animals that would brouse on its twigs. Its growth is compact and slow ; it is easily dressed, and when well cared for it forms an evergreen wall so close that small birds can hardly enter it. Its roots grow as compact as its branches, and it does not scourge the ground in its vicinity; it is rarely touched by any disease, is very hardy and permanent, hence its adaptation for green walls in gardens, nurseries, and orna-mental grounds.

It is a usual practice in some nurseries where there is a constant demand for hardy ornamental plants, to plant the English yew in ample space, and train it into a hedge figure, and by being frequently removed its roots become a compact mass of fibre, insuring its perfect safety on being transplanted, four or five feet high—a ready-made fence of the most ex-quisite colour and polish. Frequent transplanting retards the plant considerably.   Plants inserted as a hedge a foot high, and that distance apart, and allowed to remain without being disturbed, will advance twice as fast as those that are transplanted every other year.   When young, the plant only grows a few inches yearly, and the most vigorous established yew hedge seldom yields shoots a foot long.   (*See* YEW TREE.)

The Whin or Furze (*Ulex Europæus*) is seldom cultivated as a nursery plant, and when required as a hedge should be sown on the line of fence.   It is not permanent, being often killed with frost, particularly at a great altitude.   It however readily springs from the root.   Of itself it is not sufficient for a fence, but placed on the top of a mound or bank of dry soil, it forms a fence where few other plants would luxuriate.

For a whin hedge the ground should be dug over in autumn, and if limed and manured so much the better.   Well-rotted manure and ashes are fertilizing elements for the whin.

In open weather in winter, or early in spring, the seeds should be sown.   A double drill is to be recommended, as it accomplishes the object more speedily.   Stretch a line on the site of the hedge when the ground has been dug over, cleaned, manured, and pulverized, and with a draw-hoe draw a rut on each side of the line similar to that for garden pease, but only about one inch deep; into these deposit the seeds, thick or thin according to the quality.   The price of seed commonly ranges from 1s. 6d. to 2s. per lb.; which, if the seed is fresh, should be sufficient for 80 or 100 yards of a double drill; the covering is readily effected by drawing a rake along and thus closing up the drills.   The drills should be cleaned the first and second years, until the whins prevail over the neighbouring vegetation.   Pruning with the switcher is all that is afterwards required, and the hedge is made the more permanent by being pruned so frequently as to be prevented from yielding seed.

Ditching.—Where land is not so wet as to render the formation of a ditch and mound indispensably necessary before planting a hedge, these may not be formed until the paling is removed, at which time the plants will have become a fence. In forming the ditch along a straight hedge, a line should be extended about eighteen inches from the stems of the plant, which line forms the side of the ditch next the fence, and this side requires to be considerably sloped. Where the soil is dry and friable, the side of the ditch will be more oblique than where it is firm ; consequently the ditch must be wider. In some situations by road-sides, only a small trench is formed, about two and a half feet wide at top, one and a half deep, and one foot broad at bottom ; but a ditch of a common size is four and a half feet wide, two and a half feet deep, and one foot and a half wide at bottom. In forming these, the greater part of the earth is cast over to the opposite side of the hedge, which is commonly about four feet high. It is therefore necessary for the person who finishes the bottom of the ditch to cast up the earth on the surface, while another turns it over to the opposite side of the hedge, placing a part between the hedge and the ditch, and raising it to the depth of five or six inches. In this way any vacancy may be filled up by fixing the adjacent branches into the soil. The ground thus transferred is made high along the side of the hedge, and sloped down into the field, which prevents the hedge from impoverishing the soil or injuring the crops.

Fences for Protection of Hedges.—The most judicious mode of protecting young hedges depends on the resources of the district in which they are situated. It is therefore unnecessary to enlarge on this subject. Where thinnings of plantations or heavier timber abound, wooden paling should be adopted, and the figure or description of the erection will depend on the kind of wood available. The smallest thinnings of larch and fir plantations are suitable for upright fences, and larger timber for horizontal rails. The cost varies much in different localities. Where timber is scarce, recourse should be had to wire fencing.

Sunk Dikes.—In forming a hedge and sunk fence there is no method more effectual, where suitable stones are at hand, than to cut out a ditch, casting up the best soil to one side, on the line where the hedge is to be planted, facing it up with stones to the height of from two to three feet, and planting the hedge about eighteen inches back from and alongside of the mason work.   The two surfaces of good soil form a mound above the ordinary level of the adjoining ground, and, supported by the stone facing, afford ample drainage ; and these advantages of position, soil, and drainage, commonly yield an efficient fence at an early date.

Cleaning, Manuring, and Pruning.—It is well known to all who have had experience in rearing hedges, that the first and most important point is to have the ground thoroughly cleaned, and that this should be carefully attended to in so far as rooted weeds exist in the ground before the plants are inserted.   A hedge for the first three years requires to be cleaned by hoeing on each side and hand-weeding close to the plants.   During this period no agricultural crop should be allowed to grow within four feet of the plants.   After a hedge has been three years planted, it should be manured on both sides, in all parts where it is not thoroughly vigorous; the manure should be well made or decomposed, and immediately dug in on being spread.   The steel fork is more suitable for this purpose than a spade, as it is less liable to injure the roots.   The best season for manuring and digging is in autumn or in open weather throughout the winter, and the influence of manure is not more apparent on any plant than on the hawthorn.

Pruning should begin three years after planting, or if the young hedge is very vigorous, after two years' growth.   The most approved figure, and that most easily kept, is wedge-shaped—broad at the base and tapering to a point at the top. But for ornamental purposes the figures may be diversified. The autumn or the months of winter are the usual time for pruning hedges, but a second pruning at midsummer has the effect of increasing their closeness.   When young hedges are apt to get bare or thin near the ground and show a vigour at the top, it is advisable to prune the upper half only, once or twice during the summer months.   This has the effect of directing the growth towards the strengthening of the lower branches, and keeping the hedge in proper shape, and the hedge-switcher is the most speedy and efficient implement for all the fences of field or forest.

The four most important points to be attended to in the raising and managing of live-fences are well-prepared ground, well-prepared plants, thorough cleaning, and thorough fencing. These things attended to, hedges soon become an ornament, a shelter, and a sure protection.

The practice of planting timber trees in the line of hedge has a very injurious effect on the fence, and seldom fails to produce inequalities and gaps. Some trees are less ruinous to underwood than others, but plants in hedges cannot maintain a uniform vigour and regularity of growth when subjected here and there not only to the shade and drop of taller trees, but to the exhausting influence of their roots, so strong and vigorous compared to those of the plants of a pruned hedge. The kinds of hedge plants most suitable for underwood are holly, yew-tree, and privet, but the thorn (or quick) is readily injured by being overspread by any other tree.

Old thorn hedges are not unfrequently met with in a stunted state, entangled with weeds and full of gaps, so that to form a fence requires the assistance of paling. To improve this state of affairs the hedge should be sawn over near to the surface of the ground, and a yard wide of ground on each side should be forked over, carefully shaking out all the roots of weeds, after which a liberal supply of well-prepared manure should be dug in during autumn or in the winter months. Where vacant spaces exist the soil should be cast out, and a trench formed four feet wide and about two deep, into which fresh soil from the field should be deposited and well manured; a single or double line of thorns should then be inserted ; the plants should be strong and well rooted from being often transplanted. Such plants after being topped will yield strong shoots the first year. The planting of thorn should be performed in the end of the year, or in open weather in winter.

For this purpose well-nursed plants of holly or of beech are also suitable for filling up gaps in hedges. After a cut down hedge has made the growth of two summers, it should be formed into shape by being pruned; before being so, however, the hedge should be gone over, and all small vacancies which were not of such dimensions as to necessitate the insertion of young plants should be closed up by training the adjacent shoots into the vacancy and fixing them with a forked stick, or by tying them into the proper position with willow twigs. A hedge thus treated requires to be thoroughly protected, and kept free from weeds until it becomes a fence.

XVIII

THE PINE TREE.

THE genus *Pinus* is the most important of any belonging to the natural order *Coniferæ*, and perhaps the most valuable of any genus of ligneous plants. It consists of evergreen trees, natives of Europe, Asia, and America. Most of the species produce timber of a great size, abounding in resin. Their leaves are generally needle-shaped, disposed in groups of two, three, or five, enclosed by a scaly sheath around the base of the group. Pines generally flower in May and June; the male and female flowers are separate on the same tree; the cones become ripe in the end of the second year, or eighteen months after the time of flowering. In their native countries pines generally grow in masses, to the exclusion of other trees; and the genus is remarkable for yielding timber of large dimensions on poor soil in elevated situations, and frequently they are found on the extreme limits of arborescent plants. Some of the species are therefore valuable in subduing the severity of the climate in exposed districts, where they are generally planted as the forerunners to broad-leaved trees, and such as are not adapted to endure the severity of an exposed situation.

All the species have a tendency to sport by cultivation in soil and climate different from that in which they grow wild, and by this means alterations in form and foliage are produced which tend to obscure the distinctions between the ordinarily recognised species.

P. sylvestris: the Scotch Pine.—This is the most valuable timber tree of the genus in Britain, or even in Europe. It is found in great perfection in native forests in the Highlands of Scotland ; it also abounds in a wild state in many parts of Norway, Sweden, Russia, Poland, and Germany. It is almost invariably in an open or heathy soil that it springs naturally ; a grassy vegetation, or a close herbage of any kind but heath, is hostile to the growth of the young plant. In grass it seldom appears a close crop, even in the vicinity of the native forests, where the seeds are thickly dispersed ; whereas in moorland, with a short heathy cover, the seeds readily come in contact with the ground, and vegetate. The stems of the heath are a sufficient protection for the young plants, and open enough to prevent them from damping off through confinement. In collecting the seeds of Scotch fir, it is of great importance that they should be gathered from the best native forests, such as those of Abernethy, Duthil, Glenmore, and Rothiemurchus on the Spey, or Braemar on the Dee, where the trees are found in an indigenous state. There exist many varieties of Scotch pine, but these are found very rarely in the best native woods ; there the foliage and figure of the trees are nearly all alike, and their timber of the same age is uniformly red, hard, and resinous. But where the tree is removed from its native habitat, and repeatedly propagated in a different soil and climate, it runs into a number of varieties of foliage and form, and after a few generations in cultivation it becomes degenerate, short-lived, and its timber is comparatively worthless.

The first published account we have of the varieties of Scotch pine is in a *Treatise on Forest Trees*, by the Earl of Haddington, published in 1760. His Lordship says, " When I cut firs that were too near the house, there were people alive here who remembered when my father bought the seed. It was all sown together in the seed-bed, removed to a nursery, and afterwards planted out the same day. These trees I cut down, and saw some of them very white and spongy, others of them red and hard, though standing within a few yards of one another. This makes me gather my cones from the trees that bear the reddest wood, as I said before." Boutcher, in 1775, says, " It has been an old dispute whether there be more than one sort of the Scotch pine or fir," but the difference of the wood cannot be owing to the age of the tree and the quality of the soil, for he adds that he has seen many pine trees cut down, "of equal age, in the same spot, where some were white and spongy, and others red and hard."

A very minute detail of the true kind was published in 1811, by Mr. George Don, of Forfar, who describes several varieties in a planted wood near that town. In describing the cones of variety No. 1, a degenerated sort, he says, " The cones are considerably elongated, and tapering to the point, and the bark of the trunk is very rugged. This variety seems to be but short-lived, becoming soon stunted in its appearance, and it is altogether a very inferior tree. It produces its cones much more freely than variety 2d; the seed-gatherers who were to be paid only by the quantity, and not by the quality, would seize upon the former and neglect the latter." Of No. 2 he says, "Its cones are generally thicker, not so much pointed, and they are smoother than that of variety No. 1. The tree seems to be a more hardy plant, being easily reconciled to various soils and situations. It grows very freely, and quickly arrives at a very considerable size. This is the sort which, I conceive, might constitute a distinct species, and from the disposition of its branches I would be inclined to call it *Pinus horizontalis*. May I here be allowed to conjecture that the fir woods which formerly abounded in every part of Scotland, and the trees of which arrived at a great size, had been of this variety or species? I have certainly observed that the greater part of the fir woods of the present day, and which are so much complained of, are of the common variety, or variety 1st." Mr. Don adds that No. 2 retains all the good qualities ever ascribed to the Scotch fir, and accounts for the "decline" of the tree in this country from the circumstance that No. 1 produces its cones much more freely than the other, and seed-gatherers, who are paid by the quantity and not by the quality, seize upon the cones of No. 1, to the neglect of the better sort.

Mr. Don's description of the cones of No. 1 accords exactly with that of the third or fourth generations of the tree degenerated by cultivation away from its native soil and climate, in which case its cones are always long and tapering, while the more pure the tree is, as in its native locality, the shorter, rounder, and lighter in colour are the cones; so distinct in appearance that any seed-collector of experience has not the smallest difficulty in knowing at once from their appearance whether they are gathered from the native Highland forests or from degenerate plantations.   Cones are always most abundant on the latter.   Though Mr. Don's description of the trees as they appear is very correct; though he speaks of the "defect" and "decline" of the Scotch fir, and recommends the cultivation of the well-marked variety; and though his remarks have been of great use in directing notice to the tree, yet it does not appear distinctly that he attributed the defect to degeneracy, but to the cultivation of a bad variety, although he adds, "May I here be allowed to conjecture that the fir woods, which formerly abounded in every part of Scotland, and the trees which arrived at a large size, had been of this the best variety?"

Several instances are known of plantations grown from seeds during last century from the celebrated native forests on the Spey, and although they occupy soil of various qualities, the timber in all these woods has been famed for its quality, while, in several instances, adjoining woods of the same age, and on the same description of soil, grown from degenerate plantations, yielded wood very inferior; the march boundary of the lands sometimes forming the line between the good and the bad timber.

Botanists generally agree that none of the differences which the tree assumes are sufficiently distinct and permanent to constitute a specific character; but of this tree, Loudon, in the *Arboretum Britannicum*, justly remarks, page 2150, "The reason why we wish to keep every variety and sub-variety as distinct as possible is, that in the practice of arboriculture, whether for useful or ornamental purposes, a variety is often of as much importance as a species, and sometimes indeed more so; for example, in *P. sylvestris*, the Highland variety is known and acknowledged to produce timber of a superior quality to the common kind." Numerous instances of the propagation of this tree from different sources tend, in every way, to establish the fact, that this tree, by cultivation for several generations, is very apt to become degenerate; and as it not only yields cones most abundantly at a low altitude, in a district uncongenial to its best form of development, but also produces them at a much earlier age than in the Highlands, degeneracy is thereby accelerated throughout the country.

The natural law of deterioration appears to be somewhat general: the finest variety of wheat obtained from the genial climate of the south, when sown in some of the more unfavourable parts of North Britain, generally produces a good first crop, but the experience of the agriculturist tells him that the variety changes, and that it is more profitable to renew the stock than to continue to reproduce from the seed of his own growth.

The effects of soil and climate were pretty generally understood throughout the north by the rural inhabitants of the last generation, in the cultivation of flax. It being an annual crop, a change in the plant soon became manifest, and not more than the first crop of seed produced in some districts could be sown without the fibre becoming of a coarse and degenerate quality. This furnishes an illustration of a law in nature to which many of our native plants are subject. The Scotch pine, however, being a tree of great duration, and by no means in general cultivation, its character remained more obscure, and being of limited interest, its properties were more slowly recognised.

More than thirty years ago, the Highland Society of Scotland, aware of the degeneracy of the plantations throughout Scotland, on account of having been propagated from inferior stock, and with the view of improving the timber of the country, offered premiums, both for collecting the greatest quantity of seeds of the *P. sylvestris*, from the most celebrated forests in the Highlands of Scotland, with satisfactory evidence that the seeds were sown, or sold for sowing; and also for raising the greatest number of plants from seeds of this description. The offers by the Society extended over a period of upwards of ten years; and all the premiums, both for collecting the largest quantity of native seeds, and for raising the greatest number of plants, were awarded to Messrs. Grigor and Co. of the nurseries, Forres. The effect was a reform in the cultivation of the tree, and planters generally now obtain native seeds, or plants raised from such seed, taken direct from the indigenous forests. Since notice was directed to this subject by the Highland Society of Scotland, the first extensive planter of the true native pine was the late Sir George M'Pherson Grant, Baronet; but by far the most extensive planters of the tree have been the late Duke of Sutherland and the late Earl of Seafield, each of whom has formed forests composed of several millions of the plant. Large plantations of it have also been made by Lord Lovat, Mr. Ellice of Glenquoich, Mr. Dempster of Skibo, and Mr. Mactier of Durris, and, on a smaller scale, the tree has been employed in the formation of plantations in almost every county in Scotland and England. Few Scotch landowners have interested themselves more in the cultivation of forest trees than the late Sir Thomas Dick Lauder, Bart. Respecting this tree, he gives the following advice in his edition of Gilpin's *Forest Scenery*, vol. i. p. 177 :—" It should be carefully remembered by planters that sundry wretched and worthless varieties of the Scotch fir have crept into use, which in some measure accounts for the miserable appearance of the low-country planted trees. The greatest care should be taken to plant nothing but those trees raised from the seed of the true *Pinus sylvestris* of the mountains."

A succession of unfavourable seasons since 1860 has occasioned a great scarcity of Scotch pine seeds, and consequently of young plants of the native tree of this country. This gave rise to large importations of the seed from the Continent, from

which plants have been grown very extensively throughout
Scotland.   These plants are utterly worthless, except in the
most favourable situations.   In the most sheltered nursery-
ground they seldom survive the second winter without showing the influence of frost; and unless protected in early
winter, they generally become quite brown by the spring of
the year, and unsaleable.   This is a fact well known to Scotch
nurserymen, and to many in England, whose nursery grounds
stand elevated and exposed.   The difference between the
plants of the native Scotch pine and the Continental *P. sylvestris* is quite perceptible when one year old, but much more so
at the age of two years; then the two sorts brought into
view on elevated ground standing side by side could readily
be distinguished at a mile's distance, so great generally is the
contrast in colour : the foreigner has a dead and withered
appearance, while the native plant stands green and scathless.
Plants from imported seed have also the disadvantage of
forming bare roots, and are, on that account, more difficult to
transplant in safety; therefore to treat the plant skilfully, it
should be transplanted at the age of one year, which will
have the effect of giving it a more fibrous root, and of retarding its upward growth, which has the effect of diminishing
the influence of frost to some extent during its nursery
management.   The upward growth of foreign Scotch fir,
however, in good shelter and in a favourable climate, is more
rapid than that of the native plant.   This is very decided in
early life, but their girth is generally less, and, at best, the
tree assumes the tall, slender appearance which I have
observed conspicuously in the planted pine woods of Germany, even where the trees had ample space.   Some young
plantations formed in the Highlands of Scotland with plants
from foreign seed, suffered so severely by the summer frosts
of 1863 that the succeeding summers have not restored their
vigour, while the native plant stood exempt from injury.

The cones of the Scotch fir are ripe in the end of the year,
but after being exposed to the frosts of winter a less degree
of heat extracts the seeds, therefore they may be gathered in

March; after this time the heat of the weather opens their scales, and the seeds fall out.

If a large quantity of seed is required, the cones should be placed from six to eight inches deep on a kiln, laid with deal two inches broad, with a vacancy of half an inch between each, to admit heat, and also to allow the seeds to fall through on the cones being turned. The heat to be applied should not advance beyond 130° of Fahrenheit's thermometer; but a degree approaching to that may be kept up for the first four hours, after which the heat should be withdrawn for half an hour, and the cones then turned over with a spade or shovel, and the seeds which fall through into the pit underneath should be removed. It may be necessary here to remark, that the pit should be formed about a foot or two feet under the level of the fireplace, in order that, for the safety of the seed, it may be always cool, or of a comparatively low temperature. After the first turning the heat should be continued for other four hours, but the temperature rather decreased, and another cooling of the kiln should then take place, when the cones should be turned, and the seeds removed as formerly; by this time the cones are nearly all open, and their depth on the kiln is more than twice that at which they stood when the heat was first applied. A steady heat of about 110° is then kept up for a few hours longer, which completes the process of drying, and all the seeds fall from the open cones on their being riddled. It usually happens, however, that a small proportion of cones remain unopened; these should be picked out and placed along with the close cones for the next kilnful.

Brick or metal kilns are quite unsuitable for the drying of seeds, and cannot be safely employed; these materials imbibe a greater heat than wood, and the seeds which come in contact with such substances are deprived of their vegetative principle.

If only a few pounds of seeds are required, these are most conveniently extracted by exposure to the influence of sunshine. By placing the cones in a warm sunny situation, sloping to the south, in a few days they will open, and the seeds may be sifted out.

When the seeds have been extracted they should be moistened with water, which will detach them from their chaff or wings; and then, by winnowing and sifting, they are readily made clean. The influence of our sunshine in April is not powerful enough to open the cones of some of the foreign species; but the heat stated for extracting the seed is adapted to all the species of the pine, with this difference, that some sorts require the heat kept up for a longer period than others, particularly if the cones have been gathered immediately on their becoming ripe.

The usual time of sowing Scotch pine in South Britain is the middle of April; while in North Britain the last week in April or first in May is the most approved time. The soil best adapted for seedling pines is that which is well pulverized, and rather dry and sandy than otherwise; such as is not apt to get hard by alternate rain and drought; and it is not necessary that it be made rich. After being dug and smoothly raked, the ground should be marked off into beds, four feet broad, with alleys twelve or fifteen inches wide. The beds should be opened by removing the surface soil into the alleys, by the operation called *cuffing*, which is performed by a wooden-headed rake; this soil forms the cover for the seeds; one pound of good seed is sufficient for sowing ten or twelve yards in length of a bed; the cover should be from one-fourth to one-half of an inch thick; the thinnest cover is generally enough in heavy soil; but no exact depth can be fixed, as that which is required in extreme drought is too deep during a continuation of wet weather.

The beds require to be carefully protected from the ravages of birds until the young plants have been a few weeks above ground. The different sorts of linnets are generally the chief depredators; but in some districts the larks are very destructive; they not only eat the seeds while germinating, but also the cotyledons when above ground; pigeons and partridges are still more ruinous, particularly the latter during the night, but these are less apt to find out the crop, and a season sometimes passes over without a visit from them.

The seed-beds during the first and second summers require no further attention beyond that of weeding. After the second summer's growth the seedling plants are fit for being planted into bare moorland or heath. In the north of Scotland a greater number of Scotch fir plants are inserted in plantations at this age than at any other.—(See MOORLAND PLANTATIONS.)

To adapt plants for situations where they have to contend with a rank surface vegetation or any other herbage than short heath, they must be first transplanted into nursery lines; which is generally done when the plants are two years old. The lines should be eight or ten inches distant, and the plants two or three inches apart. These dimensions are sufficient for their remaining one year in the lines, but nearly twice this space is necessary for the plants if they are required to be a second year in the lines, which brings them to the greatest size and age at which the Scotch pine should be removed. While on this subject, it is worthy of remark that the plants best adapted for living in a bare and barren exposure, such as a hill-top, are those which are transplanted at the age of one year, and nursed one year in the lines, a sort rarely used, but most tenacious of life.

The price of two years' seedling native Highland pines is commonly 2s. per 1000; one-year-old seedling and one year transplanted, 3s. 6d.; two years' seedling one year transplanted, 4s. to 5s.; two-year-old seedlings two years transplanted, 8s. to 10s. per 1000; the common varieties of the tree produced from plantation woods are frequently sold at fully one-third under these prices.

There is no other tree that grows so freely, and produces timber so valuable on poor soils of very opposite qualities. It luxuriates on the dry and gravelly heath-covered moors, its roots penetrate among the fissures and débris of rocks, and support the tree in the most scanty resources of almost every formation. Stagnant water is ruinous to the tree; but as its roots generally range near to the surface of the ground, it exists on a very thin stratum above water, where trees in general perish. Of all soils common in waste lands, pure bog is most uncongenial to its growth; and although the plant is sometimes seen to live in soil composed almost wholly of this vegetable substance, yet the pine requires a mixture of inorganic matter, in order that it may produce timber.

In early life the tree rises in a formal shape, particularly in planted woods, where it indicates its age by the whorls of its branches, or their marks on the trunk. Trees fifty years of age are met with in sheltered situations, which furnish a scale of their growth during the respective years of that long period, when their entire height is often seventy feet. Some of the tallest pine trees in Scotland measure upwards of 100 feet in height, but in a wild state they are seldom found to exceed seventy feet. But although the growth of the tree when young, particularly in planted woods, is according to a regular form, yet in its native wilds it soon assumes a very different character, presenting a massive trunk, with ramifications irregular and beautiful. It is an Alpine tree, preferring an elevated situation, a northern exposure, and a cool climate. Throughout the Highland districts isolated groups of the tree arise here and there in broken and varied outline, scattered around lakes, and on the rocky knolls of an undulating surface, while single specimens stand throughout the brown heath, investing the scene with an air of grandeur and antiquity, of solemn and solitary beauty, which no tree but the Scotch pine and cedar could confer. Such scenes are of a very striking character during the heat of summer and the snows of winter. In the Highlands of Morayshire, along the roadside from Carr-bridge towards Aviemore, for several miles magnificent specimens of the native tree stand with massive trunks, broad and umbrageous heads, displaying a ramification equal to that of the oak.

Wordsworth, who had a lively perception of the picturesque grandeur of the native Scotch fir, wished to insert a few plants of them in his grounds at Rydal, which we supplied, and to show his appreciation of the tree I take the liberty to quote the following letter, which he returned on receipt of the plants :—

" RYDAL MOUNT, 20*th February* '45.

"DEAR SIR,—Your plants were received with much pleasure, and you will be glad to learn that they are not injured. My garden lies in something of a hollow, and is yet covered with snow, but they are placed in a sheltered plot in front of the house, and will be transplanted to the garden as soon as the snow will permit.

" You were quite right in inferring that the fir was a favourite tree with me, indeed, as perhaps I have told you before, I prefer it to all others except the oak, taking into consideration its beauty in winter, and by moonlight, and in the evening.

"Accept my sincere thanks for this mark of your attention, and even still more for your good wishes so feelingly expressed.—I remain, dear Sir, faithfully yours obliged,

" WM. WORDSWORTH."

Some of the best trees of the species in the kingdom stand in the county of Moray, near Forres. A tree on the estate of Brodie, lately felled, measured fourteen feet in circumference at three feet from the ground; the trunk was sound and shapely, with little or no taper to the height of about fifteen feet, where it ramified into a large head nearly seventy feet high, and fully ninety feet in diameter. The age of this tree was 150 years. It had lost its top in early life and having had ample space it ramified and spread like an oak. Its root inverted forms a prominent figure on the lawn at Blackfriars Haugh, Elgin, displaying a · solid circular body of gnarled timber, eleven feet in diameter.[1] The ground on which this

---

[1] This tree grew in the parish of Dyke, on the Hardmoor, the spot distinguished by affording the scene of the mainspring of the drama of the tragedy of Macbeth. It was on the Hardmoor, on the western side of the park of Brodie Castle, where Macbeth and Banquo, returning victorious from an expedition in the Western Isles to wait on King Duncan, then in the castle of Forres, and on a journey to Inverness, are represented to have been saluted by the weird sisterhood. Banquo, impatient, after a fatiguing journey on this blasted, and to appearance boundless waste, thinks of the termination of his journey, and asks

" How far is 't called to Forres?"

tree stood was very wet, and notwithstanding its immense size none of its roots descended more than two feet under the surface. Other fine trees, but of inferior size, stand throughout the forests of Brodie, Darnaway, and Dalvey, supposed to be remnants of the old Caledonian forests, which at one time extended from the Highlands of Perthshire to the shores of the Moray Firth.

On the banks of the Spey at Rothiemurchus, Glenmore, Abernethy, and Duthil, and along the northern slopes of the Cairngorm mountains, native Scotch pine timber is produced, in quality not inferior to the finest pine timber of any country. Throughout the present century extensive removals of timber were effected, particularly from such parts of these forests as were situated near to the roads and rivers which gave an easy transit for the wood.

For many years the forests of Rothiemurchus yielded a large revenue, often exceeding £18,000 per annum.

Glenmore forest, the property of the Duke of Richmond—

"The land of the mountain and flood,
Where the pine of the forest for ages hath stood;
Where the eagle comes forth on the wings of the storm:
And her young ones are rocked on the high Cairngorm."

In the end of last century the forest of Glenmore was considered the finest in this country. His Grace the Duke of Gordon about that time sold the principal part of the timber to Mr. Osbourne, an eminent wood-merchant in Hull, who finished felling it in 1804. The timber was floated to Speymouth, and principally employed in naval purposes. One of the finest frigates built there of this timber for his Majesty's service was named " The Glenmore."

This forest is situated in a glen, and surrounds Lochmorlich, where the water of Abernethy, the Druie, takes its rise, close on the north-west of the mountain of Cairngorm. Its length is upwards of four, and its breadth nearly three miles. When I last inspected this forest, there were still a great many fine trees, particularly on the borders of the lake, but none notable for great dimensions, except a few measuring from nine to ten feet in circumference, of little value, being knotty, bushy, and blemished. They stood at great distances, commonly from 50 to 100 yards apart, and evidently had not been considered of consequence when the intermediate ones had been felled. In other parts they were in patches, on the border of the lake, and on hill-sides; in both situations they grow rugged in figure, and of great girth.

Notwithstanding the openness of this situation, and the fertility of the soil, it seemed not congenial to the natural reproduction of pine timber; partly from the ground being depastured, but principally from the exuviæ of the old wood, for in the interior of the forest a young plant was rarely met with. On examining this place, a flat of ground at a turn of a small rivulet was pointed out as the spot where the largest trees grew, one of which was called "The Lady of the Glen," the largest in the forest. It was cut up, and a deal from its centre presented to his Grace the Duke of Gordon by Mr. Osbourne. I have seen it in the entrance hall of Gordon Castle; it is 6 feet 2 inches long, and 5 feet 5 inches broad. The annual layers of wood, from its centre to each side, number about 235, indicating that number of years. A brass plate attached to the plank bears the following inscription:—

"*In the year* 1783,

WILLIAM OSBOURNE, Esquire,

Merchant of Hull, purchased of the Duke of Gordon the forest of Glenmore, the whole of which he cut down in the space of twenty-two years, and built during that time, at the mouth of the river Spey, where never vessel was built before, forty-seven sail of ships, of upwards of 19,000 tons burthen. The largest of them 1050 tons, and three others little inferior in size, are now in the service of His Majesty and the Honourable East India Company. This undertaking was completed at the expense (of labour only) of above £70,000. To his Grace the Duke of Gordon this plank is offered as a specimen of the growth of one of the trees in the above forest, by his Grace's most obedient servant, WILLIAM OSBOURNE.

"HULL, *September* 26, 1806."

I saw many blocks of extraordinary size near the spot where this tree grew. The surface soil is composed of thin sandy peat earth ; the subsoil of rich brown clay, which feels quite soft, and forms a great part of the subsoil in the glen. Perhaps no district in Scotland is better calculated than this, so far as quality of soil goes, for growing larch, oak, or various kinds of valuable hardwood.

The herbage consists of *Calluna vulgaris, Juniperus communis, Tormentilla officinalis, Polygala vulgaris, Agrostis vulgaris, Narthecium ossifragum, Vaccinium Vitis Idæa, Erica tetralix,* and *Prunella vulgaris,* etc.

Along the outside of this forest young wood to the extent of several square miles has sprung up since the removal of the old forest.

These trees grow slowly until they reach the age of twelve, which perhaps is owing to their roots not penetrating earlier into the rich subsoil. They are of all sizes under fifty feet ; some crowded, and others quite thin. This young forest is of the usual age for bearing seed, but very few cones are to be seen ; and on examining the ground around the trees, few of those of former years are found, and those are smaller and rounder than the cones of the low-country planted trees, as is invariably the case in native forests.

The largest trees to be found in Strathspey at the time stood on the outskirts of the forest of Abernethy, the property of the Earl of Seafield, some of which were fifteen feet in circumference, but short in the bole, and bushy. But in the close parts of all these native forests throughout Strathspey, trees of great girth display their clean boles to the height, in many places, of upwards of forty feet, in figure as straight and in taper as elegant as that of a billiard cue.

Strangers to those forests are surprised at seeing the size and closeness of the trunks to each other, and must admire the value of timber contained in small space. The forests in Strathspey however have lately been greatly reduced, owing to the high price of timber, enhanced by the introduction of two railways into the district ; but although a great quantity of fine timber has disappeared, the young native woods are better protected, which, with very extensive plantations of recent formation, occupy a greater space by some thousands of acres than was covered with timber trees at any time during the present century. The following is an extract from the report on the pine forests of Scotland made by me to the Highland Society of Scotland in 1836, respecting Abernethy native forest :—

" It is one of the most ancient in Scotland, and from time immemorial has been famed for the quality of its pine timber.

" It stands on the southern extremity of Morayshire, on the south side of the Spey. The water of Nethy winds through it, and is of the greatest importance to the forest, as it supplies water-power to the saw-mills, and floats the timber to the Spey, by which it is conveyed to the sea-port of Garmouth. The timber of this forest contains a large quantity of resin, and is therefore very inflammable. In the year 1746 a great proportion of it was burned down, but a large extent has produced a new crop of excellent timber. The ground on which the forest stands is partly hilly and partly level; the smaller hills, and the sides of the larger to a considerable extent, being entirely covered with trees. The soil is of various qualities, but is principally composed of a thin sandy moss, with a subsoil of hard hazelly-coloured gravel, and in some parts it is a black mould mixed with white sand, and very stony.

" Along the banks of the Nethy I had an opportunity of seeing a great quantity of very fine timber, barked and prepared for floating; the largest of which measured 10 feet 7 inches in length, 6 in girth at the root end, and 5 feet 2 inches at the other end. The number of annual layers or rings at the root end indicated its age to be seventy-three years, and that at the upper end sixty-one years. The timber was of excellent quality, well-hearted, clean, and full of resin, and although from thirty-two to thirty-four of the last-formed rings composed the sapwood, yet it bore a comparatively small proportion to the bulk of the whole trunk, the trees having of late years made but little progress. Many of the trees were much older and smaller, consequently their wood was closer and of superior quality.

" I was directed to a part of the forest which stands about seven miles south of the Spey, as affording the best specimens of large trees, many of which I measured close above the swell of the roots, or about the height of one foot from the surface, at which the largest were from 10 to upwards of 13 feet in circumference, and at the height of eight feet from the ground from 9 to 12 feet, tapering with clean trunks to the height of from 20 to 35 feet, and shooting up to the entire height of from 40 to 65 feet. These very old trees stand on low and level ground on the side of the Nethy; but perhaps the finest tree in this forest stands on a steep hill-side adjoining, though not highly situated, which measures in circumference, at the height of one foot from the surface, 13 feet 3 inches; and at eight feet high 12 feet. It tapers to 32 feet of trunk, its whole height being about 50 feet, with a top branching like an oak, to which all the large trees in point of form bear a strong resemblance. A few yards distant from this tree one of similar dimensions had lately been felled; the stump and roots remaining to indicate its size. The annual rings of this root indicate the age of 242 years, and that of the top 224. The top lay at the distance of 27 feet from the root, and I imagine that the tree had grown about that length in eighteen years, that being the number of years intervening between the ages of the root and top. Several others had been felled of nearly the same size, which had almost attained the age of 200 years. I observed from the size of the interior layers that the trees had rapidly advanced in growth between the ages of 8 and 70, the growth having afterwards diminished, and eventually the outside layers, although distinct enough to be numbered, are very minute, and the whole timber is equally strong, hard, and red, to within less than an inch of the bark. Many of them had been thrown down by the great flood of 1829, the stumps of which still remain, and show that the roots had derived all their nourishment from the surface soil, none of them being more than one foot from the surface where the subsoil is hard and gravelly. These are discernible above ground, and each forms a rib to the height of several feet on the side of the trunk."

The soil in the best native forests is of very various qualities. The popular belief is that a poor soil yields the best pine timber, but the soil of the native forests does not generally support this theory.

The soil of the old forest of Glenmore is generally of a rich deep-brown earth, rather clayey than otherwise, and apparently adapted for any description of hardwood timber.

In Rothiemurchus the soil is very changeable, and it affects the size, and not the quality, of the timber. On the opposite side of the Spey, in the parish of Duthil, better pine timber can nowhere be found than that produced on slopes composed of rich deep-brown clay. Abernethy furnishes the greatest variety of soil. Generally speaking, it is a hazelly gravel, and parts change from a dry sand to a deep black mould; on the whole it is, perhaps, the least fertile district of soil occupied by the forests of Strathspey; yet uncultivated moorland has produced an ash-tree upwards of twenty feet in girth; and while traversing these woods, we were surprised to find a common alder, standing on low meadow ground, near Lurg, which measured upwards of fifteen feet in circumference, which is beyond the girth of any tree of this species of British growth hitherto recorded. From these illustrations of the resources of the soil in the native forests, we need hardly add, that it is at least equal to the soil on which the Scotch pine produces inferior timber throughout the country, and that the superiority of the native Highland pine ought by no means to be attributed to the timber being produced on poor soil.

The trees in the native forests are in general older than most plantations throughout Scotland, and are of greater size, even in proportion to their age. Notwithstanding this, it is very uncommon to see a single tree in a decaying state. We observed several trunks that had a few feet of timber scooped

out from the side of each to be used as *candles* by the cottars,
yet the trees continue quite green and healthy, with the
hollows overhung with turpentine *icicles* several inches in
length.

The pines grown in these districts appear to be of one
species, and differ from the great bulk of those produced in
the low counties of Scotland in the following respects :—

The Highland pine is of a more robust and shaggy appearance. In early life it grows, although crowded together, to a
greater girth; it is found to attain a greater size on very wet
ground; its wood is redder and harder, and consequently
more durable, and is found to be more inflammable. It produces very few fertile flowers or cones, and what it does
produce are uniformly found to be rounder, smaller, and
whiter, and it outlives many generations of the common
cultivated fir, and ultimately attains a larger size.

It may be difficult to ascertain the differences in plants
necessary to constitute a distinct species, but if the superiority
of the Highland pines to the common tree of the low country
should not be attributed to a difference in kind, the great
proportion of the trees in Scotland, by repeated cultivation,
must have lamentably degenerated, since it is known that
thousands of the common fir have arrived at maturity, and
thousands have died of old age, without ever producing timber
in any respect comparable to those of the districts now
attempted to be described; and they who aim not to propagate these magnificent objects of nature, overlook that analogy
which is everywhere observable in the works of creation.

In the parish of Dyke the remnant of a plantation seventy
years old, which did not form half a cover on the seventy-three acres of ground which it occupied, was lately disposed of
by public sale for £2500. The wood had always been celebrated for its fine quality, and the last of it was employed for
railway sleepers, where the timber specified was that of "the
best native Highland." It could not be distinguished from
the finest Strathspey native wood, and people were recently
alive who knew that the young plants of this wood were

taken from the native forests in Strathspey, having when they were children been employed at the formation of the plantation. An instance is given in Lawson's *Manual* of "a plantation recently cut down, which stood on the north side of the Perth and Dundee road, nearly ten miles from the former town, the seed of which was, seventy or eighty years since, received from the forest of Mar; and the timber, although grown on a poor, damp, tenacious clay, besides attaining to a great size, was found equal in quality to that for which the above natural forest is esteemed."

The durability of the red timber of the Scotch pine was supposed by the celebrated engineer Brindley to be as great as that of the oak; and Dr. Smith, in his essay on the production of timber in the *Transactions of the Highland Society of Scotland*, vol. i. p. 165, says that he has seen some Scotch pine grown in the north Highlands, which, when taken down after it had been 300 years in the roof of an old castle, was as fresh and full of resin as newly imported timber from Memel, and that part of it was actually wrought up into new furniture.

There is no timber more generally useful than that of the Scotch pine. The first thinnings of a plantation, retaining the branches and leaves, are suitable for being employed in the erection of screen fences, to protect young plantations from the influence of the sea, and from the effects of the weather in cold exposures, for sheep-flakes, and similar purposes. The next thinnings are adapted for coal props, an article in constant demand at all shipping ports. The usual prices for common props range from 1s. 9d. to 2s. per dozen; crown props sell at from 2s. 9d. to 3s. per dozen.

These articles are bought in lengths of six, twelve, and eighteen feet; seventy-two lineal feet make a dozen. Common props must stand three inches in diameter at the small end, and crown props are not under four inches.

Prop-wood is very profitable when the carriage to the shore is not expensive; and its value is not influenced so much as that of other wood is by the price of foreign timber.

Young pine wood is also much used for paling, staves for dryware casks, lath, and such purposes. When in a more matured state, it is cut into deal for flooring, and all the other purposes of house-carpentry, railway sleepers, etc. The quality of pine timber is regulated to some extent by the soil, and more by its age, but most of all by its variety, that is, the purity or absence of degeneracy in which it is produced. The plantation wood of degenerate quality sells at present for about 4d. or 6d. per cubical foot, while the native Highland variety fetches about double that rate, and its value is regulated by the price of foreign pine timber, which has of late been much reduced. In a treatise on the strength of timber, published by P. Barlow, F.R.S., showing the result of experiments on the strength of various kinds of timber, selected from the Woolwich dockyard, it is stated that pieces of Riga fir, 7 feet long by 2 inches square, broke with a weight of 422 lbs.; pieces of the same dimensions, grown in Mar forest, broke with a weight of 436 lbs., each being the mean results of three experiments. The mean results of a second trial are recorded, with pieces 6 feet long, 2 inches square. The Riga fir broke with a weight of 467 lbs., and the Mar forest fir with a weight of 561 lbs., showing a difference greatly in favour of the Scotch native timber.

The Scotch pine, as well as all the pines, are subject to the ravages of various kinds of insects; of these, several species of beetles are among the most destructive, particularly the *Hylurgus piniperda*, while the plants are in the nursery ground. This is a small beetle of the size and colour of the seed of Scotch pine, and varying in colour from a black to a light brown, as do the seeds of most kinds of the pine. The young shoots of all the kinds are commonly formed by the end of June, and it is in this month that this insect, active and on wing, generally begins its depredations, by boring into the young shoot, and perforating it in the centre like the stem of a tobacco-pipe. The more vigorous and succulent the growth the more subject it is to be attacked, especially in sheltered places, so that its ravages are chiefly confined to nursery-grounds or to foreign pines placed in rich soil and shelter. The leading shoots are often destroyed, which renders the plant comparatively worthless. The first symptoms of attack appear from the pale foliage and drooping young shoot. These should be immediately picked off, collected into a bag, and burned; many such twigs will be found to contain the insect, which shifts from twig to twig during the summer months. Its effects are seldom observed in the native forests of Scotland, or in plantations in exposed moorland, and then only on the sheltered lateral twigs. I have never seen an exposed leading shoot injured in such places, therefore its operations have only the effect of a gentle pruning or fore-shortening of the tops of the lateral branches. There are other beetles more destructive to the Scotch pine in the forest than the one described. These consist of several species, abounding in ground where fir woods have recently been cut down. They find a congenial home in the old roots, and the mossy herbage and decayed leaves protect them in the severity of winter. It is when such places are replanted that these insects are most destructive by feeding on the bark of the plants of the fir tribe. They attack the plant first at the surface of the ground, particularly if it stands closely surrounded by a mossy herbage. When the bark is broken the plant yields a flow of rich resinous sap, on which they feed, and the result, in the course of a few weeks, is the death of the plant. Field-mice sometimes commit the like depredations in similar situations, and on larger plants, but their attacks are not so frequent.

No effectual cure for these evils is yet known, further than keeping the plants clear at the surface of the ground. The precautions best adapted to avert these casualties are detailed in the article on replanting forest ground.—(*See* p. 74.)

Young plantations, commonly under thirty years of age, are subject to an attack from the larvæ of several species of moth peculiar to the pine. I have seen several acres of the Scotch pine, both in the Highlands and in the low country, stripped of every leaf, while all experiments in smoking and dusting with poisonous substances were of no avail. The larvæ descend into the surface cover on the ground, and undergo their transformation; but fortunately, the same trees, so far as I have ever known, are not infested during a succession of summers. Such repeated attacks would infallibly be fatal. An attack retards the growth of the trees during that and the following year, but I have never seen the trees killed. The first and second years after an onslaught, the trees assume a curious tufted appearance, their leaves being confined to the extremities of the branches, or to the newly-formed twig.

Squirrels were known to abound in the large forests of the Highlands from time immemorial, without any complaint on their doings till of late; but they have spread to an alarming extent throughout the north of Scotland, and their depredations, by barking the succulent stems of the pine tribe, have become very serious, as the effect is to deprive the tree of its top. On many estates gamekeepers are employed in shooting them down. In doing so, however, care should be taken that the shot do not perforate the leaders to a great extent, as that occasions a flow of turpentine, and forms a weak point, at which the tree is apt to snap during a gale of wind.

Pinus sylvestris Haguenensis (Loudon.)—This is the Continental variety of the Scotch pine, named after the great forest of Haguenau, situated in France and Germany, and extending on both sides of the Rhine to about 30,000 acres, and from which British nurserymen are often supplied with seed, not only in seasons of scarcity, but to a considerable extent every year. The same variety constitutes the plantations of *P. sylvestris* throughout France, Germany, and Prussia. When young, the plant in this country advances rapidly, rather more so than the native Scotch tree, provided the soil is good and well sheltered; but it is of no value as a timber tree in ordinary exposed plantations. We have already referred to it under Scotch pine. The seeds are imported annually at less than half the price of those of the native pine.

Pinus s. Pumilio (Hænke); *P. s. Mughus* (Loudon); *P. s.*

Montana (Baumann).—These pines are closely related to the
Scotch pine, but they are all of a very dwarfed habit of
growth, and bear a close resemblance to one another. They
are natives of high mountains, and abound on the Alps, the
Pyrenees, and other cold exposures throughout Europe, where
they become bushy shrubs; whereas, with a little more
shelter, some of them attain to the size of low bushy trees.
Their yearly shoots are very short, with the foliage thickly
set, of a dark green, and their habit being broad and spreading, adapts them for cold and windy situations. As these
sorts are often found occupying the more elevated grounds
close to the forests of *P. sylvestris*, it is believed that their
peculiarities and habits are produced from the effects of
climate and situation, by the repeated generations of a tree
susceptible of great change in the course of a long period,
under such circumstances. They all produce wood, red, hard,
and durable, and from its very resinous and inflammable
quality it is used as torch-lights by the inhabitants of the
countries where it is produced. All these sorts blossom at
an early age, and produce cones very abundantly, particularly
pumilio and *mughus*. I have seen the *pumilio* produce a
female blossom during the third summer of its growth in the
nursery, and an infertile cone the fourth summer, while the
plant was yet under one foot in height. In nurseries these
trees are only grown for ornamental purposes, or by lovers
of variety. Although they are all stemless trees, the *montana* being only a tall bush, yet there are purposes for which
they may be found very suitable. I have seen in some of the
northern islands of Scotland the *pumilio* and *mughus*, though
only about two feet high, form a cover of rich green dense
foliage, on the most exposed ground, composed of pure moss,
and so wet that water might be compressed from the soil at
every season of the year, and where any other ligneous plant
could hardly exist.

The habits and the tenacity of life in these trees in the
most inhospitable soil and climate, adapt them admirably for
being planted as underwood, associated with the native Scotch

pine, as screen-fences, with the view of intercepting snow-drift, which so frequently blocks up railways throughout Highland districts. *P. montana* is best adapted for this purpose, as it becomes a bush twenty feet high. There is very little demand for these trees in nurseries, but as their seeds are readily obtained from the foreign-tree seed-merchants at a cheap rate, the plants could be grown to order at a few shillings per thousand. There are five or six other varieties of *Pinus sylvestris* recorded in nurserymen's catalogues and in cultivation.

P. Laricio (Poir), or Corsican Pine.—This tree was introduced into England about the middle of the eighteenth century. It is a native of various parts of the south of Europe, and of the west and north of Asia. In the island of Corsica it frequently attains to the height of 140 feet, but it does not inhabit the poorest soils, nor is it found natural at a great altitude. This species, like others extending to various countries, has many varieties. It is remarkable for its rapid growth when young. In the Jardin des Plantes at Paris it attained the height of eighty feet, with a trunk seven feet in circumference, in a period of fifty-five years, and it has been known to produce a top shoot three feet long in one season. It is quite hardy, but in poor, exposed ground, the vigour of the tree, after a few years, is not greater than that of the native pine in upward growth, and its girth is somewhat less. It therefore requires good soil and a favourable situation, into which it should be removed at an early age, as, like all other pines of early vigour, it is apt to form a root small in proportion to its top, which renders it difficult to remove with safety. The price of young plants fluctuates very much, on account of the scarcity of seeds in some seasons on the Continent, from whence they are usually imported. One and two year seedling one-year transplanted plants commonly range from 10s. to 12s. per 1000. The timber of the tree, when young, is soft and easily worked, like that of the rapidly-grown Scotch pine; when old, it is said to be tough and resinous.

P. L. Austriaca (Hoss).—The black pine of Austria was introduced into this country by Messrs. P. Lawson and Son of Edinburgh, in 1835. It is propagated to a considerable extent throughout Britain from imported seed. It is perfectly hardy and of robust growth, particularly in soft soil of any quality. It carries a breadth along with its upward growth, which adapts it for exposed ground. In its native country it sometimes attains the height of 100 feet. Its mode of treatment is similar to that of the Scotch pine, but as its roots are apt to grow more bare and straggling than the roots of that tree, the best rooted plants are those that are transplanted when one-year-old seedlings, into nursery lines, where they may remain one or two years, as may be necessary to prepare them for the ground they are intended to occupy. The tree commonly yields a close compact foliage of the darkest green, and its timber is strong and resinous.

P. Pinaster (Aiton) : The Cluster Pine.—This tree is indigenous throughout the south of Europe, in the countries along the shores of the Mediterranean, and in other parts of the world. Like all pines of an extensive geographical range, the species is possessed of numerous varieties, sometimes distinguished by the length and colour of their foliage, the size of their cones, etc. The tree was introduced into England by Gerrard at the close of the sixteenth century. It is readily grown from seed, which should be sown in the end of April in beds four feet wide, allowing one pound weight of seed to every six or seven lineal yards of a bed ; the cover should be half an inch deep, and the beds, like those of all other pine crops, should be protected from the ravages of birds. The seedling plants should be transplanted at the age of one year into lines about eight inches apart, and the plants about two inches asunder. When they have been one year in lines they are fit for removal into sand, or bare exposed ground. If they are not planted out finally after being one year in lines, they should again be transplanted into wider space in the nursery, in order to keep their roots sufficiently fibrous ; for two years undisturbed, at any age, in almost any description of soil, is generally fatal to the safe removal of the pinaster.

Notwithstanding the rapid growth of the plant when young, it is not a favourite tree with planters in general throughout this country, except within the influence of the sea, where the variety *maritima* and all the other varieties of the tree grow better than most plants. In rich or wet soil it does not endure the frosts of winter; a dry, deep, sandy soil is indispensable to its profitable growth, and in the vicinity of the sea the temperature is to some extent equalized and adapted for its development. Some of the best specimens of the tree in Britain are to be found in the county of Norfolk, standing nearly 80 feet high, with trunks 12 feet in girth; many such trees stand at Westwick Park, where few other species of trees would become timber. On this property the tree has been planted at intervals during the last 150 years, and it is the principal species in a plantation of 500 acres produced from seed grown on the property. In raising any of the more tender kinds of the Coniferæ, home-grown seed should always be preferred to that grown in a warmer climate; the rate of growth may be rather shorter for the first few years, but if they are exempt from the influence of frost, as is the case with the native pine and the home-grown larch, compared with those from imported seed, there is often all the difference between a healthy plantation and a total failure. In open sands the pinaster strikes its roots to a great depth, and derives its support from soil several fathoms below the surface; it is of rapid growth, often attaining the height of 30 feet in twenty years, and a tree has sometimes been known to grow five feet in two years. In Scotland, when the tree is interspersed in plantations, it has a tendency to grow crooked and towards the side most exposed, and appears to prefer an open and airy situation. I have seen it in some of the weather-beaten islands of the Hebrides affording shelter as a hedge five or six feet high where few or no other ligneous plants could live. In a congenial soil it becomes a large handsome pyramidal tree, and all the varieties are distinguished by their light green foliage of a clustering habit of growth, occasioned by bare spaces on the branches which have produced male catkins, and which ever after are destitute of leaves, and give the foliage a singular tufted appearance. The cones are generally produced in groups which point outward in a star-like form, whence the name *pinaster* or star-pine.

In general the wood is white and soft, and fit only for the purposes of common deal. The greatest triumph recorded in connexion with the history of the pinaster is that of converting several hundred square miles of drifting sand in France into a thriving forest, which I detail under SEA-SIDE PLANTING. These forests on the *landes* of Bordeaux, and between that city and Bayonne, form the principal field for the supply of seeds of the *P. pinaster*, which seldom exceeds 2s. 6d. per lb. To nurserymen this tree is generally unprofitable. This arises partly from want of a regular demand for the young plants, and partly on account of their readily overgrowing, when they must be cleared out. Therefore, grown to order they could be furnished at a few shillings a thousand, or half the usual rate.

P. P. folis variegatis is a very ornamental tree, and is propagated by inarching on any of the common kinds of the species.

P. ponderosa (Douglas).—The heavy-wooded pine was introduced into England in 1826. It is a native of the north-west coast of America, and of California, and is found growing mostly on alluvial soils. In this country it is of a very vigorous habit of growth; when a few years old it produces leading shoots often about two feet long and one inch in diameter. Its leaves are thickly set, and eight to ten inches long (produced in threes). It is quite hardy, but not likely ever to become a useful tree in this country, as it forms a root small in proportion to its top, and is easily blown over by the wind.

P. Cembra (L.)—The Cembran or Siberian pine is a native of the Alps, of Siberia, of Switzerland, Italy, and other mountainous districts, where it is frequently found associated with the dwarf variety of the *P. sylvestris*. It was introduced into England more than a century ago by Archibald, Duke of Argyll. The tree is

quite hardy, and consists of several varieties. It has for
many years yielded crops of cones in Britain. Some of the
best trees are found at Dropmore, Kew, and Whitton. The
seeds are large and edible, and being produced abundantly in
Switzerland they form part of the food of the peasantry of
those districts in which the tree is indigenous. The seeds
should be sown fully half an inch under the surface of the
ground; they do not vegetate till the second spring, and the
plants are remarkable for their slowness of growth; they
should be transplanted when two years old into nursery lines.
This becomes an erect tree with a smooth bark, with leaves
(five in a sheath) of a fine green silvery appearance. Although
slow of growth when young, yet in after years it advances
more rapidly, and is believed to be a tree of great duration.
It retains its lateral branches down to the surface of the
ground in a very marked degree, and these are commonly
very short in proportion to the thickness of the bole. A
tree of this species planted at Croome in Worcestershire was
45 feet high at the age of thirty years. Though of formal
growth it is generally reckoned handsome, and its male
blossoms, which are of a bright purple, are the most attractive
and ornamental of all the pines. When young its root is
large and fibrous in proportion to the size of its top, which
enables the plant to be removed of a large size, and adapts it
to endure a rough exposure.

P. Strobus (L.) : The Weymouth Pine.—This tree is a
native of America, and abounds on the hill-sides from Canada
to Virginia, attaining its largest size in the State of Vermont,
where it is sometimes found upwards of 150 feet in height,
and from three to five feet in diameter. The tree began to
be cultivated in England in the beginning of the eighteenth
century, having been planted in large numbers by Lord
Weymouth on his estates in Wiltshire, and having grown
vigorously it was called the Weymouth pine. It is easily
propagated by the same treatment as the Scotch pine. In
rich soft soil it grows to a large size, provided it is protected
from the severity of the weather by hardier trees, or

reared in masses. It has a soft and delicate appearance, is of formal growth, and has silky foliage. Its timber is white and soft, and forms the white American pine of commerce, which is extensively imported into Britain. It is remarkable for being clean and free of knots ; it is easily worked, and is generally employed for inner doors, boardings, mouldings, and the furnishings of house-carpentry. One of the largest trees in Britain is at Strathfieldsaye ; it measures about 100 feet in height, with a trunk about four feet in diameter. Although the tree when young often produces top shoots two feet long in one summer, yet its average progress throughout England, of fifty or sixty years' growth, does not exceed one foot yearly. Few pines advance more rapidly during their nursery treatment ; consequently they soon overgrow it ; and the demand for the plant is very small, the tree being generally unprofitable in Britain.

P. excelsa (Wallich).—The lofty or Bhotan pine is a native of the Himalayan mountains, where it attains the height of from 80 to 100 feet. In appearance it bears a great resemblance to the Weymouth pine, but it is of a stronger and more robust habit of growth, with leaves considerably longer, and branches more drooping. It was introduced into this country in 1827. It forms a very ornamental tree, and its growth is equal to that of the Weymouth pine. Ten-year-old plants are generally 12 feet high. The timber is white, soft, and very resinous ; on the slightest incision the tree readily yields a pure and limpid turpentine. Though the tree has a rich and luxuriant appearance in soft and sheltered ground, it assumes a bare and naked aspect when exposed to the influence of wind, which prevents it from becoming a generally useful timber tree. It grows from imported seed, with the treatment adapted to the Scotch pine.

P. Lambertiana (Douglas) : The Gigantic, or Lambert's Pine. —This tree is a native of the north-west coast of South America, and was introduced into England in 1827 by Douglas, who gives the following account of it in its native country :— "One specimen which had been blown down by the wind, and which was certainly not the largest, was of the following dimensions. Its entire height was 215 feet; its circumference at three feet from the ground was 57 feet 9 inches. The trunk is usually straight, and destitute of branches about two-thirds of its height. The species covers large districts about 100 miles from the ocean, in latitude 43° N., and extends as far to the south as 40°." It is added, "It grows where the soil consists entirely of pure sand, and in appearance incapable of supporting vegetation; here it attains its greatest size, and perfects its fruit in most abundance." I have seen imported cones nearly 18 inches long, but these are generally badly filled, and I have seldom found among them a cone that yielded ten seeds fit for germination. The plants are perfectly hardy in the climate of North Britain; even when only one-year seedlings they withstand the influence of the severest winters. The young plant arises of an elegant appearance, with leaves of a grassy green about four inches long, and five in a sheath; its yearly shoot is shorter than that of the Weymouth pine, and seldom exceeds 12 inches long, and is well matured before winter; and the plant when only a few years of age becomes remarkable for its great girth in proportion to its height. The specimens of the tree which I have seen in this country in rich soil with shelter give little promise of ever attaining a great size; and I expect that if ever it becomes great in Britain, it will be when placed on dry sandy links like that on which Douglas found it indigenous. The scarcity and consequent price of young plants have caused them to be inserted in situations too rich for their development, where they are very apt to be infested with the pine-beetle. The timber is said to be white, closely grained, and full of turpentine.

XIX

SPRUCE FIR.

SPRUCE FIR—*Abies.*—This genus belongs to Monœcia Monadelphia in the Linnæan system, and to Coniferæ in the natural order of plants. It consists of several species of evergreen trees, natives of Europe, Asia, and America, which affect a soft and moist soil, with a cool climate, and are generally remarkable for their erect growth and profusion of foliage; the leaves are solitary, and the cones, which ripen during the year in which the blossoms are produced, are pendant.

A. excelsa (De Candolle).—The lofty or Norway spruce is indigenous to the north of Europe, abounding in Norway, Sweden, Lapland, Denmark, and the north of Germany. It is the commonest tree of the genus in cultivation throughout Britain, and it is believed to be the loftiest tree indigenous to Europe. It is of great beauty, of very uniform growth, assuming a conical form, and, when allowed sufficient space in a congenial soil, it retains, even at an advanced age, its branches and luxuriant foliage, down to the surface of the ground.

In some of the more fertile valleys of the native countries of the tree, it has been known to attain the height of 180 feet.

It was early introduced into Britain, but the precise period is not known. The earliest notice of it in England is given in the *Hortus Kewensis,* by Turner, in the middle of the sixteenth century. In Britain the tree blossoms in May and June, and the cones, which it produces abundantly, become ripe in the following winter.

The mode of propagation is by seed. The cones may be collected in winter or in spring, and the mode of extracting the seeds and of sowing them, is similar to that recommended for the *Pinus sylvestris*. (*See* PINE TREE.)

When the plants have grown two seasons in the seed-bed, they are generally from seven to nine inches high, and fit for being transplanted; but sometimes, from unfavourable soils or seasons, they are of weaker growth. In that case, if they stand thin, they may be safely left in the seed-beds during the third summer, and then transplanted. No other species of the Coniferæ admits so well of being kept three years in the seed-bed; but when the spruce has sufficient space, its naturally fibrous roots adapt the plant for removal, in perfect safety, at that age.

The plants are sometimes removed direct from the seed-bed into their permanent abode. In moorland, and situations where the surface herbage is not too strong for plants of this size, and particularly where the soil is of a kindly nature, the practice is quite successful. The more common mode, however, is to transplant the seedlings into nursery-lines for one or two years, according to the description of ground they are intended to occupy. One-year transplanted plants are seldom much larger than they were at the time of being transplanted from the seed-bed, but they are much hardier, their roots are more bushy, and they suffer less from transplantation the second time. If plants are to be removed after being only one year transplanted, they may be inserted in the lines much closer than when they are to remain two years. In the former case, the lines may only be about eight inches apart, and the plants about two inches asunder. In the latter case, the lines should be one foot apart, and the plants three or four inches distant. A common method practised by nurserymen is to insert the lines closely, and before the second season's growth to remove every other line, thus making the distance of the lines apart the second double that of the first year. Two-years transplanted plants are those usually employed in plantations; but if the plants are allowed ample space, they may, in most soils, from their naturally fibrous roots, be successfully grown three years in lines, after which they should be removed, either into the forest, or transplanted again in the nursery, to insure their future success.

The spruce admits of being removed at a great size, even though more seldom disturbed during its nursery growth than any other tree of the order.

Although the tree is so hardy that it is never affected by any degree of frost during winter, yet it never attains to a very great size in an exposed situation. It is, however, valuable as a nurse to other trees in young plantations, from its numerous lateral branches and dense mass of persistent foliage, which furnishes great shelter.

It derives its nourishment chiefly from the surface, and luxuriates in soil which is cool and moist, and, with a surface soil of ordinary quality, it is one of the few trees that will thrive where the subsoil is wet and retentive. Even in dry and sandy soil, unsuitable for the growth of its timber, the spruce often advances sufficiently to form a shelter for the establishing of other sorts, which ultimately become more valuable; but on approaching the size of a timber-tree in such situations, it assumes a sickly aspect, and presents a scanty foliage, compared with the luxuriance of the tree in the cool alluvial deposit of the sheltered glen, where it is developed in its most attractive form.

Perhaps the finest specimen of the species in Britain stands at Studley, Yorkshire. In April 1853 this tree measured 124 feet 6 inches high, and is now upwards of 130 feet. At a yard from the surface of the ground its circumference is about 16 feet. Its age is believed to be about 160 years.

At Monymusk, Aberdeenshire, in an avenue formed on soft alluvial soil, incumbent on gravel, watered by a mountain stream and sheltered by the surrounding hills, there are some fine specimens of Norway spruce, which were planted about the year 1720. As these trees have been frequently measured from time to time, I am enabled to give the following tabular view of the dimensions of five of them in 1841 and in 1851, showing their progress during ten years, after having attained to the age of about 120 years :—

No.	Height in 1841.		Height in 1851.		Cubic Contents in 1841.	Cubic Contents in 1851.	Cubic Progress in ten years.
	Ft.	In.	Ft.	In.	Ft.	Ft.	Ft.
1	95	6	100	4	198	252	54
2	98	0	105	6	238	306	68
3	98	0	103	3	213	240	27
4	98	3	102	4	200	254	54
5	102	0	106	9	187	233	46

These trees have not increased in height during the last ten years, and not much in girth.

From this statement it will be seen that the average yearly growth in height during the previous ten years referred to, was only from six to seven inches, but the average yearly increase of timber in each tree was about five feet. They are well furnished with branches and foliage far down on their massive trunks, each of which is from four to five feet in diameter at the surface of the ground, and their healthy appearance gives promise of a further increase in their dimensions. Their unusual vigour at this advanced age arises from their enjoying sufficient shelter, on a soft soil, rich in alluvial deposit, and from having at no period of their growth suffered from confinement by being pressed upon by other trees.

In the growth of the tree in plantations, for the ordinary purposes of timber or for profit, it is not necessary that the lateral branches should be preserved; indeed, the quality of the timber is improved by the trees pressing so closely that the branches nearest the surface become enfeebled and drop off. Pruning should never be practised on this species. Under ordinary circumstances, a plantation of it is generally fit for being felled at the age of seventy or eighty years.

In growing this tree for ornament, it is absolutely necessary that it have shelter, and at the same time sufficient
space; for the natural beauty of the species is only fully
developed when the tree stands alone.

It is of formal growth, uniformly conical, and the foliage
on a good specimen is always rich and luxuriant down to the
surface of the ground. In the absence of those bold ramifications which adorn many other species, the gentle tapering
of its lofty top adds a gracefulness to the object, adapts it the
more for standing on a soft soil and enduring the vicissitudes
of a northern climate.

One of the best specimens of the tree in Scotland stands at
Blair-Athole. Its height is 110 feet 6 inches. It is full of
luxuriant foliage down to near the surface of the ground,
where the trunk measures four feet in diameter. It stands
in a low and sheltered glen on the banks of a stream, and in
soil composed of red loamy gravel incumbent on limestone.

After being a few years established in suitable soil, this
species advances rapidly, and generally between the age of
fifteen and thirty years, its yearly top-shoots are two feet long,
and with moderate shelter it is often sixty feet high in forty
years.

In close plantations the earliest thinnings are used as prop-
wood, and sell at prices similar to that of the Scotch pine.
The timber is white and soft; it is finely grained, and free
from knots only when grown in a close plantation. The
ordinary purposes for which it is used are planks and spars
for gangways and scaffolding, masts for small vessels, boarding
for roofing, railway sleepers, water-troughs, conduits, and
sluices. In value it seldom exceeds the price of the more
degenerate sort of Scotch pine timber. The principal resinous
production of the tree is Burgundy pitch, which is the con-
gealed sap melted and clarified by boiling it in water.

A. nigra (Michaux).—The black spruce fir is a native of
North America, and was introduced into Britain by Bishop
Compton, in the end of the seventeenth century. It has
reached the height of from sixty to seventy feet at Painshill

near London ; and at Brahan Castle, Ross-shire, it has attained the height of upwards of fifty feet. The tree seldom exceeds these heights in Lower Canada, New Brunswick, Nova Scotia, and the districts of Main, where it constitutes indigenous forests of great extent.

In this country it is esteemed only as an ornamental tree, for the richness and density of its foliage, and for its hardiness. It luxuriates in a moist soil, and affords great shelter and seclusion. In favourable circumstances its lateral branches often strike root into the ground, and form a circle of young plants around the parent tree. This species seldom produces trunks of great diameter ; but it is remarkable for thriving and maintaining its ordinary girth when the trees stand unusually near to one another. It is raised from seed in the same manner as the Norway spruce. It is of much slower growth than that species, but its timber is strong and elastic.

A. alba (Michaux) : The White American Spruce Fir.— This tree was introduced at the same time as the species last described, but this in every respect is an inferior species— weaker in growth, and less ornamental. Under the most favourable circumstances it rarely attains to the height of forty feet.

The seeds of this tree are often imported from America in large quantities, at a very low price, and it is sometimes to be met with in British nurseries in lots of many hundred thousand plants, which can be sold at a very cheap rate.

When young it has a striking resemblance to, and is often mistaken for, the Norway species. When four years old, and in after life, the plants have a more twiggy appearance ; the leading shoot is generally short and slender, and destitute of that robust habit of growth that distinguishes the Norway species. But in poor soil the Norway plant generally assumes a white or pale appearance, and then the sorts are the more difficult to recognise ; yet the one species yields a trunk fit for the mast of a large ship, while that of the other seldom exceeds the dimensions required for the mast of a fishing-boat.

A landowner called on us one autumn previous to his going abroad during winter, and requested to be furnished with plants to form a plantation of a few acres, which he had enclosed from a field of arable land, for shelter and ornament. On our producing a list and marking kinds which we considered adapted for his purpose, the only sort he objected to was the Norway spruce, stating that from some peculiarity in the soil it would not grow, and although he admired the common spruce very much, yet no spruce but the *Douglasii* came to anything on his property. The Norway tree was ignored at the time, but it afterward became apparent that no Norway spruce existed on the property, and that the white American species was mistaken for it.

A. Douglasii (Lindley): Douglas's Spruce Fir.—This is a fast-growing magnificent tree, a native of the banks of the Columbia river, in North-West America. It was introduced by Douglas the celebrated Scotch collector, whose name it bears.

The first plants in England were produced from seed in 1827. One of these planted at Dropmore attained the height of nineteen feet in ten years, when it began to produce cones. At that place the tree now forms very heavy timber, where some of the best specimens stand upwards of ninety feet in height. The tree is found to be quite hardy, and is much esteemed throughout the country.

For some time the best Scotch production of this species in the far north stood at Coul House, in Ross-shire. It measured 47 feet 6 inches in height, with a trunk six feet in circumference, at the age of twenty-four years; but on account of its standing in an exposed situation it only advanced ten feet in additional height during the next ten years, when its trunk measured 8 feet 6 inches in girth. Other trees in the north, much younger, but better sheltered, are now far loftier.

In Morayshire some fine trees of this species embellish the lawn at Seapark. At Brodie Castle the first generation of this tree, grown from seed produced there, stand about twenty

N

feet high, and the yearly shoots of these are sometimes upwards of two feet long. The young plants from British grown seed are now getting so plentiful and cheap, that landowners possessed of suitable soil may insert them profitably by the thousand.

Hitherto, on account of their value, the seeds have been sown and the plants grown under glass; but the open ground treatment, adapted for the common spruce or for the Scotch fir, is found quite suitable.

According to Douglas, the trunks of this species in the forests of the North-West of America vary from two to ten feet in diameter, and from 100 to 180 feet in height. He gives the dimensions of a stump of this tree near Fort-George, on the Columbia river, which measured, exclusive of the bark, and at three feet from the ground, forty-eight feet in circumference. In this country the tree frequently forms leading shoots three feet long in one season, and notwithstanding that circumstance it generally appears bushy in proportion to its height. Its foliage is of the richest description, and bears a striking resemblance to that of a vigorous yew-tree. The flag-staff at Kew Gardens is of the timber of this tree imported from its native country, and stands 150 feet high.

It is believed, like other spruces, to thrive best in ground having a moist subsoil, unaffected by the drought of summer, particularly in sheltered ravines. Instances have occurred of the tree having grown freely on gravelly and open subsoil, and died suddenly after protracted drought; and it is to be expected, that with this, as with other species of Coniferæ, excess of drought, or adverse circumstances in any form, will be more fatal to trees of vigorous growth, than to those whose previous progress had been of a more moderate rate.

A. Canadensis (Michaux) : The Hemlock Spruce Fir.—This is a handsome ornamental tree from Canada, of a pendulous habit, but not worthy of cultivation in this country as a timber-tree on account of the slowness of its growth.

A. Smithiana (Loudon) : The Khutrow or Morinda.—This tree is known also as the Himalayan spruce fir, introduced into Britain in 1818—a plant of great beauty when in health, but it generally becomes diseased before reaching the size of a timber-tree.

There are several other species of the spruce fir in cultivation, and employed as ornamental plants, which are not of sufficient interest to call for any notice in this work.

XX.

THE SILVER FIR.

SILVER FIR.—This is the genus *Picea* of Linnæus, but the *Abies* or spruce fir of the ancient and of some of the modern botanists. It is one of the most ornamental trees belonging to the natural order Coniferæ, and consists of several species, natives of Europe, Asia, and America. Some of the kinds are only recently introduced into Britain, and are still very rare trees. All the species advance in a uniform habit of growth, and yield a straight bole, of regular taper, large in proportion to the lateral branches, which range horizontally in regular whorls, each of which presents a flat or frond-like surface of foliage. The genus bears a striking resemblance to the spruce fir, but the leaves are less numerous, and lie more flat on both sides of the branchlets, and thus form two ranks ; and the cones stand erect on the branches, and, when ripe, their scales being deciduous fall away, while the spruce fir cones are pendant, and their scales are persistent.

Picea pectinata (Loudon).—The common silver fir is found indigenous in the north of Africa, and throughout central Europe, abounding on the lower slopes of the mountains, and in glens in France, Germany, Italy, and Spain. On the Carpathian mountains and on the Alps it occupies an elevation between 3000 and 4000 feet. In narrow valleys in the south of Germany, between the Swiss mountains and the Black Forest, on rich, friable, loamy soil, it attains the height of 150 feet, with a trunk sixteen to twenty feet in girth. From this district it was introduced into England in the year 1603, by Serjeant Newdigate, who then planted two two-year-old seedlings at his residence at Harefield Park, Middlesex. According to Evelyn, these in 1679 had become "goodly masts," the largest being eighty-one feet high and thirteen feet in circumference. The other tree was of a less size, in consequence of its having been scorched on one side by the burning of the house adjoining to the spot where it stood.

The silver fir, when young, is one of the most tender timber-tree plants in British cultivation, and although its progress is tardy during the first ten or fifteen years of its age, yet, after it gets a little established in a suitable soil, and attains the height of eight or ten feet, it advances with a rapidity both in height and girth which is seldom surpassed by any species of the Coniferæ.

The cones of this tree become ripe in the beginning of winter, and the scales fall away, allowing the seeds to be extracted without the aid of artificial heat. They should be sown in April, in a well-pulverized soil. The fertility of the seeds, which varies exceedingly in different seasons, and is indicated by their being full and plump, should, in sowing, regulate their closeness, or the quantity which should be deposited in a given space. The distance of two inches is the space most suitable for the young plants to stand asunder in the seed-bed. The mode of laying down the crop described in the article PINE-TREE is suitable for the silver fir, and the soil should form a cover on the seed about half-an-inch deep. The seed-beds should be protected from birds until the end of May, when the crop of young plants generally appears above the surface of the ground.

When the plants appear above ground earlier in May, it is sometimes necessary to protect them, as the slightest touch of frost destroys them. The protection may consist of a cover of broom, or of the branches of any evergreen trees, which may either be spread flat on the surface, or the smaller branchlets may be stuck into the ground, close enough to protect the plants, and yet so open as to admit enough of air. Another mode, which is sometimes the most convenient, and protects the beds both from the ravages of birds and the influence of frost, is to cover the beds immediately after they are sown with straight-drawn straw, which forms a complete protection, and is easily regulated in closeness or openness to suit the health of the young plants. In situations where such a covering is apt to be disturbed by the wind, a straw rope should be stretched along the centre of the bed, across the covering, and fixed by pegs to the ground. Some seasons, however, and particularly when the plants are late in rising through the surface, they are much less subject to injury during the first year than in the succeeding years of their nursery treatment. In the month of May, and in early seasons in the end of April, the slightest touch of frost destroys the newly expanded foliage and young growth, when its influence is hardly perceptible on any other tree; therefore twigs of evergreens or some other slight covering is often required to preserve the top shoots of the young plants.

When the second year's growths are matured, the plants should be transplanted into lines. But in the case of being frost-bitten, the usual method is to allow them to remain the third year in the seed-bed, that they may form tops, which they do more readily before than after being disturbed. In transplanting into lines, a moderately sheltered and shaded situation should be preferred, such being more likely to produce healthy and well-topped plants of silver fir than an open sunny exposure. The lines should be a foot apart, and the plants a few inches asunder. After remaining in the lines two summers they should be carefully lifted, and either removed to the forest or replanted into lines to become a larger size before being finally planted out. So slow is the early growth of this tree, that plants thus treated, at the age of six years, seldom exceed one foot in height; but their roots are always found to be large and bulky compared to the size of their tops, and the girth of the stems is uniformly great in proportion to the height of the plant.

The soil most suitable for producing silver fir timber is a

rich deep loam, not absolutely wet, but that which is cool and moist, rather than dry, such as usually occurs in valley ground, ravines, and along the slopes of Scottish moorland, at no great altitude. But the tree also thrives in heavy clay, and in soils of very opposite qualities, that are not apt to be much affected by severe drought.

It is valuable for filling up vacancies in woods, and for being inserted among timber which is intended soon to be removed. The excess of shelter in such places, which is fatal to many kinds of young trees, tends to promote the growth of the silver fir, and preserves it from the effects of late spring frosts, to which, in this country, it is so liable while a plant. This tree is altogether unfit for the bare and rough elevations adapted to the native pine and larch. In countries where the tree is indigenous, it is frequently associated with the oak; and it is uniformly an inhabitant of a region less elevated than that occupied by the *P. sylvestris*.

The yearly growths of the young plants of silver firs are always very short, and from the circumstance of the buds of the top shoots being generally unfolded simultaneously with those of the lateral branches, and being exceedingly tender, and sensitive to the slightest degree of frost, the plant generally suffers severely in the opening up of the season. When it becomes more established, however, and has reached the height of ten or fifteen feet, it assumes a more robust habit, and shoots vigorously; then the buds on the older and lower branches become first developed, and those of the top shoot at a later period, when they are generally exempt from injury.

The tree has seldom been planted in masses by itself, but where such has been the case in suitable soil, it has been found that it admits of standing very close on the ground, which tends to produce tall clean wood of a fine grain, and notwithstanding the unusual closeness of the trees to one another, they are found to attain a great girth, so that in suitable soil, the cubical contents of the timber, per acre, in seventy years is seldom equalled by any other tree. Hitherto

the species has been employed chiefly for ornamental effect.
In Devonshire, Sussex, Bedfordshire, Hampshire, Wiltshire,
Derbyshire, and Northumberlandshire, its height is considerably above the ordinary size of timber trees, and the best
specimens range from 110 to 120 feet high, many of which
measure 300 cubical feet of timber; and several instances
are recorded both in England and in Scotland, of a single tree
having produced upwards of 200 cubical feet of timber at the
age of seventy years. In the latter country the tree is of
more recent introduction, but specimens nearly a century old
are found interspersed throughout plantations, and ornament-
ing park scenery adjoining many of the principal residences
in North Britain. At Duff House, Cullen House, Ballindalloch Castle, Balnagown Castle, it has long since become
conspicuous by the regularity of its pyramidal head, towering above all other trees, and generally ranging from 90
to 100 feet high; and instances seldom occur where a
tree of this species, of advanced age, in ordinarily favourable soil, is not very far beyond the size of other trees of
equal age.

The wood of the silver fir is generally of a pale yellow, and
the texture of its grain is commonly very irregular, even in
different parts of the same tree, if it has had space to become
of great girth. This arises from its unequal growth at different
periods of its existence. A transverse section of the trunk,
cut near to the surface of the ground, generally exhibits fifteen
to twenty circles, very closely arranged around the pith; after
which every year's circle or growth of wood usually becomes
greater in diameter, to the number of twenty, thirty, or forty,
according to the quality of the soil. In ordinary cases, after
that period the layers are formed of smaller diameter, and the
timber is consequently of finer grain. Grown into large timber,
it is commonly formed into deal for the purposes of flooring;
for which it is very suitable from its never showing any tendency to warp. When grown in masses the timber is generally
finer, more equal, and adapted for masts, joists, rafters, and all
the ordinary purposes of carpentry, and it is sometimes used

in wood-engraving. It usually sells at the price of common pine timber.

This, like most of the other species of this tree, is very productive of resin. It yields the Strasburg turpentine of commerce, so named from the extensive produce of the forests contiguous to that town. This resinous fluid is collected from the small blisters or tumours formed under the outer bark or *epidermis*, and from other exudations of the tree. It is the only tree of the tribe that yields turpentine employed in the preparation of clear varnishes and artists' colours ; and the resinous juice of the tree is manufactured into several articles of great efficacy and importance in medicine and in farriery. The essential oil of turpentine is the produce of this tree, and is much esteemed in cases of strains and bruises.

None of the Coniferæ spring from the roots when felled, but roots of the silver fir have been known, after the removal of the tree, to produce annual circles of ligneous matter, increasing the diameter of the stump, and forming yearly deposits, which have sometimes continued for many years. Around the outside of a stump about two feet in diameter I have counted eight very distinct circles of this woody substance, of good average breadth, corresponding to the number of summers which had elapsed since the tree was felled. The substance has a creamy white appearance, as compressible as cork, and apparently impervious to moisture. Of the instances of this curious formation, the most remarkable is that recorded by M. Dutrochet, of a stump of silver fir, felled in the Jura forests in 1743, which was still full of life when examined at the end of the year 1836. This formation consisted of ninety-two layers of woody matter, formed during that number of years by the roots, deprived of their trunk and leaves, and the wood which composed the stump at the time the tree was felled had in 1836 entirely disappeared.

P. balsamea (Loudon) : The Balm of Gilead Silver Fir.— This tree was introduced into Britain from America at the close of the seventeenth century by Bishop Compton. During the second and third decades of the present century this was the most common silver fir, both in the Scotch and English nurseries. It has now become comparatively rare, having died off, and disappeared throughout the forests. I observe that in seedsmen's lists the seed is now offered by the ounce, at a price which would have bought a few pounds forty years ago. This arises partly from the scarcity of the tree now in this country, and partly from its nature being better understood, and grown only on a small scale, and that not for its timber. It yields cones abundantly in this country, and it is readily grown from seed by the mode of treatment recommended for 'the *P. pectinata*, and in the nursery it is of rapid growth, being at the age of five or six years twice the size of the common silver fir at the same age, and at the age of eight or ten years about three times its height ; but its vigour is of short duration, and it often ceases to grow before it is twenty years of age. In the most congenial soil, deep, moist, and well sheltered, it seldom lives beyond the age of thirty or forty years, and it is an unusually fine specimen of the tree that reaches forty feet in height, or contains twenty cubical feet of timber. Notwithstanding this it is a useful tree, on account of its early and rapid growth, shelter, and ornamental effect. At the age of ten years it is commonly more than that number of feet in height. Its foliage is closely set, shorter, and more dense than that of the common silver fir. The leaves are of a dark, shining green above, and silvery underneath. It advances in a pyramidal form, and gives a richness and shelter to all newly-formed plantations into which it is introduced, and no other tree is better fitted for shelter and ornament in shrubberies and beltings, until plants of greater duration become established. But in all light gravelly soil, where the subsoil becomes affected by the drought of summer, it assumes a sickly aspect and dies, while in a deep rich soil of ordinary moisture its foliage is found of the darkest and most luxuriant description. Since "Christmas trees" have become so fashionable in this country, the plant may be recommended as possessed of sufficient elegance for that purpose.

Its timber is used for the purposes of common deal. The resinous extract of this tree in its native country is known under the name of Balm of Gilead, or Canadian balsam, and a medicinal preparation of its turpentine is said to possess great efficacy in certain stages of consumption. The tree yields its resinous substance more profusely than any other of the Coniferæ. The bark, buds, and cones are often literally encased with turpentine, which is of the most penetrating taste, and highly odoriferous. It exudes copiously on the slightest incision, and the species appears to inherit an early decay from an overflow of resinous fluid.

P. nobilis (Loudon): The Noble Silver Fir.—This is a most beautiful species, but of very slow growth when young. It was introduced into England by seed in 1831, and in October 1837 the largest plant in England did not exceed the height of from one to two feet. At that time the price of plants was three guineas each. Douglas, who introduced it, says it is a majestic tree, forming vast forests on the mountains of northern California, and producing timber of excellent quality. He adds, "I spent three weeks in a forest of this tree, and day by day could not cease to admire it." Of the original introduction, some of the plants came to the north of Scotland from the Royal Horticultural Society. One at Coul House, Ross-shire, has attained the height of 43 feet 3 inches, and the girth of 4 feet 6 inches above the swell of the roots, the growth of thirty-three years. This species, on account of its beautiful form and foliage, will no doubt be always esteemed as an ornament. Seedling plants are now getting plentiful throughout the country. The produce of home-grown seed and plants may be had at a few shillings each; but although instances are recorded of plants, somewhat advanced, having yielded top-shoots three feet in length in one season, yet, on the whole, the growth of the tree is inferior to that of the common silver fir. The mode of cultivation bestowed on the common silver fir is suitable for this tree, and it is said that the timber of both species is very similar.

There are several other species of silver fir, natives of California, esteemed in ornamental collections, such as *Grandis*, *Amabilis*, etc., still too rare and costly for ordinary plantations.

P. Pindrow (Royle) : The Upright Indian Silver Fir.—This tree is found on the Himalayas and other mountains in India, at an altitude of from 8000 to 12,000 feet, where it is sometimes found to attain to the height of 150 feet, and 20 feet in circumference.

In this country it often forms a beautiful tree, with dark green foliage. It grows rapidly, and stands any degree of frost during winter, but, like most silver firs, it is apt to be retarded by late spring frosts, after its leaves are expanded. Plants one to two feet sell from 1s. 6d. to 2s. 6d. each.

P. Webbiana (Loudon): Captain Webb's Silver Fir.—This tree is also a native of the mountains of India, and is found at the same elevation as the *Pindrow*, but it does not attain to so great a size.

In this country, in all but shady situations, it is very subject to the casualty to which all silver firs are liable—unseasonable frosts. In health this tree is perhaps the most ornamental of the genus. The upper surface of its leaves is of a dark green, with white or silvery stripes distinctly marked underneath. The cones are large and prominent, of a beautiful purple colour. This species now yields seed in Britain, where no doubt repeated generations of the tree will become more and more hardy. Some of the finest specimens in England stand at Dropmore, and at Theydon Grove, near Epping.

P. Cephalonica (Loudon): The Mount-Enos Fir.—This tree is indigenous to the mountains of Greece, where it flourishes at an altitude of 4000 or 5000 feet, and forms a handsome tree sixty feet high, with a trunk from nine to ten feet in circumference. When its upward growth subsides, it forms a broad spreading top, increasing the diameter of its bole. It is quite hardy in North Britain, and grows with considerable vigour. Plants one foot high are sold at one shilling each.

P. Nordmanniana (Loudon): Nordmann's Silver Fir.—This tree was discovered by Professor Nordmann, of Odessa, on the Crimean mountains, at an elevation of about 6000 feet, where it attains the height of nearly 100 feet. The timber in its native country has a high reputation; and in this country the young plant is late in expanding its leaves in spring, whereby it is generally exempt from injury. It is reckoned the hardiest silver fir yet introduced, but it does not attain to the size of the common tree of the genus.

XXI.

THE LARCH.

THE LARCH.—*Larix Europæa* (Dec.)—This is the only genus of the natural order Coniferæ, or cone-bearing tribe, that is composed of deciduous trees.

This, the common species, is a native of the Alps of France and Switzerland, and of the Apennines in Italy. It is also indigenous on all the elevated and rocky situations in the Tyrol, and on other mountains in Germany.

It appears that the tree was introduced into England during the early part of the seventeenth century. In 1596, Gerard, a physician in London, published a catalogue enumerating upwards of a thousand sorts of plants of foreign and domestic growth, in which he gives an accurate description of the tree. But the first account we have of its being introduced into England is given by Parkinson, an apothecary in London, who wrote in 1629. In his *Paradisus* he notices the larch to be "a rare tree, and nursed up with but few, and those only lovers of variety." Evelyn, in 1664, mentions a larch, a flourishing and ample tree, growing at Chelmsford in Essex. Miller, in the first number of his *Gardener's Dictionary*, published in 1731, states that the tree was common in English gardens, and that some large trees at Wimbledon produced annually a great number of cones. Miller again, in the seventh edition of the *Dictionary*, published in 1759, states that the larch had become plentiful and common in most of the English nurseries, and that of late years great numbers of the tree have been planted, adding, that "those which had been planted in the worst soil and situations had thriven best." From this period the nature of the tree became better known; and as none of the tribe, after having begun to yield cones, seeds more abundantly, the tree spread rapidly. The fertile plains of England, however, were not adapted for its full development, as the elements most conducive to its prosperity were wanting, namely, an elevated, open subsoil, affording a ready discharge of moisture, and a clear and open atmosphere. Generally speaking, all the trees of the species in Britain, remarkable for their size, are produced in the vicinity of high mountains and running streams.

From the accounts we have of the introduction of the larch into Scotland, some state that it was first planted in 1725 at Dalwick in Tweeddale, and several years afterwards at Dunkeld, Monzie, and Blair; but the dates of the various reports do not exactly correspond.

According to a report by the Rev. James Headrick, minister of the parish of Dunnichen, published in 1813, the larch first reached Scotland shortly after the middle of the seventeenth century. In his *General View of the Agriculture of the County of Angus or Forfarshire, drawn up for the consideration of the Board of Agriculture,* he says, " It is generally supposed that larches were first brought to this country by one of the Dukes of Atholl about eighty or ninety years ago. But I saw three larch trees of extraordinary size and age, in the garden near the mansion-house of Lockhart of Lee, on the northern banks of the Clyde, a few miles below Lanark. The stems and branches were so much covered with lichens that they hardly exhibited any signs of life or vegetation. The account I had of them was, that they had been brought there by the celebrated Lockhart of Lee, who had been ambassador to Oliver Cromwell at the Court of France, soon after the restoration of Charles II. After Cromwell's death, thinking himself unsafe on account of having served a usurper, he retired some time into the territories of Venice. He there observed the great use the Venetians made of larches in shipbuilding, in piles for buildings, and other purposes, and when he returned home he brought a number of larch plants in pots, with a view to try if they could gradually be made to endure the climate of Scotland. He nursed his plants in hothouses and a greenhouse, sheltered from the cold, till they all died except the three alluded to. These, in desperation, he planted in the warmest and best sheltered part of his garden, where they attained an extraordinary height and girth."

An account published in the *Transactions* of the Highland Society, under the authority of the late Duke of Atholl's trustees, states that the first larches at that place were brought from London by Mr. Menzies of Migeny, in 1738; that five small plants were left at Dunkeld, and eleven at Blair in Atholl, as presents to the Duke of Atholl. The five were planted in the lawn at Dunkeld, in alluvial gravelly soil, abounding with round stones, in a sheltered situation, elevated forty feet above the Tay, and 130 above the level of the sea. Three of the five were cut down, two of which were felled in 1809, of which one measured 147 cubical feet, and the other 168 cubical feet. The last-mentioned was sold on the spot to a shipbuilding company in Leith for 3s. a foot, or £25, 4s. the tree. The other two larches are of immense size, and continue to grow on the lawn at Dunkeld. The popular account, that the trees at Dunkeld were the first larches introduced into Scotland, and that they were imported from Italy with other exotics, and nursed in a hothouse, does not appear to be correct; but as the trees at Dunkeld are now among the oldest and largest in the country, it is by no means surprising that, in the absence of the first imported trees, the tradition respecting their treatment should be engrafted on the celebrated tree on the banks of the Tay at Dunkeld.

Of the larches planted at Blair, one, 106 feet high, was cut down, from which the coffin was made of the celebrated Duke of Atholl who planted the tree so extensively. Between the years 1730 and 1740, the larch plants were in request by many of the Scottish landowners, who planted them to a small extent, and generally ruined them by inserting them in soil too rich and cultivated for their future success. Frequently, too, the mistake occurred of inserting the plants too close to the residences, and in places where the tree had not space to become full grown, as is often the case at the present time in fixing the situation for the *Cedrus deodara*, or Indian cedar, which has been almost as common in Scotland during the early part of the present as the larch was in the early part of last century.

It is reasonable to believe that the larches planted in Scotland during the early part of last century were the produce of seeds ripened in England, since many of the trees had, in that country, yielded seeds at least fifty years before the time of their insertion at Dunkeld. Whatever may have been the size or age attained by the imported plants cultivated by Mr. Lockhart on the banks of the Clyde, it is certain that they were of no use to Scotland, in a national point of view; for it was not from them that the tree was increased. Here it is interesting to observe the more fortunate fate of the seedling plants of British growth, introduced by Mr. Menzies, and planted in Perthshire. They formed the great source from which sprang all the finest larch plantations to be found throughout Scotland during the end of the last and early part of the present century, pointing at once to the importance of acclimatation. A separate article will be found on that subject, p. 24.

From an abstract of a report on the Atholl plantations, I find that James Duke of Atholl planted—

In 1738, ...	16 larch plants.
Between 1740 and 1750, . . .	350 „
Between 1750 and 1759, . . .	1575 „
Total, . .	1941

John Duke of Atholl succeeded Duke James in 1764. He planted 11,400 larches. He had great difficulty in obtaining plants to a great extent. Although the earliest planted trees on the estates had begun to yield cones in his time to a small extent, the greatest number of plants that they produced yearly did not exceed 1000.

At the date of the accession of the celebrated planter, Duke

John, in 1774, the extent of the plantations of all sorts of plants was only about 1000 Scotch acres. In 1829 the extent of the whole of the plantations amounted to 13,378 Scotch acres, of which 8604 were of larch alone, besides those mixed with other plants.

The following table shows the divisions of the forests and the extent of the kinds planted, etc., to 1829 :—

ABSTRACT OF THE DUKE OF ATHOLL'S WOODS AND FORESTS, 1829.

Divisions	Oak		Larch		Spruce		Scots Fir		Mixed		Birch		Total		Observations, and Nos. of Plan.
	ac.	dec.	ac.	dec.	ac.	dec.	ac.	dec.	ac.	dec.	ac.	dec.	Scots acres.	dec.	No. in Plan-book.
Dunkeld, ...	11	763	279	997	..	..	207	983	195	432	..	..	695	175	I.
Fungarth, ...	117	958	..	..	..	..	5	663	1	834	..	..	125	455	II.
East Boat, etc.,	23	067	175	476	..	..	38	..	20	308	..	..	256	851	III.
Drumbuie, ..	132	..	464	486	11	335	..	..	100	877	..	..	708	698	III.
Letter and Lows,	53	709	291	501	..	..	28	770	2	..	2	684	378	664	IV.
St. Columbus,.	37	106	530	139	..	..	..	..	257	544	..	..	825	789	VII.
Guay, etc., ..	88	510	95	200	..	..	..	..	11	006	..	..	194	716	X.
Loch Ordie, ..	20	..	2791	122	150	..	..	..	..	..	..	..	2,961	122	XVII.[1]
Loch Hoisher,.	..	..	2031	870	200	..	..	..	..	..	..	..	2,231	870	XVIII.[2]
Inver, etc., ..	104	196	116	103	15	474	25	036	250	284	21	613	632	706	VIII.
Daluaruoch,..	118	814	713	795	..	..	34	..	84	715	..	..	951	324	IX.
Laighwood, ..	24	768	..	..	..	..	..	..	..	..	..	..	24	768	V.
Kincraigie, ..	1	493	1	250	..	..	..	..	..	..	..	..	2	173	XI.
Kinnaird, ...	..	..	..	..	..	..	..	..	1	186	..	..	91	186	XII.
Dalcapon, ...	110	158	..	..	..	..	..	..	94	000	..	..	114	158	XIX.
Tulliemett, ..	78	353	161	987	..	..	..	..	..	..	..	..	240	640	XVI.
Edradour, ...	42	142	99	433	..	..	..	..	11	910	..	..	153	485	XXI.
Balnamoor, ..	18	450	..	500	..	..	..	..	..	..	..	..	18	950	XIII.
Balnaguard,..	..	..	..	..	..	..	9	500	..	..	..	..	9	500	XIV.
Logierait, ...	80	..	253	200	..	..	..	..	..	..	..	..	333	200	XV.
Killiehangie, .	28	263	299	503	..	..	..	..	..	..	..	..	327	766	XX.[3]
Around Blair,.	..	..	300	..	..	..	..	..	1000	..	..	..	1,300	000	
Strathord, ...	..	..	..	..	..	..	..	..	800	..	..	..	800	000	
Total, ..	1070	750	8604	542	376	809	348	952	2932	096	24	297	13,378	466	

[1] Planted 1815 to 1818.   [2] Planted 1825 and 1826.   [3] Planted 1817 to 1819.

NOTE.—On the Duke's accession in 1774, the total number of acres planted was about 1000. Consequently he has planted about 12,378 Scots acres, or 15,473 English statute acres. Allowing 2000 plants in each Scots acre when planted, makes the number of trees 24,756,000, but which in reality has been considerably more; and 10 per cent. may be allowed for making good, which makes the total number of trees planted 27,231,600.

As the ages of the respective plantations in each estate vary considerably, they could not be marked on this abstract; but they are distinctly enumerated in the Book of Plans, in which the numbers correspond with those inserted here opposite each division.

These plantations formed the source from whence the numerous tree seed merchants of Perthshire derived their supply early in the present century, and before the Continental larch seed became an article of commerce.

Since the larch has become a cultivated tree on the Continent in very warm situations, the imported seeds are found to become more and more tender; for seed collectors, finding an ample supply at their doors, are not likely, with the great competition in the trade, to ascend the rugged heights where the tree is more hardy and indigenous for a supply of cones, and for many years Continental seed has formed the chief supply of the British nurseries, yielding not less than twenty millions of plants yearly; and during the unfavourable seasons since 1860 for the produce of Scotch larch seed, the plants raised from Continental seed must have amounted yearly to a much greater number.

In appearance the larch is an elegant object, of a beautiful conical form, and in favourable situations it is of the utmost regularity of figure at every stage of its growth. Its trunk is straight, and if allowed proper space it becomes massive. It produces a great number of branches, commonly horizontal, and when old of a graceful pendant figure, aspiring somewhat toward the extremities, with branchlets which proceed from the main shoot of the branch, of a drooping habit of growth, The leaves, which are of a lively grassy green, are produced in the form of bundles, except on the young shoots, where they stand individually. The tree yields its male and female flowers in April and May, and arrayed in its summer's dress it forms a very graceful and engaging object. The common larch is the only tree of the genus that is worthy of cultivation for timber; compared to it all the other species appear feeble and ungainly, and how interesting soever they may appear in collections, from the peculiarities and habits of their growth, they are, in point of ornament, inferior to the common tree. It consists of many slight varieties, of which some yield red, and some white, blossoms or female catkins. The former are the more common in Scotland. Trees are also frequently met with of a delicate pink blossom, and of all the intermediate shades between the red and white. The unripe cones and the young wood of the trees exhibit to some extent the same colours, but the red and the white flowering kinds cannot be distinguished in winter. The red is decidedly the hardiest tree, and best adapted for a late climate. The white, or *L. Europœa flore-albo*, is imported from the Tyrol. I have repeatedly found that the seedling plants of this variety produced from seed imported from the Tyrol have failed to ripen their tops so as to resist the early frosts of autumn, and even when grown from Scotch seed they are more tender than the red variety. All the finest specimens of the tree throughout the country are of the red larch.[1] The structure of the branches and the rate of growth vary considerably in different trees of the species. But although the most distinct colours of the blossoms can be propagated from seeds with little or no variation, yet the more important properties of structure and growth are unfixed or accidental, and are not so uniformly inherited by the seedling plants. It often happens that late frosts in the end of April and in May, which is the time of the blossoms of the larch, destroy the crop of seed. Unlike the pine tree, the larch blossoms and ripens the cones during the same year. The cones are ripe in the beginning of winter, and in collecting them it is of importance to make choice of

[1] In a recent work on forest management the following statement is given : —" There are two varieties of larch generally found in cultivation in the plantations of Scotland, namely, the white and the red. The white is the variety which attains the greatest dimensions of timber, and is the sort most generally cultivated, although the two kinds are often seen growing together in the same plantation, and that by mere accident. It is said that upon the Atholl estates the red larch does not attain to more than one-third of the cubic contents to which the white larch does ; and this is observable in every plantation where the two varieties are found growing together."

The foregoing statement is very apt to mislead the inexperienced planter. The white larch when young is of more vigorous growth than the red, but as it is far more tender, its vigour soon subsides ; and it ought to be avoided as far as possible. I once grew it purely, alongside the red variety, in the nursery, but found it extremely tender ; its top shoots seldom ripened, and in spring they drooped, retaining the foliage of the previous year, and became of no value. Fortunately the larch plantations in the north of Scotland contain only a very small proportion of the white variety, probably not exceeding 5 per cent., and in grown woods I have never found the white the largest, but the reverse. I know that the cultivated trees at Dunkeld and throughout the Atholl plantations and at Monymusk are of the red variety.

those produced by healthy free growing trees, as such yield the largest cones, the largest seeds, and the strongest seedling plants.

Various methods of extracting the seeds are recommended by different authors. That of boring and splitting the cones, although well suited for manufacturing a few handfuls, and well adapted to the circumstance of the times a century ago, is now far too tedious. The present prices of larch plants cannot afford the time and labour of dissecting the cones individually. Another method recommended in some books on this subject is to "scatter the cones on the prepared seed-beds, to let them lie there till natural chemical action makes them discharge a sufficient number of seeds, and then to rake them off, either for dispersion on other seed-beds, or for preservation till another season." Plausible though this method may appear, it is not practicable in the climate of Britain, and can never yield a crop. The influence of our sunshine does not extract the seeds from the larch cones with any degree of regularity even at the hottest season of the year, and that is a period at which the crop does not admit of being sown.

To extract the seeds of larch safely and speedily, the cones should be placed on a timber kiln. Brick or metal covers are unsafe, and the cones should be heated on wood. They should be laid on about six inches thick. The temperature should be raised to 100° Fah., but not above 110°; with this heat the cones will be quite dry in about ten hours, during which time they should be turned twice, and the seeds which fall out at each turning should be immediately removed. The heat may be kept at a lower degree, but in that case the time for drying will require to be extended. The vegetative powers of the larch seed cannot be destroyed by a degree of heat considerably higher, but even at a higher temperature very few of the seeds escape from the cones. The object therefore of drying the cones is to render them brittle, preparatory to their being thrashed. When dry they should be removed from the kiln in a warm state, and laid about six inches deep, on a floor formed of a fine stone causeway, where they should be thrashed to pieces with flails. The operation is facilitated by the dust and seeds being sifted out from among the cones every hour or so as they accumulate. After sifting, the cones should be replaced on the floor with an additional quantity warm from the kiln, which should undergo the same operation, and so on till all the seeds are obtained. While in a mill of any ordinary construction the seeds are invariably injured, by thrashing they are protected by the dust and pieces of cones surrounding them; even when the stroke falls on the spot where the cones and seeds are thin, the interstices between the causeway stones preserve the seeds from injury. After thrashing, the seeds should be dressed with a common barn fanner, which prepares them for being sown. The price of seed is commonly from 2s. to 3s. per lb., and the expense of manufacturing seldom exceeds 6d. per lb., whilst that of boring and cutting up cones individually with a knife exceeds the value of the seeds.

In the climate of London, the larch should be sown about the middle of April, and in the climate of Scotland during the last week of that month. If the ground requires to be enriched at the time of sowing, it should be with well-rotted manure, decayed leaves, or vegetable mould; but the best method is to make the larch crop follow after a crop which has been well manured a year previously. The ground should be light, well pulverized, and smoothly raked. The seeds should be moistened for a few days, and then sown in beds four feet wide. One pound-weight of seed should occupy about four lineal yards of a bed, and the covering should be one-fourth of an inch deep. The plants generally appear above ground in a fortnight or three weeks, according to the state of the weather. The crop is subject to many casualties. It must be protected from the attacks of birds until the plants have all been a week or two above the ground, and situations should be avoided which are subject to the ravages of the grub-worm.[1]

---

[1] It is now more than twenty years, and shortly after the introduction of the Italian rye-grass, that I occasionally grew in the nursery a plot of this

The beds should be carefully weeded throughout the season. The young plants will have completed their first year's growth by the end of September, when a good crop should cover the ground, and stand from four to seven inches high. A full crop is too close for remaining another year. It is therefore a common practice to loosen the plants with a fork, and to thin them out to the extent of one-half during the ensuing winter or spring. Those that are picked out should be transplanted into lines about one foot apart, the plants standing a few inches asunder. Here they remain one year, when they, as well as the plants left in the seed-beds, which have become two years old, are fit for being planted into bare moorland. Larches should never remain beyond two years in the seed-bed. When stout plants are required, such as those best adapted to overcome and extirpate a cover of furze, or other rank herbage, the plants should be allowed to remain two years in the nursery lines. A common practice is to remove every other line in the lot of one-year transplanted plants which affords the lines of two-year transplanted plants during the last year of their nursery growth a space of two feet asunder. They should never be allowed to remain more

grass for the sake of its seed crops. On one occasion, a crop of seedling larch was made to follow the Italian rye-grass ; the soil was of a fine friable mould, with a considerable mixture of sand and peat, well adapted for forming fibrous roots, at the same time well fitted for the development of the larvæ of the *Tipula* or crane-fly. I do not recollect in what month the grass plot was broken up, but the surface sward was regularly placed under a stamp of clean earth, and the ground well pulverized. The larch crop was laid down about the last week of April, in beds four feet wide, with alleys one foot ; about the last week of May the crop had appeared above ground quite regular, and had a very promising appearance ; and it was, I think, in the first week of June, after an absence of a few days from home, that late one evening, after a heavy shower, I was surprised on looking at the crop to find that the plants had disappeared entirely in five or six beds near one side of the lot. A similar space adjoining contained an irregular half crop ; onwards the beds stood much more complete, but at least 600,000 to 700,000 plants had disappeared. On disturbing the surface and examining the ground, the enemy was discovered to be the grub-worm in considerable numbers. As it was on a Saturday night, the men, women, and boys who were in the habit of working on the grounds were immediately collected, and provided with dibbles ; they went over nearly half an acre of ground that evening, forming a division between the full crop and the blank land; this space they completely perforated with holes—throughout the beds and alleys in every blank space, where no plants than two years in nursery lines without being transplanted. At that age they commonly stand from 2 to 2½ feet high, a height which is suitable for the roughest description of forest ground.

Sometimes two-year-old seedlings are put into nursery lines, and nursed for two years, when they generally attain the height of three feet. In forest planting the tree does not appear to be well adapted for removal beyond this age and size without being put back by the operation. Notwithstanding its being deciduous, no plants can be employed so successfully as healthy larches closely planted to subdue and extirpate a rough cover of herbage. The quality of nursery plants varies exceedingly. The usual prices for larches of the best description are—

One-year seedlings,	.	.	1s. 6d. to 2s. per 1000.
Two-year seedlings,	.	.	3s. 0d. to 4s.	„
One-year transplanted,	.	5s. 0d. to 6s.	„
Two-year transplanted,	.	8s. 0d. to 12s.	„

Protracted drought during the months of May and June sometimes proves fatal to the larch crop, but a scarcity of

stood; the holes were made to the average depth of about four inches. A deep cut or incision was also made with a spade along some of the alleys three or four inches wide at the surface, with the view of intercepting the progress of the enemy. On visiting the ground on Sunday morning the space was black with crows—a place where they were not usually seen, and never since at one time have I seen so many within the grounds. While they betook themselves to some trees in the neighbourhood, without disturbing them long, I saw the holes were generally empty, but some of them I found to contain two, and some three grubs. This operation had the desired effect, and no further injury was observable by the larvæ that season. The assistance of crows was unexpected, as they had not frequented the place previously. My idea was that the holes would form the earth somewhat compact in the bottom, so that the grub when entrapped would be unable to escape, at least for some time, and that by a second application of the dibble into the same holes, they would be destroyed or sunk, so as to render them harmless for the future; this second operation, however, was found quite unnecessary. I need hardly state that ever since I have avoided laying down larch, or any of the fir or pine tribe, after a grass crop. Nevertheless, occasionally, and during the last two summers in particular, grubs were sometimes detected in the ground by the people weeding the one-year seedling crops, and small blanks here and there indicate the place of their abode, when a few dibble holes around the spot is found a sure remedy against further ravages.

plants more frequently occurs from a want of seed, occasioned by the late frosts in spring destroying the blossoms.

The growth of every kind of plant is interrupted to some extent by being transplanted, but no kind will undergo the operation more successfully or take to the ground more readily than a healthy larch, provided it is not of an age beyond that recommended. It flourishes in soils of very opposite qualities, from the dry and sandy to that which is wet and clayey; but in the presence of springs or moisture in the soil, it is absolutely necessary that the water have a free exit. Although frequently fine specimens of the tree are to be met with on flats of sandy loam or clayey gravel, and on various other qualities of soil, yet it is on the declivities, along the slopes of ravines, on the shattered *débris*, and the disturbed soil of the landslip or avalanche, that the tree is found to luxuriate in its greatest vigour. In stagnant moisture, or on a substratum of till, or ferruginous gravel, it becomes stunted and full of disease. But a ready discharge of moisture in the soil must always be accompanied with a free circulation of the atmosphere. It affects a cool and elevated situation, with a cloudless sky; therefore in all low and sheltered situations the tree should have sufficient space for the full development of its foliage.

On the Alps and Apennines it luxuriates at a great height, and some scattered specimens of it are to be met with near to the highest range of vegetation. In many of the Highland districts of Scotland it may be observed filling the straths with massy timber, and ascending the mountain sides associated with the native pine, and, with the exception of that tree, perhaps no other plant, native or foreign, was ever spread over so great a space in so short a period.

I think the larch is inferior to the native pine for producing timber at a great altitude; on this point I must differ from some published opinions of high authority. The failure of larch, however, in many such situations, not unfrequently arises from the surface-soil at great altitudes throughout the country being composed of too pure a peat soil, which is less adapted for the larch than for the native tree. No degree of cold can injure the larch during winter, provided the summer has ripened it, but few trees are more sensitive of the slightest touch of frost while in a growing state. The leaves being remarkably fine and tender, are readily affected by a sudden change of weather. On southern exposures, along the warm slopes of steep mountains, the larch is a very precarious tree. The fine weather which frequently occurs during April and early in May readily brings trees in such situations into full leaf to be blighted by a succeeding frost, from the effects of which they are always very slow in recovering, and sometimes die. One of the largest plantations formed in the north lately died, to some extent from this cause, along the southern slopes, a few years after the plants were inserted; while on the flats or level grounds of the same plantation the casualty was only perceptible, and on the northern declivities the plants were exempt from injury. For the sake of appearance, as well as of economy, the larches in such places should always be interspersed with some other tree.

A plantation consisting of larches purely has a bare and uninteresting aspect during a great portion of the year, but, under any circumstances, the tree is generally found to thrive better in a mixed plantation than in a plantation purely of the species. It grows well when associated with the Scotch pine, but the best specimens of rapidly grown larch timber are generally to be met with interspersed with oak, which is late in the development of its foliage, and drawing its nourishment from a considerable depth in the soil, affords advantages to the larch beyond any other tree.

The distances at which larches and other trees should be inserted in a plantation must always be regulated by the situation. In planting bleak and exposed moorland in the Highlands of Scotland with larch and Scotch pine, 4000 plants per imperial acre are usually allowed, while on low sheltered situations 3000 plants are quite sufficient; and the proportion of larch usually ranges from one-fourth to three-fourths, according to the suitableness of the soil for the growth of the tree, and the value of early thinnings in the district. By having the kinds regularly interspersed, a choice is afforded in thinning the plantation, so that ultimately it may be continued mixed, or made to consist either of the one species or of the other, as may be found desirable. For a description of plants adapted to various situations, and the mode of inserting them, see the article on that subject. The larch, when young, is of quicker growth than any other coniferous tree, and for that reason it is best adapted for extirpating furze and rank herbage. For this purpose, as already stated, two-year transplanted plants are employed, and they should be planted immediately after the furze or whins are cut down. When closely planted, from three to four feet apart, they generally overtop and suffocate the furze, so that the expense of frequent clearings for the growth of the plants, even where there exists a mixture of other trees, is rendered unnecessary. In such cases, however, the larches require early and gradual thinnings, as soon as the herbage is subdued.

One-year-old seedling larches, when produced from five to eight inches high, are adapted for being finally planted into moorland, where the soil is favourable and the heath short. Some of the most healthy and rapidly grown plantations in the north of Scotland were formed of larches of this description; but two-year-old plants are most frequently employed in planting moorland when their average height is about twelve inches, and sufficient to overtop the herbage on the surface of the ground.

In favourable situations in the north of Scotland, the tree has been known to attain a height of upwards of forty feet in twenty years, but thirty feet is the ordinary height in that period. In Morayshire, along the banks of the Findhorn, the tree shoots up with great vigour. I have known larch thinnings removed from a plantation composed of larch and oak of the age of forty years, on the estate of Relugas, average twenty cubical feet each tree, and sold on the spot, eleven miles from the shore, at 1s. per foot. About forty trees per acre have been removed at a time from some parts, and twice that number left interspersed with oak. This growth of measurable timber may be stated as the average produce of the larch in the best situations on the banks of the Findhorn, in the forests of Darnaway, Altyre, and Logie, after a growth of forty years, and where ordinary care has been bestowed in thinning the plantations.

One of the finest clumps of larch ever produced in Scotland was grown at Brahan Castle, on the banks of the Conon, in Ross-shire. This clump originally occupied about one imperial acre. When I last saw it, the remaining trees were about eighty years of age, and averaged about ninety feet in height. This space must have yielded, besides small thinnings, at least 15,000 measurable feet of timber, a portion of which still remained on the ground; the trees standing on an average about twenty feet apart, many of them containing 100 cubical feet each. On the opposite side of the river the larches likewise advance with great vigour. In a mixed plantation, having a northern aspect, on the lands of Conon House, some trees, not exceeding forty-two years of age, were lately thinned out, and their cubical contents ranged from twenty-two to twenty-eight feet each, a size about twice that of the hardwood with which they were associated.

At Ballindalloch Castle, on the banks of the Spey, there lately stood some large specimens of the tree which were planted in 1767. The following table shows the girths of four of these trees at the age of seventy, and their progress up to January 1851, after having been planted eighty-three years :—

	GIRTHS IN AUGUST 1837.				GIRTHS IN JANUARY 1851.			
	At 1 foot.	At 6 feet.	At 12 feet.	At 18 feet.	At 1 foot.	At 6 feet.	At 12 feet.	At 18 feet.
	Ft. In.	Ft. In.	Ft. In.	Ft. In.	Ft. In.	Ft. In.	Ft. In.	Ft. In.
No. 1, .	9 6½	8 5	8 4¾	6 6½	11 8	9 6¼	9 3	7 7
No. 2, .	8 7¼	7 1	6 4	6 0	9 10	9 0	7 5	7 1
No. 3, .	10 6	8 4	7 1	6 6½	13 5	9 8	8 2	7 8
No. 4, .	9 1	7 3	6 5	6 4½	10 6¾	8 4	7 7	7 5

The height of these trees then ranged from 75 to 80 feet. The soil is dry and sandy, with a subsoil of sand, and, in some parts, of sandy gravel and clayey gravel, elevated from 30 to 40 feet above the Spey, and about 400 feet above the sea.

The larch was introduced at an early period into Aberdeen-shire. At Monymusk, Westhall, Tillyfour, and other mansions in that county, venerable specimens are to be met with, which have stood since the tree was a rare plant in Scotland. At Monymusk House, the oldest tree is said to have been imported from Switzerland at an early period in last century, but its dimensions are far exceeded by younger trees at Paradise on the same property, situated on the banks of the Don. This beautiful Alpine spot is composed of soft alluvial soil, incumbent on gravel, and drained by a mountain stream near to its confluence with the Don. It stands 340 feet above the level of the sea, and at the distance of fifteen miles inland. The surface of the ground is level, richly wooded, and embosomed with high mountains. Benachie, with its rocky crest rising 1000 feet above its base, towers aloft and guards it on the north. In this plain stands the finest specimens of the larch I have ever seen. They form an avenue, and are interspersed with Norway spruce firs. All these trees are of great uniformity in size and shape; and as many of them have been measured from time to time by the forester on the estate, I am enabled to record the dimensions of two of the larches, and their progress, after having attained a hundred years of age :—

	MEASUREMENT IN 1841.				MEASUREMENT IN 1851.			
	Girth at surface.	Girth at 60 feet.	Height.	Cubical contents.	Girth at surface.	Girth at 60 feet.	Height.	Cubical contents.
	Ft. In.	Ft. In.	Ft. In.	Feet.	Ft. In.	Ft. In.	Ft. In.	Feet.
No. 1, .	12 6	4 4	101 3	233	15 0	4 9	102 4	276
No. 2, .	12 0	4 2	106 0	203	14 0	5 0	99 6[1]	269

[1] Top broken.

These two trees are only fair specimens of a few dozens
standing in the same avenue. The table shows their dimensions at the age of one hundred years, and their progress
during ten years thereafter; but at the end of that period
their vigour began to cease, and in 1861, when I applied to
ascertain their progress during the ten preceding years, I was
informed by the intelligent forester, Mr. Taylor, that, at the
height of twenty feet, in many instances their girth had only
increased one, two, and three inches beyond that of 1851, and
during that period they had added little or nothing to their
height. There was but one exception to this general state of
affairs; it was a larch known as, or marked, No. E. The
general appearance of this tree indicates more vigour than
that of the others. It stands quite clear of other trees, and
has a river frontage, and is sufficiently near to draw nourishment from the running water, but perfectly free from stagnant
water. Its girth at the height of twenty feet was, in 1841,
8 feet 1 inch; in 1851, 8 feet 6 inches; and in 1861, 8 feet
11 inches. A gale in 1860 blew down one of the larches in
the line, and it was found that decay had begun, but only in
a small spot of no importance near the root, and the quality
of the timber was found very superior. When I first saw
these remarkable trees (in 1851), they were 110 years of
age. At that time I observed that the top shoots of many of
them were quite visible above the extremities of their lateral
branches, but very soon after their vigour subsided, and the
yearly growths of the tops disappeared, and now they present a more round-headed form, and of late they in general
have a more open appearance, arising from the foliage having
become more thin and scanty.

The Norway spruce firs associated with the larches in the
avenue at Paradise are said to have been planted in the year
1720, and, notwithstanding the advance of about twenty years
in their favour, the larches are on an average the largest in
cubical contents, although not quite so lofty as their asso-
ciates. The avenue in which these trees stand is proportionally wide, and they have long luxuriated in ample space.

Planted in straight lines, their massive stems arise like the shafts of columns, and taper to the height of upwards of eighty feet of measurable timber. Those who have seen the larches at Paradise generally reckon them the finest specimens in Britain.

I observe from the *Court Circular* that Her Majesty the Queen has just paid a visit to Paradise (October 1866), returning to Balmoral on the evening of the same day, after accomplishing a journey of upwards of ninety miles through the hills of Aberdeenshire.

The larch, in a congenial soil, and under proper treatment, has been found to produce a greater quantity of valuable timber than any other tree. The great mistake which has hitherto attended its cultivation has been a complete disregard of the acclimatation of the plant : hence the diversified state of the larch plantations throughout Scotland. Some are vigorous and profitable, others blighted and half dead, though standing in the same description of soil and climate. Indeed, acclimatation has hitherto been looked on as a chimera, and has been stated to be so by some recent writers on arboriculture, who have supported their opinions by quotations from the writings of distinguished men who have fallen into error on this subject. They probably never had the opportunity of comparing or testing the hardiness of trees of the same species, grown from seeds produced by trees in a warm country, alongside of those produced from trees inured to a far colder climate. (*See the article under the head* ACCLIMATATION.) This subject is of special importance in the growth of larch. Hardiness forms the sure basis of success, and the best means of avoiding many of the diseases and casualties to which the tree is subject.

The late King of the Belgians took a lively interest in the success of the Belgian larch forests. I have frequently heard that during his tour through the north of Scotland, about the year 1818 (then Prince Leopold), his remarks on the larch plantations in the Highlands awakened among the landowners an interest in the tree. His Majesty was aware of the neces-

sity of acclimatation.   In forming a large forest of larch in
Belgium, he gave special orders to have the plants from the
nurseries of Messrs. John Grigor and Co., Forres, and sent the
orders twelve months before requiring the plants each year,
during the three years required to complete the forest.   All
these plants were the produce of Scotch-grown larch seed, and
formed at least part of the third or fourth generation of the
tree in Scotland.

There can be no doubt that the reason why acclimatized
plants of the larch are so seldom sought after is, that the
great difference between plants from Scotch seed and those
from foreign seed is not known.   It is seldom that the contrast in hardiness is brought under the planter's notice.   If
he has once seen the hardiness of the plant inured to the
climate, it forms a subject not to be forgotten ; and then a
judicious planter would no more allow the plants from foreign
seed to be inserted in his plantation than he would any other
species of half-hardy tree.

I am not prepared to say that the influence of acclimatation
will extend so far as to make the tree more and more hardy
in proportion to the number of successive generations that are
produced by seed-bearing.   It may be that a few generations
of the tree in this country will render it as hardy as its nature
will admit of becoming.

The plantations in my neighbourhood are generally very
healthy, and probably are composed of the third, fourth, and
fifth generations from the Atholl larch forests.   Where bare
moorland adjoins some of them, the young seedling plants
may be seen in acres, from self-sown seed, like to the indigenous
growth of young pines beside the native forests.

A person unacquainted with the nature of the trees and the
influence of seasons would be apt to suppose that if the plants
were too tender the severity of winter would clear the ground
of them in early life, so that there would be no chance of their
remaining to die by frost at a more advanced state in the
forest.   This, however, is not the case so uniformly as might
be expected.   Many are severely injured, but plants in nur-

sery lines often renew their tops or leading shoots. Removal by transplantation into nursery lines, or into the forest, has a tendency to stop a luxuriant growth, and to harden and sustain the plant for a short time against the severity of frost. The period of the larch's greatest vigour is that at which it is most likely to suffer; that occurs near the time it is about to assume a timber size, or from twelve to twenty feet in height—more or less in some plantations,—dependent, to some extent, on the nature of the soil, and on the seasons. The period of the greatest vigour of the larch is readily ascertained by inspecting the concentric circles on the stump of a felled tree in the same description of soil.

The failure of the tree at this period of its growth in many districts throughout Scotland arises purely from its being too tender; and its display of dead wood, with a few young shoots emerging here and there from its stem, are just the symptoms which all half-hardy plants present when frost-bitten.

Another prevailing mistake in larch management throughout Scotland is the neglect of early thinning. The confinement to which it is subjected just at the period it attains the size of a timber tree, when commonly from ten to twenty years of age, and frequently afterwards, is an immense loss to landowners and to the country at large. No other tree is so speedily ruined for want of sufficient space. The larch leaves are tender and minute, they present only a small surface to the influences of the atmosphere, and therefore, in a crowded plantation formed of trees of the same size, and situated on a level surface, the leaves fail to elaborate the sap necessary for the formation of timber, and the trunk becomes, and continues to be, bark-bound, bare, and stunted. On its native Alps it enjoys an inequality of surface, which furnishes more space for its foliage, and a better exposure to light; there also it possesses an advantage which always accompanies indigenous forests—that of the trees being very unequal in size and in age. Close planting is certainly necessary in rearing larch timber in bleak and exposed situations. It prevents the play of the biting winds on the surface of the ground, particularly

P

at the opening of the season, when the tree is most likely to suffer; and although the top of the plant, an important part, is least sheltered, yet it least requires shelter, for the leaves of a healthy young plant are always first expanded near to the surface of the ground, and some time before the higher ones become exposed to the influence of the weather. Thus a wise economy in nature develops the top bud at a time when shelter is least required and safety is most certain. But the benefit of shelter is not the only advantage which arises from close planting. It is found that a close cover of young plants of any kind, with their branches and foliage in direct contact with the surface, destroys the native herbage, converts the decayed stems and roots into a manure, and directs the whole energies of the soil to the growth of wood. The progress of a plantation is always marked by a rapid growth immediately after the trees have formed a cover, and suppressed the surface vegetation. It is at this stage of growth that thinning is absolutely necessary, but much neglected, in larch plantations. The tree should have space in proportion to its growth. Numerous instances are met with of broad-leaved trees having been neglected, and rendered unshapely by confinement, which have yet, by care, been restored to health and to a proper figure; but there is no instance in which the larch, after having been stinted by confinement, or reduced to a scanty supply of foliage, has ever regained its vigour, or become a healthy, well-proportioned tree. This neglect is so general in larch plantations in some quarters, that it has given rise to the idea that it is a tree of slow growth after the age of forty.

The larch is subject to many casualties. Of the insects which infest it, the *Coccus laricis* is the chief. It was observed in the end of last century throughout the larch plantations of Scotland and England. It is rarely met with extensively on trees of vigorous growth; but in confined woods and low situations, where there exists a great humidity of the atmosphere, it often prevails to a great extent. Wet seasons, or whatever tends to decrease the vigour of the tree, favour this insect. But its effects are most fatal immediately after frost late in spring, or early in summer, such as that which occurred in May 1861. On the 10th of that month the thermometer exposed reached 80°, while on the following night or morning of the 11th it descended to 19°, being 13° below the freezing point. This had a very damaging effect on the larch plantations throughout the country. Such a visitation is usually followed by a severe attack of the *Coccus laricis*, which multiplies with amazing rapidity,[1] and not unfrequently overspreads the surface of the young branches, consuming every leaf as it emerges from the bud, and several weeks sometimes elapse before the tree is able to renew its foliage. The sap appears to exude by the numerous perforations, and to form a black coating resembling the effects of smoke, and the top shoots of such trees are uniformly short and feeble.

This state of affairs shows the advantage of a northern aspect and a cool situation, where the tree may not come early into leaf. There are some dry, gravelly knolls exposed to sunshine, which early in spring imbibe heat, raising the temperature of the soil, and expanding the leaf very early. In such soil the larch is a very precarious crop. Ample space for the tree increases the extent of foliage, and is always favourable to the tree when attacked. The *Coccus* is least injurious where there is a free circulation of the atmosphere, and where the plantation does not consist purely of larches. Plantations formed of plants free from disease, and thoroughly isolated from infected trees, have been known to

---

[1] Respecting some kinds of this insect, Curtis states the startling fact, that the spring broods are all females, and do not require any intercourse of the sexes to render them prolific. They are pregnant at their birth, and if the nit (as it is termed) brought forth by the fly in the spring be taken and kept entirely excluded from its companions, it will be able to produce young; and if one of these be treated with the same precaution, it will yet be found to retain the same powers of conception; and thus one may proceed for twenty or thirty generations. This will explain their otherwise marvellous multiplication; and the warmer the weather the more rapidly families increase; so that it has been calculated by an eminent naturalist, that from *one* egg 729,000,000 of plant-lice might be produced in seven generations; admitting forty to be the maximum, and twenty the minimum, the average would be thirty, and the generations from the spring to the autumn amount from sixteen to twenty, or upwards.

remain free from the insect for many years. From the extent
and nature of the field in which the insect prevails, no practicable cure is likely to be discovered beyond that of using
the best means to promote the health of the plantation.
When the soil is congenial, and the plants acclimatized (that
is, the produce of Scotch seed), there is little to fear from this
insect.

The larch is subject to an atmospheric blight, which is
often observed to some extent, but it sometimes falls with
great severity. It occurs at various periods when the tree is
in leaf, but most frequently about the end of July or the
beginning of August. It blasts the foliage, giving it a
yellowish tinge or ripened appearance, particularly the oldest
leaves, those recently expanded appearing only slightly
affected, except in severe cases. Its effects are frequently
ascribed to frost, but it commonly occurs during warm weather, when the thermometer indicates a high temperature.
Throughout some districts it attracted little notice until its
effects the following summer, by the want of foliage, exhibited
numerous dead twigs throughout the tree. During the summer of 1850, this blight severely injured the larch plantations
throughout Scotland; other trees also, in some districts, suffered, but in a less degree, from the same casualty. In a
plantation composed of larch and Scotch pine, twenty-four
years old, situated in Inverness-shire, upwards of 1000 feet
above the level of the sea, where the progress of some of the
trees was marked yearly, the larches were tallest, but previously to 1850 their girths, on an average, were equal.
During that year, however, on account of this casualty, the
marked larches barely made any perceptible increase, while
the Scotch pine made an increase of nearly three inches,
namely, from $27\frac{1}{4}$ to 30 inches in girth.

The rot in larch is a disease which has been noticed by
various writers on larch since the beginning of the present
century. In some parts it is known by the term *pumping*,
doubtless from the resemblance which the hollow trunk bears
to a wooden pipe or pump. It is generally believed that the

disease begins in the root, and its progress is upwards. It is not confined to the centre of the trunk. Old trees, when cut down, are frequently found to be extensively diseased, where the centre and many of the first-formed circles of wood are perfectly sound. This, however, is not uniformly the case, but as the circumstance frequently occurs, it shows that the disease had not existed in the young plants, nor arisen in consequence of the treatment bestowed on it during the first stages of its growth.

This disease has always been found to prevail most in larch plantations formed on ground which had previously yielded timber. The decaying roots, and the fungi which accompany them, tend to injure the roots of several sorts of forest trees. In this respect the larch is by far the most sensitive; but instances have occurred of two-year-old oaks having been killed by the white fibrous matter, or spawn, from rotting wood, cutting off the tap-roots of the plants. I have known a few cartloads of manure composed chiefly of sawdust to create mycelium, and to kill several hundred thousand seedling larch plants in one season, and to render the ground unsuitable for that crop for many years. Every description of fresh manure is injurious to the health of the young plant. Where this disease is likely to occur, plants as young as possible should be preferred in all cases where they are suitable. The natural figure of the larch tree is admirably adapted for being supported by a small root, and its root is generally found smaller than in any other genus, in proportion to the cubical contents of the tree. The natural figure of the tree, however, is changed in all neglected plantations, the expanse of branches in the body of the tree is destroyed by confinement, and when a plantation is newly opened by thinning, the trees which possess top branches only, seldom escape having their roots injured by the influence of wind. In such cases, the movements perceptible on the surface during a hurricane indicate the strain to which the roots are subjected. The injury they thus sustain is followed by decay, and no doubt this is an extensive source of rot in larch timber.

Another source, we believe, arises from the roots resting on a retentive subsoil. A ferruginous crust or subsoil which retains water is generally found to yield diseased larch timber. During winter, when the tree is in a dormant state, its small fibres retain life in stagnant water for a considerable time ; but in the months of August and September, the season at which floods frequently occur, the spongioles are in an active state, and stagnant water has been known to rot the weakest fibres in the course of two days. Excessive moisture is thus calculated to commence the disease in woods standing on retentive subsoil, where it is imperfectly drained. The rot in larch no doubt arises at first from a decayed or injured root, and whatever tends to damage a fibre should be avoided.

It is generally admitted by vegetable physiologists that a perfect root with its spongioles in a healthy state has the power, at least to a great extent, of selecting the food adapted to the plant; but let the spongioles become mutilated or decayed, and they possess that power no longer, but absorb any fluid with which they come in contact.

The rot is sometimes met with in young woods in the best description of larch soil. It appears in thinning, when the trees are no stronger than prop-wood ; this is commonly the result of planting very large plants with mutilated roots.

I lately observed, on some fine specimens of the larch, about forty-five years' growth, being removed from a mixed plantation of larch and oak, that the timber was generally quite sound and healthy, pumping was of very rare occurrence, not more than one or two instances of it in an acre on the average, and where it did occur it was generally in the centre of the trunk. I saw, however, on one of the largest roots near the outside, which yielded one of the heaviest trunks on the ground, that the disease had taken possession of one side only, forming a bulge or swelling which enclosed the vacuity in the timber ; the stump left in the ground showed that the ailment arose purely from one main root. To find out if possible the source of the disease, I had the root laid open by excavating the ground for the distance of a few yards, where it was traced to a decayed post or stump, the remains of an old paling which had protected a cottar's garden. The extremity of this root had been cut off in trenching the garden ground, which brought the wound on the root in immediate contact with the mycelium which overspread the decayed post, and to which the disease in the root and trunk of the larch was clearly traceable.

The Larch Blister.—This is a disease which appears on the stems or trunks of the tree, with a discharge of the sap, commonly on one side of the trunk, while on the other side the ascent of the sap deprives the trunk of its regular shape, giving it sometimes a flat or a bulged appearance. It destroys the timber at the point of attack, which commonly extends to twelve or eighteen inches in length. I find it prevails to a very serious extent in many larch plantations throughout the country; in others it is comparatively rare, such as in the larch plantations in my neighbourhood. My experience of this disease has therefore been very limited, and I cannot speak positively respecting its origin. I may state, however, that the plantations in which I have seen it most prevalent are situated either at a great altitude, or in places where frost is likely to fall heavily. Often finding a great uniformity in the position or height of the blisters on the respective trees, I am inclined to think that it arises from a severe frost in summer, or while the tree is in a growing state, not sufficient to kill the shoot, but to interrupt the free circulation of the sap, and to form a weak point on the trunk. On examining the leading growth of a vigorous larch, the produce of a summer in which a severe frost has occurred, the casualty is often indicated by an immature or weak point marking the length of the shoot at the time of the occurrence. A summer frost or blight sometimes marks very distinctly the vigorous shoots of the osier, forming a black spot or blemish, which in the operation of weaving occasions it to snap at the precise point.

I expect the safety of the larch from this, as well as from other diseases to which the tree is subject, is to be found chiefly in having the plants of worthy extraction, the produce of trees which have been acclimatized to the country.

Favourably situated, no tree becomes so valuable in so short a time as the larch. In many districts the first thinnings are useful for rustic paling, props for plants, sheepflakes, and for similar purposes. It is much esteemed for hop poles, and it is in constant demand for coal props at all shipping ports, at from 2s. to 3s. per dozen, of seventy-two lineal feet, where it usually forms the return cargoes in coal vessels. As an agricultural timber it is particularly valuable, being durable as posts in building, in palings, and in all structures where it comes in contact with the ground. Healthy trees at the age of fifty years have been sawn into half-inch laths, and employed in cladding fences, gates, etc., where it has been known to remain sound for twenty years without paint. It is adapted for lintels, rafters, joists, and the main timbers in buildings ; but from its propensity to warp, and from its being more difficult to plane than common deal, it is seldom employed in flooring or in the lighter purposes of finishing. In some districts it is manufactured into carts, which are found to be very durable.

It is constantly employed for railway sleepers, for mill axles, and in shipbuilding. For the last-mentioned purpose the main roots, in connexion with part of the timber of the trunk, form knees of great toughness, strength, and durability. Since the late reduction of the duty on timber, larch has sold at from 1s. to 1s. 4d. per cubical foot. I am induced to believe that the value of larch wood in the manufacture of furniture is not yet fully appreciated. In thinning a larch plantation, of thirty-two years of age, of my own planting, I had a tree felled and cut up into plank and deal, and after the timber was well seasoned, it was sent to a cabinetmaker, who was instructed to make it into a writing-table, with drawers. The table was produced purely from the timber furnished. It was varnished, had a fine polish, and a beautiful, clean, yellow colour. It has now been upwards of four years in use, and has become darker in colour, being now of a rich brownish yellow, retaining its fine polished silky gloss. Although the timber was rapidly grown, and some of the deals about a foot broad, there is not the slightest appearance of warping, and the joints are all close and perfect—so much so that it is difficult to find out the points of junction. Few trees of British growth could furnish anything superior.

Larch bark is sometimes preserved for the purpose of tanning. It generally sells at half the price of oak bark, the value of which is very fluctuating. As larch bark is twice the bulk of that of oak, in proportion to its weight, it is seldom profitably manufactured.

The Venice turpentine of commerce is produced by the larch. The full-grown tree in its native districts is pierced to the centre with an auger. The turpentine is conducted by a tube into a trough, and it requires no other preparation to render it fit for sale than straining it through a coarse haircloth. A healthy tree, when tapped, is said to yield seven or eight pounds weight yearly for forty or fifty years. The turpentine flows from May to September, but the timber is rendered of no value.

No tree is so valuable as the larch in its fertilizing effects, arising from the richness of the foliage, which it sheds annually. In a healthy wood the yearly deposit is very great; the leaves remain and consume on the spot where they drop, and where the influence of the air is admitted, the space becomes clothed in a vivid green, with many of the finest kinds of natural grasses, the pasture of which is highly reputed in dairy management. And in cases where woodland has been brought under grain crops, the roots have been found less difficult to remove than those of other trees, and the soil has been rendered more fertile than that which follows any other description of timber.

The other species of larch sometimes met with in collections are less vigorous than the common species, and cannot be recommended as timber trees.

XXII

THE CEDAR.

CEDAR.—*Abies Cedrus* (Lindley); *Cedrus Libani* (Barrelier). —The cedar of Lebanon is an evergreen tree, indigenous to Mount Lebanon in Syria, and to Mount Atlas in the north of Africa. It is the most magnificent and picturesque of the Coniferæ, or cone-bearing tribe. When allowed sufficient space, it is more remarkable for the size of its bole and the enormous expanse and number of its branches than for its height. The elegance and grandeur of its appearance are frequently extolled in sacred history beyond that of any other tree. In the poetical style of the Hebrew prophets, it is used to symbolize the spiritual prosperity of the righteous; and whatever was beautiful and comely in the human countenance, or commanded the admiration and respect of the beholder, whatever was fearful even in the execution of the judgments of Heaven, or peaceful and prosperous in the fulfilment of its prophecies, found a striking similitude in Lebanon and its cedars.

Some suppose, from the soft wood yielded by the cedar when grown in this country, that the cypress, or some other tree, must have produced the timber of Biblical times; but none else indigenous to Syria is found to suit the habit and character described by Ezekiel; and its failure, under the unnatural circumstances of climate in which it is placed here, is not extraordinary, for others of the Coniferæ, such as the Highland Scotch pine, are known to degenerate the more in the quality of their timber the oftener they are propagated in an unsuitable clime, at a distance from their native country.

The modern history of Lebanon and its cedars is ample.

Almost every traveller who has visited Syria has ascended Mount Lebanon. Belon is perhaps the first; he records his visit in 1550. He says, "At a considerable height up the mountain, the traveller arrives at the monastery of the Virgin Mary, which is situated in a valley. Thence proceeding four miles farther up the mountain, he will arrive at the cedars— the Maronites, or the monks, acting as guides. The cedars stand in a valley, and not on the top of the mountain, and they are supposed to amount to twenty-eight in number, though it is difficult to count them, they being distant from each other a few paces. These the Archbishop of Damascus has endeavoured to prove to be the same that Solomon planted with his own hands, in the quincunx manner as they now stand. No other tree grows in the valley in which they are situated; and it is generally so covered with snow as to be only accessible in summer."

We have the reports of successive travellers, at dates down to the present time. Ranwolf in 1570, Thevenot in 1655, Maundrell in 1696, Bruyer in 1702, La Roque in 1722, Pococke in 1745, and others, bear witness to the existence of the remains of an ancient forest in the highest part of the mountain, which ultimately became reduced to a single clump of very old and large cedar trees.

Lamartine, who visited the cedars in 1832, says, "We alighted, and sat down under a rock to contemplate them. These trees are the most renowned natural monuments in the universe. Religion, poetry, and history have all equally celebrated them. The Arabs of all sects entertain a traditional veneration for these trees. They attribute to them not only a vegetative power which enables them to live eternally, but also an intelligence which causes them to manifest signs of wisdom and foresight, similar to those of instinct and reason in man. They are said to understand the changes of seasons; they stir their vast branches as if they were limbs; they spread out or contract their boughs, inclining them towards heaven or towards earth, according as the snow prepares to fall or to melt. These trees diminish in every succeeding

age. Travellers formerly counted thirty or forty; more recently, seventeen; more recently still, only twelve. There are
now but seven. These, however, from their size and general
appearance, may be fairly presumed to have existed in Biblical
times. Around these ancient witnesses of ages long since
past, there still remains a little grove of yellower cedars, appearing to me to form a group of from 400 to 500 trees." In
September 1836, M. Laure, an officer in the French Marine,
in company with Prince de Joinville, visited Mount Lebanon,
and ascended to the cedars, which stand on an almost level
space or plain, entirely surrounded by the steep peaks of the
mountain. They found fifteen of the sixteen old cedars mentioned by Maundrell; but all more or less in a state of decay.
One of the healthiest of the old trees, but perhaps the smallest,
measured 35 feet 9 inches in circumference. All the trees
are much furrowed by lightning, which seems to strike them
more or less every year. Surrounded by the old trees, are
about forty comparatively young trees, the smallest of which
have trunks from ten to twelve feet in circumference. They
add that there was not one young plant of cedar in all the
wood of El-Herze, and that the soil of the forest of Lebanon,
on which there was not a single blade of grass growing in
September 1836, is covered to the thickness of half-a-foot
with the fallen leaves, cones, and scales of the cedar, so that it
is almost impossible for the seeds of the trees to reach the
ground and germinate.

The cedar was introduced into England after the middle of
the seventeenth century; and here it still possesses much of
that grandeur and majesty for which the tree is so celebrated.
At Claremont, in the neighbourhood of London, it stands 100
feet high, with a bole of 5 feet 6 inches in diameter. The
loftiest cedar in England is supposed to be one at Strathfieldsaye. It is 108 feet high, with a trunk upwards of
3 feet, and a head 74 feet in diameter. But perhaps the
largest, as well as the handsomest, specimen in Britain, is a
tree at Sion House, about 80 feet high, with a trunk 8 feet in
diameter at three feet from the ground, and a head 117 feet

in diameter. In Scotland and Ireland the tree is of more recent introduction. At Hopetoun House, near Edinburgh, are several fine cedars, about 120 years old, having been planted in 1748. The following shows the circumference and the rate of growth of the two largest, which stand each about 70 feet high, with tops of nearly 100 feet in diameter :—

Year, . . .	1801.		1820.		1825.		1833.		1848.	
	Ft.	In.	Ft.	In.	Ft.	In.	Ft.	In.	Ft.	In.
First Cedar, . .	10	0	13	1¼	14	0	15	1	17	1½
Second Cedar, .	8	6	10	9½	11	4	12	3	14	2

The largest cedars in the Highlands of Scotland are at Beaufort Castle, the residence of Lord Lovat, in Inverness-shire. These were planted in 1783, and allowing them to be four years old at that time, their age (1867) would now be 88 years. The largest measures 18 feet in circumference, with a top upwards of 60 feet in diameter, and about 50 feet high. From measurement made, one of these trees, although considerably broken by a severe storm, increased eighteen cubical feet of timber in four years, and another twenty-eight cubical feet during that time. The largest cedar at Beaufort Castle contains upwards of 250 cubical feet of timber ; and the tree rises with all that boldness of outline and permanence of aspect which rendered it the glory of Lebanon and the boast of Palestine.

The cedar seldom yields cones before it is forty, and sometimes not before it is one hundred years of age, and it is not until the tree has produced several crops that the seeds can be depended on to vegetate. The catkins appear in autumn, and the cones require two years to come to maturity. Perhaps no other tree is naturally so destitute of the means of increasing its species. The cones, when ripe, do not drop

from the tree, nor do they discharge their seed, like other
species of the Coniferæ, through the influence of sunshine.
Unless they are gathered, they remain attached to the tree
for many years, when the scales give way and the seeds are
shed, leaving the axis of the cone adhering to the branch.

The cones should be gathered before the beginning of
April, when the seeds should be extracted and immediately
sown. It is from cones, the growth of the Levant and of
England, that the British nurserymen are supplied with seed,
and no other method of propagation is in practice. Seeds
are with difficulty extracted from fresh cones. The operation
is performed by sawing off about half an inch of the bottom
of each cone, which portion contains no seeds. A hole is
then bored into the axis, which is split up by the driving of
an iron wedge or spike. To facilitate the operation, the cone
may be steeped in water for a few days, or buried in the
ground for a few weeks, before they are manufactured. The
number of sound seeds in a cone varies much in different
trees, and is much influenced by the season. About fifty are
of frequent occurrence ; and although one person can safely
extract several thousand seeds in a day, yet nature appears to
have made no provision for the rapid increase of the tree, to
re-establish the forest overrun by King Solomon's fourscore
thousand hewers. The seeds should be sown in April, in
light friable ground, and only placed about an inch apart.
The covering should be about half an inch deep in the open
ground. The young plants appear in breaking through the
ground in about six weeks, when the advantage of close sowing is often apparent, from the ease with which the cotyledons disturb and rise through the surface ; whereas, in thin
sowing, when the ground becomes even slightly crusted or
caked, a greater effort is necessary in the individual plant to
get through, and the strongest only prevail.

It is a common practice in nurseries to sow the seeds in
heat under glass, and to pot off the young plants in June,
when their cotyledons only are full grown. Others remove
the plants when one year old into small pots, and shift them

into larger as they advance. By cultivating the plants in pots their removal in safety is rendered certain, and it enables nurserymen to dispose of them at all seasons. It is attended, however, with great trouble and expense when the tree is grown in thousands, and, but for the reasons stated, that process is not necessary ; for it is found that the treatment usually bestowed on the larch and Scotch pine is adapted to the cedar; that is, transplanting the seedling plants at the age of either one or two years, into nursery lines, from which, in other two or three years, they should be removed into their final destination. Like all other kinds of Coniferæ, when allowed to remain more than two, or at most three years, without being disturbed, their roots are apt to take a wide range, and to become destitute of that fibrous bushiness which is indispensable to their successful removal. In many situations, weak plants, if left unprotected, are apt to be destroyed by vermin ; but much of the prejudice against the cedar, in consequence of its slow growth when young, may be traced to the plants being stunted in their early growth, and not removed soon enough into their permanent situations.

The summit of Lebanon is nearly 10,000 feet high, and part of the loftiest peaks are at all seasons covered with snow. Near the bottom of the mountain is Beyrout, one of the hottest towns in all Syria, where the thermometer frequently stands at from 90° to 100° Fahr. during the night. The snow of Lebanon there forms an article of merchandise, and is used in cooling wines and liquors drunk by the passing traveller. From the immense height of the mountain a different temperature prevails at various stages, as is thus described by the Arabian poet :—" Lebanon bears winter on his head, spring upon his shoulders, and autumn in his bosom, while summer lies sleeping at his feet."

It is at the bottom of the highest peak of the mountain, and at an altitude of nearly 8000 feet, that the cedars are found. The tree is exceedingly hardy. A rich loam or sandy clay soil is very suitable ; but it is found to thrive on soil of various descriptions with an open subsoil, sufficiently above the rise of water, which must, however, be so near that the roots may be able to reach the aqueduct. The majesty and beauty of the cedar in Scripture is ascribed to a similar cause : " The waters' made him great," etc. The celebrated cedars at Chelsea have become greatly decayed since the removal of a pond from which their roots derived nourishment. When Lebanon was the glory of Syria, Ezekiel, evidently a close observer of nature, had the most ample field for observation, and no doubt, on the banks of the streams, which were sustained by the melting snows at the hottest season of the year, he had observed the cedars rise to a height and magnificence which were not attained by those differently situated. " Thus was he fair in his greatness ; in the length of his branches, for his roots were by great waters." The nature of the tree remains unaltered, and all modern experience points out a similar situation as best adapted to its full development. The tree does not admit of being pruned in root or branch. It has a strong propensity to produce lateral branches of a great size, and the removal of these is found to yield no accession to the growth of the top-shoot. Deprived of its top the plant is not apt to form a leader, but becomes a gigantic bush of the wildest grandeur. When the top is preserved, the tree commonly rises in a broad conical figure, and when vigorous, will advance about two feet yearly. After it has attained its height, the lateral branches continue to extend, until the full-grown tree presents a head with a broad, flat surface, and then, as Ezekiel expresses it, " his top is among the thick boughs." When planted closely in masses, it rises like any other species of the pine tribe, with a straight naked trunk, and it scarcely differs in appearance from the larch, except in being evergreen. There are several seedling varieties of the tree ; some have leaves of a silvery, glaucous hue, others of a deep green ; some with branches upright, while others are of pendulous habit. The timber of British growth is of a brown colour, open in the grain and soft, and it does not support the reputation of the cedar either for strength or durability.

C. Deodara (the *Indian Cedar*) is the most celebrated coniferous tree of the Himalayas, where it attains the height of 150 feet, with a bole of thirty feet in girth ; and it seldom falls far short of these dimensions. It is accounted sacred by the Hindus, and is generally met with in the neighbourhood of their ancient temples. It is found in Nepal, Kamaon, and Cashmere, at elevations ranging from 7000 to 12,000 feet. It was introduced into Britain in 1822. At Relick, in Inverness-shire, a plant is now forty-five years of age, having vegetated there in the year the tree was introduced to this country. At the age of sixteen this plant stood seventeen feet high, and at one foot above the surface of the ground the girth of its trunk was two feet nine inches. Its present height is twenty-seven feet, and its girth five feet. It lost its top many years ago, and its upward growth has ever since been very slow. It is to be regretted that this casualty seems to prevail throughout the country with singular uniformity. The fine specimens of the tree in the grounds of Norman Macleod, Esquire, Dalvey, have to some extent suffered in the same way, by some of them being deprived of their tops at the time the tree attained to the height of from fifteen to twenty feet. This bears a striking resemblance to the failure which often seizes some plantations of the larch, by the destruction of the top and a display of dead twigs just at the time when the tree assumes a timber size, and has begun to put on its greatest vigour. In every species of the Coniferæ there is a period in its growth at which it is more particularly tender and susceptible of injury by frost than at other times. This usually manifests itself in imported tender trees, such as the larch from foreign seed grown in a warm country, while those grown from British seed are exempt from the casualty ; thus showing in the following generation the influence of acclimatization. We may therefore expect the next generation of the deodara from home-grown seed to be much hardier, and exempt from the casualty referred to.

Few seeds of the deodara have yet been produced in Britain. Its cones are of the size and shape of those of the cedar of

Lebanon; but they are deciduous, and fall to scales immediately on becoming ripe, in October or November.

The seeds should be sown in April, and the mode of rearing
them is exactly that recommended for the cedar of Lebanon.
The seeds are generally found to lose their vitality if kept
beyond seven or eight months. Plants are sometimes propagated by cuttings, and by grafting and inarching on the
larch; but seedlings are far preferable. The plant bears a
strong resemblance to the cedar of Lebanon; but it grows
more freely, and with pendulous spray, resembling the green
waters of a playing fountain, and the diameter of the space
occupied by its branches, in open situations, is generally
equal to its height. Like the cedar of Lebanon, it is admirably adapted for a lawn tree; but the great beauty of the
young plant frequently occasions its insertion in a spot far too
limited for its full development, such as the parterre, or
too close to the residence, where ultimately it must lose its
effect in embellishment; whereas, if properly situated, with
ample space, and a congenial soil, it will rise in its true character, combining with its beauty a majesty and a grandeur
unsurpassed by any other tree, native or foreign.

The timber of the deodara is compact and resinous; it emits
a refreshing odour, is susceptible of a high polish, and a piece
of it has been compared to a slab of brown agate. It has
been found perfectly sound in the roofs of temples which
must have stood at least 200 years.

In the *Quarterly Review* for January 1838, in reviewing
Moorcroft's *Travels in Ladaka, Kashmir,* etc., an account is
given of the excursion of Captain Johnston, in August 1827,
to penetrate the Himalaya to the source of the Jumna, and
thence to the confines of Chinese Tartary. They traced the
course of the river up to Jumnotree. Cursola, a small village
in the very heart of the chasm, is described as an isolated
cluster of about twenty-five houses, 9000 feet above the sea,
with three or four small temples, having excellent roofs of
carved deodar wood. The glen from this village to Jumnotree
was gloomy, and the peaks were completely hidden by forests

of the gigantic deodar. The Brooang Pass was only accessible over a bed of snow ; and on their descent from it on the northern side they measured a deodar cedar, and found it thirty-three feet in circumference, and from sixty feet to seventy feet high, without a branch. " On the mountains that enclose the valley of Kashmir, Moorcroft tells us, are immense forests of deodar, the timber of which is extensively used in their temples, mosques, and buildings in general. Such, says Moorcroft, is its durability, that in none of the 384 columns of the great mosque of Jana Musjid was there any vestige of decay either from exposure or insects, although they had been erected above a century and a half. Most of the bridges are of this timber ; and some pieces in one were found very little decayed, though they had been exposed to the action of water for 400 years."—*Quarterly Review*, vol. cxi. No. 121, p. 118.

In every particular the tree accords with the description given of the cedar in Holy Writ, but it has never been found on or near Mount Lebanon. Like many other species of coniferæ cultivated away from its native country, it has given rise to several varieties, none of which appear preferable to the common tree. Young plants from seed are sold at 1s. each, and plants more advanced, standing three to four feet high, sell at about £10 per 100.

Cryptomeria Japonica (Don) : The Japan Cedar.—This forms a beautiful evergreen tree with leaves five-rowed, without any footstalks, short pointed, very close together, incurved, bright green, and about three quarters of an inch in length. Branches horizontal, spreading, dividing alternately into numerous branchlets, thickly clothed with leaves. Male and female flowers on the same tree. Cones of the size of a hazelnut. Scales numerous, of a brownish red colour. Seeds ripe in November.

Dr. Siebold, in his *Flora Japonica*, says it is a majestic tree, well deserving the name of cedar ; that it grows from 60 to 100 feet high, and four or five feet in diameter, with a pyramidal head ; that it occurs in great abundance on the three great isles of Japan, and most probably on the smaller ones ; that a tenth part of the forests which cover the skirts of the mountains, between 500 and 1200 feet of elevation, is composed of this tree.

It was introduced into England in 1844, and a plant grown from a cutting in 1845, the first introduced into Scotland, stands on the lawn at Dalvey, Morayshire, 24 feet high and 3 feet 4 inches in diameter above the swell of the roots. Late frosts in the opening up of the season sometimes discolour its foliage for a short period, but its branches unprotected survive the severest winters. It yields cones abundantly, and it is to be hoped that the following generations of plants from seeds produced in the north of Scotland will be found more vigorous, and better suited to our climate.

XXIII

THE CYPRESS.

THIS genus of coniferous trees probably derives its name from the Isle of Cyprus, where one of the first species prevailed extensively.

Cupressus sempervirens (Linnæus).—The common or evergreen cypress is a native of the islands of the Archipelago of Greece, Turkey, and Asia Minor, where it grows to a great size. Notwithstanding its abundance in Italy, it is said to have been introduced into that country from Greece. In Britain, where it is cultivated as an ornamental tree, it seldom attains a height exceeding forty feet. It is remarkable for its close, upright habit of growth, uniformly rising in a tapering or flame-shaped figure, resembling that of the Lombardy poplar. Its erect figure has a striking effect under any circumstances, but particularly when it rises from among a clump of round-headed trees. It has been so frequently planted as an embellishment to tombs and cemeteries, that it has become, in the language of the poets, the symbol of the last residence of man ; and the ancients considered the tree as an emblem of immortality. Perhaps the best specimen in Britain stands at Stretton Rectory, Suffolk ; it measures sixty-three feet high, with a trunk two feet in diameter. One of the oldest trees in Europe of which there is any record is the cypress of Somma, in Lombardy, respecting which, it is said, there is an ancient chronicle extant at Milan, which proves that it was a tree in the time of Julius Cæsar. Its height is 121 feet, and at a foot from the ground it measures 23 feet in circumference. So respected was it by Napoleon, that, when laying down his plan of the great road over the Simplon, he diverged from the straight line to avoid injuring it.

There are many remarkable instances recorded of the durability of cypress timber. The doors of St. Peter's, at Rome, the predecessor of the present edifice, which had lasted from the time of Constantine to that of Eugene IV. (above 1100 years), were of cypress, and were found, when removed by Pope Eugene, to be perfectly sound. Pliny says, respecting the durability of this wood, that the statue of Jupiter in the Capitol, formed of cypress, had existed above 600 years, without showing the slightest symptom of decay; and that the cypress doors of the Temple of Diana at Ephesus, when 400 years old, had the appearance of being quite new. The timber of this tree is supposed to be the "gopher wood" of which Noah's Ark was constructed. It possesses all the qualities of durability usually ascribed to the cedar; and some believe that the cedar-wood of Biblical times must have been the timber of this tree. It is not found of sufficient size and numbers for the wood to be generally known in this country, but it is in common use in Candia and Malta, and it is always employed as the inner coffin or shell for burying Popes.

The cones of the tree become ripe early in spring, and they are opened by a heat of 100° Fahr., and the seeds fall out. They should be sown in April, in early, friable soil; the plants will appear before the end of May. They should be protected from the severity of the first winter; for unless they are grown in very dry ground, and have the benefit of a warm and ripening summer, their tops do not become sufficiently matured to withstand frost. After the first year's growth is completed, the young plants should be transplanted into nursery lines; and that their roots may be kept bushy and fibrous, they must be disturbed or removed every second year until the plants are finally situated. An excellent method in practice in nurseries, is to insert the plants into pots at the age of one or two years, and to shift them into larger pots, according to their growth. The cypress requires a situation moderately sheltered, and does not admit of a great elevation. Unfavourable seasons of late, particularly the severe winter of 1860-61, have rendered this a very scarce tree throughout Britain, more especially in low, moist situations; while it is observable that trees situated in dry sandy soil, where they ripened or matured their growth, have escaped injury even in places much exposed. Beautiful examples of this tree are to be found in the cemetery on the Cluny Hill, Forres, where they luxuriate in vigour.

The Branching Cypress (*C. horizontalis*), according to some, is a distinct species; others consider it only a seedling variety. It differs in the branches, taking a more horizontal range than the common species; and it is said to yield the best timber.

C. Lawsoniana (Murray) : Lawson's Cypress.—This tree is of more recent introduction than any of the preceding kinds, and from all that is yet known of it, it promises, in Britain, to become the leading tree of the genus. It was introduced in the year 1854, from northern California, where it grows wild in valleys, particularly along the margin of streams, where it sometimes attains the height of 100 feet, with a girth of six or seven feet, in lat. 40° to 42°; and, according to Murray, who introduced it, it formed the handsomest tree seen by him in the whole expedition.

It is found to be quite hardy, yielding cones and maturing seed in this country at an early age. Being easily raised from seed, and readily taking to the ground when transplanted, being also a tree of rapid growth and of extreme beauty, it cannot fail to become soon one of our most common evergreen trees. When only a few years old, it frequently produces shoots about two feet in length, of a drooping habit, like the deodar, which take an upright form the second year. It is raised from seed like the common species, and is apt to sport into varieties, which are already numerous, of all shades of colour, from a grassy green to a glaucous hue, while some are variegated. All have a graceful and interesting appearance, and at no distant period, in good soil and shelter, this plant may be expected to be a companion to the larch and native Highland pine, adapted to the same treatment in Scottish moorland.

The timber is said to be very good, easily worked, and possessed of a strong odour.

C. Nootkatensis (Loudon) ; syn. *C. Nutkaensis* (Lambert) ;
Thuiopsis Borealis (Fischer) ; *Abies Aromatica* (Rafinesque.)—
This is a tall evergreen tree, of recent introduction, from
the north-west coast of North America, where it attains the
height of 100 feet, with spreading or curved flexible branches,
which, when old, are covered with small blisters filled with a
fine aromatic balsam.   The tree abounds at Nootka Sound, in
Observatory Inlet, and on the island of Sitcha.   It is found
to be quite hardy in the north of Scotland, vigorous in
growth, of great beauty, and its foliage is seldom discoloured
by the severest winter.

C. macrocarpa (Hartwig) ; syn. *C. Lambertiana* (Gordon.)—
This tree was introduced into this country about the year
1840, from the wooded heights of Monterey, in Upper California, where it forms a tree sixty feet high, with a stem nine
feet in circumference, with wide-spreading branches and flat
top, like a full-grown cedar of Lebanon, which, when old, it
very much resembles.   While young, and until its upward
growth subsides, it advances with a pyramidal-shaped head,
with foliage of a bright grassy green of great beauty, and it
grows very rapidly in rich soil.   In Scotland, it is generally
reckoned quite hardy, but during the spring of 1867 it was
cut down by frost in some hollow situations, where the *C.
Nootkatensis* and *C. Lawsoniana* stood uninjured.   It should
therefore have an open and airy situation, moderately elevated, on dry ground, and all the better if the severity of
frosts is mitigated by the influence of the sea.

There are many other species of the cypress, such as *Thyo-
ides*, or white cedar, a native of North America ; *C. Torulosa*,
from the Himalayas ; but they suffer from frost, or do not
attain to the size of timber trees in North Britain.

The mode of growing the common cypress from seed is
adapted to the growth of all the species.   In the absence of
seed, propagation by cuttings is practised on all newly introduced and rare kinds, but plants from these are inferior to
seedlings.

XXIV.

THE JUNIPER.

THIS is a genus of evergreen shrubs and low trees, natives of all quarters of the globe. It belongs to the natural order *Coniferæ*, and to *Diœcia Monadelphia* in the Linnæan system. It comprehends upwards of twenty species, some of which run into an endless number of varieties.

Plants are generally propagated by the berries or cones, which lie dormant in the ground for one and sometimes for two years before the seeds vegetate.

Juniperus communis (Linnæus), the Common Juniper, is a well-known spreading evergreen shrub, which grows wild at great elevations and in ground of very opposite qualities, but prefers a deep, dry, loamy soil, partially shaded with higher trees. Thus situated, it rises to the height of a tall shrub, or dwarf tree; while on poor gravel, and in wet situations, it only becomes a low spreading shrub. In the forest at Cawdor Castle, North Britain, the best specimens measure upwards of twenty feet in height, with trunks from eighteen inches to two feet in circumference.

The plant forms an excellent underwood, and shelter for game in general, and its cover is the favourite resort of the woodcock. The berries remain on the plant for two years, and are generally found at many stages of maturity; hence the name of the shrub, from *junior*, younger, and *pario*, to produce; because, while some of the fruit is ripe, a *younger* crop is still in course of being *produced*.

The berries form a powerful diuretic, and are much used in medicine; but their principal use is in the manufacture of gin. From the plant an oil is extracted which is much esteemed in the preparation of varnish. The spray is used as fuel for smoking hams. The wood is fragrant, of a light brown colour, and susceptible of a high polish; and it is generally used in forming the smaller articles of ornamental turnery. In a loose, open soil the plants yield long, tough, fibrous roots, which in the Highlands of Scotland are split up and woven into creels and hampers, which are used at farm-steadings for carrying potatoes, turnips, and for similar purposes. These articles, thus manufactured, are remarkable for their durability, when exposed to alternate moisture and drought.

J. Virginiana (Linnæus), the Virginian Juniper, commonly known as the Red American Cedar, is the tallest hardy tree of the genus. In its native country it is frequently met with sixty feet high, with a trunk two feet in diameter. It was introduced into Britain in the early part of the seventeenth century, where it attains the dimensions which it acquires in North America. In British nurseries this species is more extensively cultivated than any other. Its progress from seed, during the first four or five years, is equal to that of the Scotch pine; and no plant is more apt to produce seedling varieties, which, while they differ in the size and shape of their foliage, in the structure of their branches, and in their habit of growth, are all of them handsome. It grows best in rich, deep soil, and endures a more than ordinary degree of moisture, which has the effect of forming it into a broad and spreading tree; but it generally rises a beautifully formed evergreen, of a conical figure, affording a great shelter. It is thoroughly hardy, and is much esteemed in the shrubbery, as a nurse or shelter for more tender plants.

The quality of the timber varies considerably in the different varieties. It is generally of a glossy brown colour, and of a fine grain, susceptible of a high polish, and used in the manufactures of the turner and cabinetmaker. Seedling plants should be transplanted when they are one, or at most two years old, and afterwards removed every second year, until they are permanently situated.

J. thurifera (Linnæus), the Incense-bearing or Spanish Juniper, was introduced into this country more than a hundred years ago. It is now in general cultivation. It forms a beautiful, low, evergreen tree, with pyramidal head, seldom exceeding thirty feet in height. It yields large berries, which, when ripe, become black, and form a beautiful contrast to the foliage, which is commonly of a vivid green. As an embellishment to the lawn, or to add richness of foliage and shelter to the shrubbery, few trees are more ornamental.

J. Bermudiana (Linnæus), popularly known as Bermudas Cedar, is in its native country a tall tree, and yields timber which is much esteemed for its fragrance. It is employed in inlaying, being a preventive of moth and other insects. It is also used in the manufacture of black-lead pencils. Although it is nearly two centuries since its introduction, it is not yet generally cultivated, being too tender for the climate or Britain, and is seldom met with except in collections of the genus. In the most favoured situations, it seldom exceeds the stature of a shrub.

XXV

THE WELLINGTONIA.

Wellingtonia gigantea (Lindley); *Washingtonia gigantea* (of the Americans); *Sequoia gigantea* (Edlicher) : The Mammoth Tree.—This new genus of Coniferæ, named in compliment to the Duke of Wellington, was discovered in June 1850, in a grove in Calaveras County, in the Sierra Nevada, near the source of the Stanislas in Upper California, and about 225 miles from San Francisco. The tree having become known to the California settlers, it is said they directed Mr. Lobb, the botanist and plant-collector, to the Calaveras grove, who discovered and introduced the tree into Britain in 1853,—the "manuscript notice" of former explorers not counting as publication. Besides the Calaveras grove, there are several other groves of the tree known to exist throughout the country. An interesting description of some of them is given by Mr. Hutchings, in his *Scenes of California.* The Calaveras grove has been often described, and the public are familiar with the names of these remarkable trees,—as the Father of the Forest, the Mother of the Forest, Hercules, the Two Guardsmen, the Three Graces, etc. etc.

The name of J. M. Wooster, June 1850, is cut on the "Hercules" tree, one of the party who gives the credit of first discovering the trees, Mr. Hutchings tells us, to W. Whitehead, Esq., who "while tying his shoe looked casually round and saw the trees" (June 1850).

Mr. Hutchings gives an account of the Mariposa grove, and the girth of 132 of the trees, nearly half the number the group contains ; only three of these reach or exceed 100 feet in circumference, two are between 90 and 100 feet, and the others range from 28 to 90 feet. He reckons that when a tree reaches 90 feet in girth, it will be 300 feet high ; and if 60 feet, it may be about 250 feet high, and so on. The following quotation relates to the Mariposa grove :—" One mighty tree that had fallen by fire, and burned out, into which we walked for a long distance, we found to be the abode of the grizzly bear. Another tree, measuring 80 feet, and standing aloof, was called ' The Lone Giant ;' it went heavenward some 300 feet. One monster tree that had fallen and been burned hollow has been recently tried by a party of our friends, riding, as they fashionably do, in the saddle, through the tunnel of the tree. These friends rode through this tree a distance of 153 feet. The tree had been long fallen, and measured, ere its bark was gone and its sides charred, over 100 feet in circumference, and probably 350 feet in height.

" The mightiest tree that has yet been found now lies upon the ground, and fallen as it lies, it is a wonder still ; it is charred, and time has stripped it of its heavy bark, and yet across the butt of the tree, as it lay upturned, it measured 33 feet without its bark. There can be no question that in its vigour, with its bark on, it was 40 feet in diameter, or 120 in circumference. Only about 150 feet of the trunk remains, yet the cavity where it fell is still a large hollow beyond the portion burned off; and upon pacing it, measuring from the root 120 paces, and estimating the branches, this tree must have been 400 feet high. We believe it to be the largest tree yet discovered."—(P. 148.)

It is reported that there are several groves at distances varying from six to ten miles eastward of Mariposa.

Of the Frezno grove Mr. Hutchings says :—

" This grove consists of about 500 trees of the *Taxodium* family, on about as many acres of dense forest land, gently undulating ; the two largest we could find measured 81 feet each in circumference, well formed, and straight from the ground to the top. The others, equally sound and straight, were from 51 feet to 75 feet in circumference." It is added, " It ought here to be remarked that Mr. L. A. Holmes and Judge Fitzhugh saw an extensive grove of much larger trees than these on the head waters of the San Joaquim river, about twelve miles east of those on the Frezno ; but it has never been explored."—(P. 151.)

Every one interested in the growth of plantations in this country must have observed the great advantages derived by trees situated in glens and ravines, standing in a congenial soil and undisturbed by the influence of "far-fetched" and biting winds. An inequality of surface, even on a small scale, is conducive to rapid growth and lofty dimensions; something unusual therefore might naturally be expected in the vegetation of a country where the configuration of ground forms land-locked valleys, on the most magnificent scale yet known ; a waterfall is recorded 2550 feet in height, but being broken into three, the greatest actual height of one unbroken fall is 1500 feet, and nearly twice the height of Staubbach, the highest cascade in Europe. The shelter and seclusion afforded by the country for the production of timber is apparent from the following :—

"Tu-toch-ah-nulah lifts up his square granite forehead 3090 feet above the grassy plain at his feet, a rounded curving cliff, as smooth, as symmetrical to the eye, and absolutely as vertical for the upper 1500 feet as any Corinthian pillar on earth. I have seen the stupendous declivity of the Italian side of Monte Rosa, a steep continuous precipice of 9000 feet, but it is nothing like Tu-toch-ah-nulah, being nowhere absolutely perpendicular."

Previous to the disastrous fire at the Sydenham Crystal Palace, part of the bark from one of these enormous trees was exhibited on a prepared frame 116 feet high, showing the gigantic size of the tree from which it was peeled, which, though dead, still stands, it is said, 363 feet high, 140 feet to the first limb, 15 feet in diameter at 100 feet high, with its trunk 93 feet in circumference at the ground level, measuring outside the bark, which was eighteen inches thick. The soil which has yielded these magnificent objects is said to be a deep, rich black loam. The groves form deep valleys in the mountains, and are about 5000 feet above the level of the sea. This is an elevation considerably greater than any of the hill-tops of Britain, but it is only about one-third of the height of some parts of the Sierra Nevada or snowy range of Upper California. So far as I can yet judge of this tree, it is superior to most of those recently introduced, and is destined to occupy a conspicuous place among the hardy trees of this country. I have grown it from seed sown in the open ground, and find that the young seedling plants stand during the severest winter. At one year old the young plant is sometimes as small as a one year Scotch fir. The two-year-old seedling is as tall as that plant at that age, and more broadly ramified. It should be transplanted from the seed-bed when one year old, to give it sufficient room and to increase its root fibres, which are naturally long and bare. Transplanting has the effect of retarding its upward growth. The plant is remarkable in early life for acquiring breadth in the spread of its branches and the thickness of its stem.

In the nursery it requires frequent transplanting or pot culture to keep it in a safe state for removal. A plant on my lawn furnishes a good idea of the progress of the tree. It was sown in the open ground in April 1860. It stood the severe winter of 1860-61, when a one-year seedling, one inch, having a cover of about two inches of snow. It was transplanted in the nursery in April 1862. In the spring of 1864 it was placed into its present situation; it has now finished the growth of seven years, and stands 73 inches high; 32 inches of which were the measure of last summer's growth (1865); the spread of its branches, or diameter, is 52 inches. The plant would have been much taller had it not been disturbed by being twice transplanted, which has had the effect of considerably retarding its progress, as will be seen from the following figures, recording its progress in inches during the seven summers respectively:—1, 4, 4, 9, 8, 15, 32 = 73 inches. It will be observed that its growth last summer was six inches greater than that of its first five years from seed. A vigorous plant has the propensity of continuing its growth late in the season ; even in October I have known it to add an inch or two to its height, and yet withstand the frosts of winter; though it is subject to become brown by frost, and assume a rusty appearance, yet I have never observed it killed down to any extent. Unlike other evergreens, its autumnal growth prevents it from being safely removed from the ground, if vigorous, at that season.

In this country the Wellingtonia grows an exceedingly symmetrical and handsome tree, with small heath-like leaves lying nearly close to the branchlets, of a bluish pale green colour. The stem is very robust, and plentifully furnished with branches spread out with great regularity, forming a base of unusual breadth near the surface of the ground.

The tree is readily increased by cuttings. I have seen some of these growing rapidly in a congenial soil, and almost of a timber size, well furnished with branches, and altogether of the most shapely figure, but, nevertheless, I would prefer seedling plants.

It is supposed by some that as the Wellingtonia has already yielded cones in this country, and fertile seeds at an age not exceeding ten years, it cannot become a large tree in the climate of Britain ; but from our experience of other large trees this does not necessarily follow. The larch, for instance, has been known to yield cones at an earlier period of its growth, and yet the species attains the height of 100 feet.

The propagation of the Wellingtonia from cuttings, on account of the difficulty in obtaining seed, has no doubt a tendency to early fructification. This practice, however, is not now likely to be long continued, and we may reasonably expect soon to find seedling plants sold at a few shillings per 100. From the nature of the tree, it is likely we shall have several generations of it in the country during the present century ; and judging from climatic influence clearly impressed on some other kinds of Coniferæ, it is reasonable to believe that British-grown seed will produce plants better adapted to our climate than those of foreign growth. The seeds are remarkably small, and such as would not readily exhaust a tree of vigour.

From the minute parcels of these seeds which I have had in hand, I expect an ounce of them would number more grains than a pound weight of the seeds of many kinds of pine. In this respect the seed of the Wellingtonia resembles that of the Scripture mustard, which is among "the least of all seeds, but when it is grown it is the greatest among herbs."

From the straggling nature of the roots of the plant when it is only a few years old, it gives no promise that it will ever transplant and admit of being notched into moorland with the facility with which the young plants of Scotch pine and larch can be planted. In woodlands, however, where the soil is of good quality, and possessed of moderate shelter, such as in glens and ravines, and in places where timber stands too thin, and where there is protection, prepared spots at considerable distances apart should be formed, a few feet in diameter, by trenching, or by loosening the soil with a tramp pick in autumn. Into these prepared spots Wellingtonias from one to two feet high should be inserted in spring, and all the better if the plants have been transplanted in the nursery ground a year previously, unless they are taken from pots. The tree luxuriates in soil of very diversified quality, rather inclining to moisture than otherwise, and does not object to a great proportion of bog or peat soil.[1] It is expected that on the properties in which this tree is early interspersed, it will, within 100 years of the date of its introduction, have emerged above the height of the ordinary trees of the forest, and will be seen towering aloft, like the spires of a city.

Of this tree, the quality of the timber of British growth has not yet been well ascertained, although some of the largest specimens are said to be upwards of twenty-three feet high, with trunks four feet nine inches in circumference. The wood

[1] In the *Gardener's Chronicle* of the 9th December 1865, Mr. Day of Theydon Grove, Epping, Essex, gives an account of some of the best grown trees of Wellingtonia to be found in this country. He says,—" Of several plants in the park here, one is planted in a very wet place—in short, its roots are literally covered with water during the winter months. On measuring its growth the other day, I was surprised to find that it had grown 3 feet 4 inches in height this season."

in its native country is said to be soft and easily worked, not
apt to warp ; of a reddish colour like cedar, but scentless. It
was reported that a tree cut down a few years ago in its native
country was found to contain upwards of 3000 annual circles
of timber, indicating that number of years of age, and that
the timber all round the exterior of the trunk which had
been last formed, was very hard compared to that nearer the
centre, although the trunk was quite sound. The wonder is
that the timber held together so long. It may be possessed
of incorruptible qualities, though not strong. In a far shorter
period the heart of oak moulders into decay ; for instance, the
Great Oak of Cowthorpe, in Yorkshire, the largest I ever saw,
whose age is unknown, has been hollow for many generations,
its interior forming a vacant space equal to the size of an
ordinary dwelling apartment, yet it continues to vegetate and
form wood externally.

In extreme old age our hardwood trees, generally, are more
subject to heart-rot than the various species of Coniferæ.

XXVI.

THE ARAUCARIA.

Araucaria imbricata (Pavon).—This tree is a native of Chili. On the Cordilleras in that country, it is found at various altitudes, and in some instances it approaches not far from the line of perpetual snow. It has been known in this country since the end of last century. In 1795, Captain Vancouver touched at the coast of Chili, and Mr. Menzies, who accompanied the expedition, procured cones, seeds from which he sowed on board ship, and brought home living plants, which he presented to Sir Joseph Banks, who planted one of them in his own garden at Spring Grove, and sent the others to Kew. From this circumstance the tree was called at first, in England, Sir Joseph Banks's Pine. The tree is diœcious, and the male is said to attain in its native country to only a small size compared to the female, which reaches to 150 feet in height.

Dr. Pœppig, in *The Companion to the Botanic Magazine*, gives a detailed account of the araucaria forests. He says,—"The araucaria is the palm of those Indians who inhabit the Chilian Andes, from latitude 37° to 48°, yielding to these nomade nations a vegetable substance that is found in the greater plenty the more they recede from the whites, and the more difficult they find it to obtain corn by commerce. Such is the extent of the araucarian forests *(pinares)*, and the amazing quantity of nutritious seeds that each full-grown tree produces, that the Indians are ever secure from want; and even the discord that prevails frequently among the different hordes does not prevent the quiet collection of this kind of harvest. A single fruit (*cabeza*, a head) contains between 200 and 300 kernels, and there are frequently twenty or thirty fruits on one stem ; and as even a hearty eater among the Indians, except he should be wholly deprived of every other kind of sustenance, cannot consume more than 200 nuts in a day, it is obvious that eighteen araucarias will maintain a single person for a whole year."

Respecting the timber, he says,—"The wood of the araucaria is red where it has been affected by the forest fires ; but otherwise it is white, and towards the centre of the stem bright yellow.   It yields to none in hardness and solidity, and might prove valuable for many uses, if the places of growth of the tree were less inaccessible.   For shipbuilding it would be useful ; but it is much too heavy for masts.   If a branch be scratched, or the scales of an unripe fruit be broken, a thick milky juice immediately exudes, that soon changes to a yellowish resin, of which the smell is agreeable, and which is considered by the Chilians as possessing such medicinal virtues that it cures the most violent rheumatic headaches, when applied to the spot where the pain is felt."   The scales of the cones are deciduous, and are shed with the seed about the end of March.

In this country the plant is more susceptible of frost in low, damp, and confined situations than it is in more upland exposures.   In such in the north of Scotland, it generally stands scathless during the severest winters ; but the plants raised in this country vary in the degree of frost they can endure ; no doubt according to the elevation at which the seeds have been produced in their native country.   It is not particular as to the quality of soil, provided it is well drained. In Britain its rate of growth is too slow for that of a timber tree, seldom exceeding fourteen or fifteen feet in twenty years. One of the first introduced trees, that at Kew, is the largest I have seen, but it is small for its years.   At the time I saw it, it bore a large cone of a globular figure, and apparently seven or eight inches in diameter.

The growth of the tree is uniform, stiff, and formal, and as yet it is cultivated purely for ornament.   It forms a conspicuous object in the pleasure-ground, from the peculiarity of its construction; but dangerous if run against, as its numerous leaves are stiff, and lanceolate, and readily pierce deeply; hence arises the name in its native country, "monkey puzzle."

With this tree there is no shedding of leaves, at least for many years. The largest trees I have met with retain the leaves that clothed the young plants twenty or thirty years ago, now situated near or far asunder according to the thickness of the stem. The branches are produced in whorls; but these do not indicate the age of the tree as the number of whorls do that of the pine.

In the north of Scotland some fine specimens of the tree are to be met with. It grows vigorously at Dunrobin Castle, in Sutherlandshire. In Ross-shire, at Rosehaugh House, it forms a handsome tree; and at Conon House, in the same county, it stands from thirty to thirty-five feet high, with a girth of four feet near the surface of the ground.

Seedling plants from imported seed abound in nurseries, and are sold from one shilling upwards, according to size.

In the *Gardener's Chronicle* of 1st June 1867, I observe a report by Mr James Barnes, Bicton, Devonshire, on the splendid araucarias at that place, both male and female trees, where fertilized seeds have been produced, and promising young plants raised and planted out. Mr. Barnes has had much experience in the coniferæ, and he says it is impossible in any way to distinguish the difference between the male and female trees about 30 feet high, until they yield blossom. He gives no credit to the differences detailed by travellers of the size and shape of the sexes in their native country, unless these are developed at a more advanced stage of their growth. It may be expected that plants grown from seed produced in Britain will be found far better adapted to our climate than those raised from imported seed.

XXVII

THE OAK.

OAK.—The oak belongs to the genus *Quercus*, of *Monœcia Polyandria*, in the Linnæan system, and to *Corylaceœ* or *Cupuliferœ* in the natural order of plants. It is a well-known and valuable tree, and a native of all quarters of the world. It is adopted as an emblem of strength and durability, and from the most remote antiquity it has been invested with a superiority among trees. Transmitted through many ages, with a character so striking and pre-eminent, we justly recognise it as "the monarch of the wood." The genus is one of the most extensive and varied that is to be found in the vegetable kingdom. It comprehends about 150 species, inhabiting chiefly the temperate parts of the northern hemisphere. About a hundred species have been introduced into Britain. These branch into an endless number of varieties. The species are, generally, trees of great magnitude; but some are of a middle size, others do not attain to these dimensions, while some are mere shrubs, not exceeding the height of two feet. Some species are evergreen, some sub-evergreen, but the greater number are deciduous trees. The fruit in all the species is the well-known acorn, remarkable for its uniformity in size and shape. For the most part the oak attains to the age of at least twenty years before it yields fruit, which is usually the case with trees of long duration; but Loudon states that he has seen in a pot a plant of *Quercus lanata*, a native of Nepaul, bearing acorns at the age of three or four years.

As a useful timber tree the oak is equalled only by the pine or fir tribe; and in furnishing bark for tanning it has no superior. One of the evergreen species, *Q. suber*, a native of Spain, yields the cork of commerce. The common deciduous species possess the property, when cut down, of springing freely from the surface of the ground, which renders them valuable in coppice. In this article it is necessary that we confine our remarks to a few of the more prominent species; and that deserving of precedence is the British oak.

Q. Robur (Linnæus) : The British Oak.—This is the most celebrated tree of the tribe. It is indigenous to Britain, and throughout many parts of Europe. It comprehends many varieties, two of which are so distinct that botanists frequently rank them as species, namely, *Q. R. pedunculata* and *Q. R. sessiliflora.* The former yields the acorns on fruit-stalks ; in the latter the acorns are sessile—it yields flowers and acorns close to the branches without fruit-stalks. The first is the most approved tree, as it produces the best timber. The timber of *sessiliflora* bears a strong resemblance to that of the Spanish chestnut, and the tree is more apt to retain its withered leaves during the winter. Its greatest recommendation is, that it grows more freely than the other sorts when young, particularly in an inferior soil and situation ; but afterwards the difference in the growth of the trees, and in their ultimate size, is hardly perceptible. *Pedunculata* is the more common tree, both in natural and planted woods throughout Britain ; but intermediate varieties of the tree are often met with.

From the earliest accounts we have of British oaks, it appears that the forests were chiefly valued on account of the acorns they produced, which were generally consumed in feeding swine and other domestic animals, while in years of scarcity they were used as human food. The year of great famine, 1116, is described in the *Saxon Chronicles* as " a very heavy timed, vexatious, and destructive year ;" and the failure of mast is thus recorded :—" This year, also, was so deficient in mast, that there was never heard such in all this land or in Wales."

The British oak is invariably propagated by seed ; and, in

forming plantations, the acorns are sometimes deposited in prepared ground. By this method the young plants are subjected to many casualties, and they seldom succeed so uniformly as those that have been raised in nurseries, prepared for the forest, and planted out at the age of three, four, or five years. In collecting seed, it is of great importance to select acorns from the most approved trees. In the laying down of an annual crop, the agriculturist is aware that the result of a single season will depend greatly on the quality or variety of the seed employed. In no case, therefore, should that principle, which nature has everywhere established throughout the vegetable kingdom, be more carefully observed than in depositing those seeds which are to occupy the soil for many generations, it may be for many centuries. Time cannot develop the peculiar excellencies of a tree unless the elements of these are contained in the seed.

Acorns generally become ripe and drop from the tree about the end of autumn ; and they may be sown any time from that period to the beginning of March. Although acorns ripen in Scotland, they are not so good in quality, and seldom produce seedlings so stout as English acorns. They will grow in any soil, but that which is light and friable is best suited for the young plants. As the smallest acorns produce plants which for some years continue of a weak and feeble growth, it is usual, when the supply of seed is abundant, to separate them by passing the small ones through a riddle of a size adapted to retain the largest nuts, which only are used. In nurseries they are generally sown in beds four feet wide, and one bushel of sound acorns is sufficient for a bed twenty-five yards long of that width. The nuts should be rolled or beaten down with the back of a spade. The alleys between the beds are commonly made fourteen inches broad. The surface soil is removed from these alleys to form a cover on the seed, which should in heavy soil be only half an inch deep ; in soil light and friable it should be nearly one inch deep. In sowing during winter, or early in spring, the soil from the alleys should be spread roughly, taking care that the seeds are all hid. In this rough state the covering should remain until April, when in dry weather it should be raked and equalized on the surface of the beds. By this process the covering is exposed to the pulverizing influence of the frost, the seeds of weeds will have vegetated, and will be readily destroyed by the operation of the rake ; and early in May the oaks will rise through the newly-disturbed and consequently soft surface. At this period they are very tender, and in North Britain are sometimes injured by late frosts. Therefore a slight covering of leaves, litter, twigs of evergreens, or of any light open substance, continued to the end of May, is often found to be a useful protection. No further care is needful, but that of keeping the beds free of weeds throughout the season. The seedling plants are sometimes removed into nursery lines at the age of one year, but more frequently when two years old. This operation may be performed any time in open weather during winter or spring. In removing the plants from the seed-beds, they should be carefully loosened, so that their lateral fibres may not be injured. The extremities of their tap-roots should be cut off.

They should be transplanted into nursery lines, sixteen or eighteen inches apart, and the plants about six inches asunder. After being two years in lines they are generally from two to three feet high, and fit for being planted into the forest. If plants are required of a larger size for planting, they should be again transplanted into a greater space, where they may remain for two or three years, according to their progress. Such plants are commonly fit for hedgerows, or for situations where smaller plants are subject to injury by vermin. When the oak remains more than two years without being transplanted, it begins to strike deep into the soil, and to lose that bushiness of root which frequent transplanting is so useful in producing. Therefore, plants which remain four or five years in lines without being disturbed, may appear vigorous and well-grown, yet, for want of fibrous roots, they will be found of little use in forest planting. The oak affects a strong deep soil, elevated considerably above stagnant water ;

but it attains to a good size in sandy or gravelly ground, particularly when it is composed of a clayey mixture; and it is found to luxuriate on ground of opposite qualities, and on that too poor to produce ash or elm timber. The roots of the oak penetrate to a greater depth than those perhaps of any other tree, and therefore its prosperity is the more dependent on the under stratifications of the soil. In rich sheltered glens and valleys, when associated with trees, it becomes lofty, with a tall trunk, while on the bare and exposed upland it is found dwarfed and bushy. Although it expands into leaf at a late period of the season, it is frequently blighted by late frosts, the slightest touch of which becomes visible on its foliage. A young oak plantation should, therefore, in exposed situations, be reared with other trees of faster growth; and for this purpose, the larch, beech, Scotch fir, balm of Gilead, and spruce fir, are well adapted. In some bleak exposures it is necessary to have firs planted a few years previously, and three or four feet high at the time the oak plants are inserted. With shelter the young plant soon takes a powerful hold of the soil, and being tenacious of life it is rarely killed by confinement. When relieved, after being pretty closely sheltered, it often advances rapidly, and frequently produces summer and autumn shoots. These autumn or Lammas growths are peculiar to the oak and a few other timber trees, while young and vigorous.

In plantations or close woods, the tree, during the early stages of its growth, is generally erect and pliant; but by and by its habits become altered. Just before its top reaches its full height, which is regulated much by soil and exposure, its ramifications become more marked and dignified. But the form and outline of a tree in a close plantation are always influenced by physical circumstances. Its natural habit of growth can only be known when it stands alone. Then the monarchical character of the British oak becomes fully developed in the boldness of its outline; its roots are seen to form a spreading basis on the surface, and to mark their sure foundation in the soil; the magnitude and massive compactness of the trunk, surmounted by ponderous horizontal limbs, become conspicuous ; and in winter, the gnarled bifurcations, firmly built together, display an elegance in the twistings of the numerous branchlets never exhibited by any other tree. In summer, while the ramifications are less visible, the glossy foliage, which, in a good specimen, generally towers into galleries or tiers of light and shade, invests the whole object with grandeur. From the most remote antiquity, the oak, together with the cedar, has possessed a celebrity among vegetable productions. On occasions of festivity, when branches of trees were presented to the gods and goddesses of heathen mythology, the boughs of the oak were offered to Jupiter.

Most of the counties in England possess some remarkable trees of this species, and generally they are "full of story, and haunted by the recollections of the great spirits of past ages." In Norfolk stands the celebrated Winfarthing Oak, for the most part a ruin, but still producing foliage and acorns. Attached to the tree is a brass plate giving its dimensions thus :—"This oak, in circumference at the extremity of the roots, is seventy feet; in the middle, forty feet : 1820." It is said to have been called " The Old Oak" in the time of William the Conqueror, and its age is believed to be 1500 years. At Bixley Park, the seat of the Earl of Rosebery, in the county last named, is the Bixley Oak, seventeen feet in circumference at five feet from the ground. It is calculated to contain twelve loads of timber, and during the war £120 was repeatedly offered for it for the use of the navy. In the same county, the great oak at Thorpemarket is one of the finest trees in England, and in full vigour. At a foot from the ground its girth is twenty-two feet ; its trunk is forty-two feet long, and the tree is seventy feet high.

" The King Oak," at Windsor Forest, is said to have been a favourite tree of William the Conqueror; it measures twenty-six feet in circumference at three feet from the ground. It is the largest and oldest in Windsor Forest, and has stood upwards of 1000 years. " The Majesty Oak," at Fredville, in Kent, at eight feet from the ground, is upwards of twenty-eight

feet in girth, and contains 1400 feet of timber.   In Notting-
hamshire, "The Parliament Oak," in Clipstone Park, the
property of the Duke of Portland, derives its name from a
parliament having been held under it by Edward I., in 1290.
The girth of this tree is twenty-eight and a half feet.   This
park was formed before the Conquest, and was seized by
William, who made it a royal demesne.   "The Shelton Oak"
still stands near Shrewsbury, and is about twenty-six feet in
girth at breast height.   This venerable oak is celebrated from
having been mounted by Owen Glendower, on the 21st June
1403, that he might obtain a view of the battle of Shrewsbury,
, on his arrival with 12,000 men.   In Bagot's Park, Staffordshire, is an immense oak, twenty-eight feet in circumference
at five feet from the ground.   *Lauder's Gilpin* says of it,—
" The branches extend forty-eight feet from the trunk in
every direction ; it contains 877 cubical feet of timber,
which, including the bark, would have produced, according
to a price offered for it in 1812, the sum of £202, 14s. 9d."
This tree is quite fresh, vigorous, and beautiful.

In the West Riding of York, near the village church of
Cowthorpe, stands an oak, perhaps the largest in England.
Close to the ground it measures seventy-eight feet in circumference, and three feet higher its girth is forty-eight feet.   I
was residing in Yorkshire during part of the summer of 1853,
and the celebrity of this tree made me anxious to see it.   I
found it about an hour's drive east from Harrowgate, at the
upper end of the village of Cowthorpe, on ground slightly
elevated.   As it stands pretty much in a line with the street
or road that passes along between two lines of thatched houses,
a person has a good view of the oak before getting close beside
it.   At a little way off it presented, as I thought, only a very
moderate appearance ; on getting near to it, however, I saw
that its girth was very far beyond that of any species of living
tree I had ever seen.   I found it a great ruin.   There were
two entrances into its interior.   The principal one was of ample
dimensions to admit cattle, and thither those in the field resorted for shade and shelter.   The interior of the tree was

literally a byre. I paced within the tree in one straight line upwards of five yards. The innkeeper at Cowthorpe told us that the day previous to our visit being a holiday at a neigh-bouring school, the tree was visited by the scholars, their teachers and friends, and that eighty-four persons stood within the tree, and that it could have contained a considerable number more.

The head of the tree presented a display of dead branches ; a great many were also strewn on the ground around, the effects of a recent hurricane. The head also exhibited some widely-spread living boughs, one of which, the nearest to the ground, extended forty-eight feet from the trunk in a horizon-tal line about eight feet from the ground. I could discover no young shoots, but the live branches were well furnished with foliage of a very healthy appearance. Nothing is known of the age of this tree ; but looking at its enormous size in com-parison to trees the growth of centuries, whose ages are ascer-tained, one is led to believe that thousands of years must have elapsed since this monarch of the forest was a small acorn in the ground, and nursed among the weeds of a season. We left the scene with mingled feelings of wonder and regret ; regret that so remarkable an object should be so little cared for.

Perhaps the largest and most valuable tree ever produced in Britain was that called the " Gelonos Oak," which grew a few miles from Newport, in Monmouthshire, and was felled in 1810. It is stated in the *Gentleman's Magazine* for 1817 to have been sold for one hundred guineas, under the apprehension of its being unsound, and re-sold while still standing for £405. The expense of con-verting it was £82, making in all £487 ; and subsequently it was sold for £675. It is said to have contained 2426 cubical feet of timber, and its bark was estimated at the weight of six tons.

Some districts of Scotland contain a considerable extent of valuable oak forests ; but the dimensions of the best specimens fall far short of those of English growth. In Roxburghshire, two fine trees stand on the property of the Marquis of Lothian; one called the " King of the Woods" stands seventy-three feet high, and having had shelter in its youth, it has a clean and branchless trunk, forty-three feet long, and seventeen feet in circumference above the swell of its roots. The other oak, called the " Capon Tree," possesses quite a different character ; it has grown in a situation more exposed, and its figure is consequently more bushy. The circumference of its trunk, two feet from the ground, is twenty-six feet. The height of the tree is fifty-six feet, and the space occupied by the spread of its boughs is nearly 100 feet in diameter. Several centuries ago, when Scotland was more closely covered with timber, specimens of British oak were pro- duced in it far finer than those which flourish at the present time. Of this fact the gigantic trunks and pon- derous limbs which are frequently discovered imbedded in the soil throughout the country are the infallible historians, and indicate by their age and figure that the trees enjoyed ample space and a sufficient shelter,—elements indispensable to the growth of large oak timber.

One of the finest oak forests in Scotland is that at Darnaway, in Morayshire. Between the years 1830 and 1840 the sales of timber and bark ranged from £4000 to £5000 yearly. The oak timber usually sold at from 2s. to 3s. per cubical foot, and bark varied from £6 to £9 per ton. The age of the timber ranged from thirty to eighty years ; and, after paying every expense during the growth of the timber, the revenue of the forest per acre was double that of the finest arable land in the country.

Although the oak does not while young advance so rapidly as some other hard-wooded trees, yet it is by no means an unprofitable tree. During the early part of its growth in the forest it is seldom or never planted by itself, but interspersed and nursed with other sorts, which generally come to be of some value on being removed ; and after the oak is about twenty years of age, it generally grows as fast as most other sorts of hard-wooded trees, and its bark is always found to be of great value. Another great advantage arising from the oak is, that a plantation of it never requires to be replanted after the timber is felled. The roots readily spring, and form a rapid growth, generally for the first ten years, twice as fast as that of the most thriving plantation newly formed. This advantage is greatly enhanced from the difficulty often arising in immediately establishing a plantation on ground which has recently produced timber.

Since the late reduction of the duty on foreign timber, the cultivation of oak coppice is reckoned much more profitable in many situations than that of heavy timber. Indeed, coppice has always been found most profitable in situations destitute of a cheap conveyance to the market; the carriage of bark being always small compared with the value of the commodity. In felling coppice-wood the ground is cleared, and the value correctly ascertained per acre. (*See* COPPICE; and for the mode of " barking," see an article on that subject.) In oak forests where large trees are grown, the heaviest, or such timbers as are adapted for particular purposes, are selected; and where the trees are of vigorous growth, the vacancies thus occasioned readily disappear. An oak forest, therefore, affords a constant succession, the sizes of the trees are very diversified, and it is never exhausted. There is no tree better adapted for coming to maturity, interspersed with oaks, or more profitable in a congenial soil, than the larch. From its upright growth it occupies but a small space compared to its value; it readily overtops the oak, and sometimes affords a desirable shelter in the opening up of the season. Its roots ramify and draw their nourishment chiefly from the surface soil; whereas the oak strikes its roots deep and wide through various strata, generally embracing a large body of inorganic matter; and thus, since the trees depend on separate resources, they are more profitably associated than most other kinds. The oak is an approved tree for hedgerow timber; its roots are not apt to rob the surrounding crops, which are generally well advanced before the tree comes into leaf, and for that reason its shade is not so enfeebling as that of many other trees.

Oak timber has long since acquired a just celebrity; and there is scarcely any purpose to which it is not applicable. If strength or durability, or even elegance, is required, it is resorted to; and it is prized in times of peace as well as war. It is esteemed in our manufactories of household furniture and of spirit casks, in architecture, in the peaceable pursuits of agriculture, but above all in shipbuilding. Shakespeare, with reference to the compact texture and knotty character of the tree, describes it as " the unwedgeable and gnarled oak." The contorted figure in which the boughs are frequently found enhances their value in some parts of the country where there is a demand for shipbuilding timber. For this reason the means generally employed in the raising of straight timber, namely, the closeness of trees to one another, or pruning in order to direct the energies of the tree to the top, are departed from in the cultivation of naval timber. The oak should be allowed a considerable space to ramify. When it has advanced to the height of ten or twelve feet, if it is inclined to a straight figure, the leading shoot should be cut off, and two or three of the lateral branches which take a horizontal range in a direction where there is sufficient space for their growth should be left, and at the same time the next strongest branches should be shortened in order to impede their progress. This method generally occasions crooked trunks, which, of whatever bend they be, are always for shipbuilding more valuable than straight ones. In some cases the shoots left will incline to the perpendicular, and not in a crooked figure; but of trees thus treated a far greater number will be found of a superior mould for naval purposes than of those whose straight leaders have not been removed. In the absence of a demand, however, for shipbuilding timber, straight, clean oak timber has lately been sought after, at from 2s. to 3s. per cubical foot.

The British oak sometimes produces a great profusion of oak-galls, which is believed to detract from the growth of timber in the tree. The chief products of galls are turmeric and gallic acid. They are much used in the manufacture of ink, and in dyeing; and in medicine they form the most powerful of all astringents. Galls are produced on various species of the oak by insects of the genus *Cynipidæ.* The galls of commerce grow on the *Quercus infectoria,* an oak indigenous to Asia Minor, Syria, and Persia ; a shrubby plant which sheds its leaves, and seldom exceeds the height of six feet. The gall-flies puncture the tender leaves or shoots and deposit their eggs, around which the gall accumulates. The most remarkable galls are those formed on the male blossom of the British oak. These catkins appear in May, from one to two inches long, and having shed their pollen, become deciduous and drop from the tree in June ; but if they have been seized on by the insects and punctured while in a growing state, they remain attached to the tree until the galls are perfected, presenting the appearance of the fruit-stalks and berries of a bunch of unripe currants.

Q. Cerris (Linnæus) : The Mossy-Cupped or Turkey Oak.— This tree is a native of the middle and south of Europe, and the west of Asia. It was introduced into Britain in 1735. It is a tree of an elegant appearance, as hardy as the common oak, of much faster growth, and it grows vigorously even in very poor soil. It ripens its acorns like the British oak, and the mode of propagation detailed for that tree is alike suitable for this species, with this difference, that, as Turkey oak seedlings are generally taller than the other, it is better to transplant them at the age of one year than to allow them to remain two years in the seed-beds. This species is remarkable for producing a great number of varieties from seed, which differ greatly in the size and shape of their leaves. It appears particularly apt to hybridize with the evergreen oak ; and frequently in seed-beds of young plants a considerable number of sub-evergreen plants may be selected. The leaves of the tree are of a glossy green above, inclining to white underneath. They are lobed and sinuated or dentated irregularly. They die in autumn ; but, like those of the young beech, they often adhere to the tree throughout the winter. In good soil the tree attains the height of forty feet in twenty or twenty-five years, and a proportionable girth. The tree generally advances with a straight trunk, and, like that of the larch, large in proportion to its lateral branches. It is altogether destitute of that grandeur of ramification peculiar to the British oak. Some of the oldest specimens of the species produced in the south of England stand upwards of 100 feet high. The timber takes an excellent polish, and is beautifully veined. It is less durable than that of the common oak.

Q. C. Fulhamensis (Loudon).—The Fulham Oak is a valuable tree. It is a fine, broad-leaved sub-evergreen, and is supposed to be a hybrid between the Turkey oak and the cork tree, *Q. suber.* The original tree of the kind, a magnificent specimen, stands in the nursery ground of Messrs. Osbourne, at Fulham. It is a lofty tree, about eighty feet high, round-headed, and the girth of the trunk, a foot above the ground, is thirteen feet. It is readily increased by being grafted on stocks of the common or Turkey oak. The acorns of the original tree, which are produced abundantly, have yielded many interesting varieties, which are also propagated by grafts; but seedlings from the original tree are found to sport or vary so much that they cannot be sold for the true Fulham oak. Grafting, therefore, is the only sure mode of increasing the tree in its purity. Grafts spring freely, and on stocks of ordinary vigour frequently attain the height of three to four feet the first summer.

Q. C. Lucombeana (Loudon).—The Lucombe oak is a hybrid produced at Exeter, Devonshire, between the Turkey oak and cork tree. It is a sub-evergreen, and noted for its rapidity of growth. The original plant was produced upwards of one hundred years ago, and the south of England is furnished with many fine specimens of the tree. Its foliage has a great resemblance to that of the Fulham oak, but the figure of the tree is very different. It shoots up like a spire with a narrow and pointed top, while the figure of the Fulham oak is round-headed.

Another hybrid worthy of notice is Turner's Evergreen Oak, raised by a person of that name in Essex, at the close of the last century. It is a hybrid between the common British and evergreen oak. In summer the tree has very much the appearance of the British oak; in autumn its foliage appears more massive, dark green, and glossy. It is the fastest growing broad-leaved evergreen tree that we know adapted to the climate of Scotland. Like other hybrids, it is propagated by grafts, which, when inserted on healthy stocks of the common species, attain the height of four or five feet in two years; and afterwards its progress is fully equal to that of the common oak till it is twenty years old. We have never heard of its having yielded acorns; and we know that several specimens forty years old were never seen in blossom, though at that age they were forty feet high. It is believed that the tree will not attain to a great size, and that it is of rapid growth only when young; but as healthy specimens are found to retain the leaves of the former year throughout the summer, it is a better evergreen than any other hybrid, and its deep green and glossy foliage will always recommend it as an ornamental tree.

Each of the red, white, and black American oaks comprehends a great number of species, some of which become large spreading trees; but as their timber is generally soft and porous, they are not profitably cultivated. The first-named division is exceedingly ornamental, conspicuous among which is *Q. coccinea*, or Scarlet oak, indigenous to New Jersey, Pennsylvania, and Georgia. It was introduced into Britain about the end of the seventeenth century, where it grows in many instances more rapidly than the common oak. The midland counties of England contain many fine specimens of this tree. One at Croome, in Worcestershire, which stands about 100 feet high, attained the height of ninety feet in seventy-five years. At Strathfieldsaye, in Hampshire, a tree of this species is from ninety-five to 100 feet high, and the diameter of the trunk is three and a half feet. The species produces acorns in Britain; but it is principally propagated from American seeds, which must be sown immediately on their arrival. The confinement to which the acorns are subjected during the voyage is apt to occasion fermentation, the slightest touch of which creates vegetation, and unless the acorns are speedily committed to the ground they immediately lose their vitality. Their subsequent treatment is similar to that of the other oaks ; but as they have a tendency to get more bare in the roots than most other sorts, which occasions a stuntedness in the plants on being transplanted, they should be removed early into nursery lines, and planted into their ultimate situations before their roots acquire a great strength. The leaves of the tree are of a beautiful shining green, oblong, and deeply sinuated. They vary exceedingly in size and shape on different trees, and on the same tree at different stages of its growth. All are remarkably handsome, and are produced on long leaf-stalks. A luxuriant tree sometimes yields leaves upwards of a foot in length, and six to seven inches broad. The leaves are deciduous, and the first frost of autumn generally changes them to a beautiful yellow and red, which ripens into a crimson or scarlet of the brightest intensity. In forest scenery, scarlet oaks, with several of the red oaks of America, although of little value as timber, will always be esteemed for their richness of effect, when planted along the margins of woods, as lawn or park trees, and for being grouped in irregular masses throughout plantations, occupying declivities, and other conspicuous situations.

Q. Ilex (Linnæus).—The Common Evergreen Oak, like every other species of the genus, consists of an endless number of varieties. It is a native of the south of Europe and north of Africa. It is the commonest evergreen in the neighbourhood of Rome and Florence ; and it has been cultivated in Britain from a very remote period. It commonly rises to the height of about forty feet ; but many trees favourably placed as to soil and shelter have attained nearly twice that height. It is a tree of slow growth, but of great duration. It is somewhat tender when young, but when once established, and advanced to the height of a few feet, it grows in common well-drained soil, in the climate of the north of Scotland, uninjured by the severest winter.

It is propagated from acorns, which ripen abundantly in

England. As seedling plants when sown in beds are apt to form bare tap-roots, which are difficult to remove with safety, this species of oak is generally grown in nurseries by being inserted in small flower-pots, an acorn in each. The young plants are removed into larger pots in proportion to their growth, and are benefited by a slight protection from frost during the first three or four winters. Thus treated, they take to the ground immediately, and grow freely on being set in their final destination.

The tree forms a very handsome evergreen. The foliage is abundant, of a rich dark green, having a fine polish, glossy above, with a downy tinge underneath. The tree blossoms in May and June, yielding male flowers or catkins, from one to two inches long, on the shoots of the former year, while the female flowers are produced on the newly-formed twigs; and the acorns come to maturity during the second year. This species endures the sea exposure better than any other European oak. It is also better suited for embellishment in large towns, as it grows in a smoky atmosphere better than most evergreens. The progress of different trees of the same age in different situations is very remarkable. This no doubt depends much on soil and situation, but very much also on the variety, for some kinds are known to vary as much in luxuriance of growth as they do in the appearance of their foliage; and all the varieties are not equally hardy. When the tree is allowed sufficient space, it commonly forms a gigantic bush, and conceals its trunk with its foliage, which it retains down to the surface of the ground.

Its timber is tough and strong, and very heavy. In Spain it has a high reputation for knee-timber in shipbuilding : and it furnishes the best and most lasting charcoal. The finest specimens of the tree in South Britain are at Mamhead, where one measures eighty-five feet in height, and eleven feet in circumference; another stands fifty-five feet high, and twenty-two feet round. In Scotland, at Newbattle Abbey, there is a tree forty-five feet high, with a trunk upwards of thirteen feet in circumference. The species is believed to live in vigour for 1000 years, and it is not subject to disease.

Q. suber (Linnæus) : The Cork Tree.—This species is rather less hardy than the common evergreen oak, although it is a native of the same countries, where it abounds on dry, hilly situations. It seldom attains a height beyond that of forty feet. Its mode of propagation in this country is exactly the same as that of the evergreen oak.

The outer bark of this tree is the cork of commerce, and is by far its most important product. When the tree is young, the trunk is cleared of branches to the height of eight or ten feet; and when it is from twenty to thirty years of age, the outer coating of its bark is a formation of coarse, porous cork, interspersed with woody matter. On its being cleared off, which operation is performed in July or August, it generally takes eight or ten years to form another cover, which is of a superior quality to the first. It is not, however, till after another interval of eight or ten years, that the cork is pro-duced in its purity, and of a proper thickness; but this generally occurs at the third disbarking. The trees are ever afterwards disbarked after the lapse of a similar period, care being taken in the operation not to cut in to the inner bark or wood of the trees. The operation by no means impedes the growth of the tree, but has rather a contrary effect. The quality of the cork is said to improve as the tree grows old, and its duration extends for many centuries. Numerous parts of Britain are well suited for the growth of cork, although few specimens of the tree are to be met with. The largest cork tree in Britain, and, Loudon says, "perhaps the largest in the world," is at Mamhead in Devonshire. It is about sixty feet high, with a trunk twelve feet in circumference above the swell of the roots. It stands alone, three miles from the sea, and about 450 feet above its level. The soil is a fine, rich, red loam, on a substratum of red-stone con-glomerate. Loudon remarks, "The head of the tree is oval and compact, and its grand massive branches, each of which would form a tree of noble dimensions, are covered with rugged, corky bark, resembling richly-chased frosted silver, which is finely contrasted with the dark-green luxuriant foliage." In the same grounds there stands another tree of the same species, which does not fall far short of the dimensions stated. In Fulham nursery, near London, close by the King's Road, between Parson's Green and Fulham, stands a handsome cork tree, thirty-five feet high, with a trunk nine feet in girth at one foot above the ground, and overspread with cork to the depth of a few inches.

XXVIII

THE BEECH.

THE Beech—*Fagus sylvatica* (L.)—belongs to the natural order *Corylaceæ*, and is one of our largest deciduous timber trees. It is a native of some of the southern counties in Britain, and it is found indigenous on hills of dry calcareous soil throughout Europe. In favourable situations the tree generally attains the height of seventy or eighty feet, with a trunk five or six feet in diameter. Instances are recorded of the tree, when drawn up in a close plantation, attaining to a height exceeding 100 feet, and when allowed to stand alone, with ample space, the diameter of its trunk, as well as the spread of its branches, are at least equal to that of any other forest tree grown in this country.

The tree is propagated by seed. It blossoms in the month of May, yielding male and female flowers, and ripens the seed, which is called beech-mast, the following October or November, when it drops from the tree, and where the ground is bare is readily picked or swept up. Sometimes the seeds are sown immediately after being collected. In that case, the crop appears above ground in April, and is apt to suffer from spring frosts. It is therefore advisable to keep the seeds spread on a floor, thin enough to prevent fermentation and moulding, until the month of March, and then sow them. By this means the crop appears aboveground a month later, and generally escapes the frosts of spring, and also the ravages of mice in the ground during winter, which, near to hedges, or in sheltered situations, are sometimes very great. The ground for a beech crop should be dry and friable; if otherwise, the surface is apt to cake, confine and injure the young plants as they rise, and most so if the crop is laid‑ down in rainy weather. One bushel of good seed should, extend over forty yards of a bed four feet wide. After sow‑ ing the seeds regularly, they are readily fixed in their places by drawing a light roller over them. The cuffing should then be drawn on, covering the seeds half an inch to an inch. deep. Where the cuffing is not a sufficient cover, additional soil should be cast on with a spade from the alleys, where such is required. The beds should be left without any further raking or polish till the seeds are just about to emerge from the ground; then, in a dry day, a careful hand should go over the beds with a rake, and smooth and equalize the surface. This will occasion the soil to be loose and pul‑ verized at the time that is most required, and at same time have the effect of a weeding, by destroying the small seed‑ ling weeds. A few days thereafter, the crop is up, and gene‑ rally has a very interesting display of cotyledons; but so sensitive are the young plants at this stage to the slightest touch of frost, that some question whether the beech tree is really a native of Britain. A slight cover of straw, or fern, or twigs of spruce, fir, or other evergreens, stuck into the beds, close enough to shelter and shade the young crop, is the easiest means of affording protection, which may be removed about midsummer.

One‑year's seedling plants generally range from five to eight inches high, and may be transplanted the following spring. Or, if the plants are not too close in the seed‑beds, they may be transplanted into nursery lines after the growth of two summers. Generally the seedlings of one year make the best plants for hedges, becoming most branchy near the surface of the ground, when transplanted at that age; and for forest planting, those removed at the age of two years are, in figure, best adapted for that purpose. The nursery lines should be about a foot or fourteen inches apart, and the plants in the lines about four or five inches asunder. The plants should seldom remain more than two years in the nursery lines without being disturbed, as their roots are apt to get bare, which unfit them for successful removal afterwards. In some soils, however, the roots keep fibrous enough till the end of the third year. After being two or three years in lines, the plants are generally fit for their final destination, but if required of a greater size they should be transplanted, giving additional space according to their size. The beech admits of removal at a great size, provided its roots are kept in a bushy state, by being frequently disturbed. A few years ago I removed several hundred yards of beech hedge six feet high, to make way for the Highland Railway through my nursery ground. The hedge had been about twenty years planted, and had been occasionally dug and root-pruned, to prevent the spread of its fibres. It was trenched out in lengths of three or four feet, and carted to a new situation, where it was inserted, and at once became a complete fence along a road side. Hedges might thus be reared which would at once become fences in most places, with little or no protection. *See* HEDGES.

Few trees in a living state are so valuable as the beech for a shelter or screen fence in prepared ground. It succeeds everywhere, particularly in calcareous soil. It is well adapted as an agricultural tree in sheltering fields in bare and exposed situations, where the surface soil is apt to drift in rough weather.

I know a fine screen-fence, planted twenty-five years, thirty-five feet high at the present time. The top has not been trimmed for the last twenty years. To the height of eight feet it has been regularly under the shears, with an occasional dressing at the root of lime and manure, at intervals of six or eight years. It retains its foliage in winter, and is a close and compact shelter down to the surface of the ground, affording April weather in March, and the usual warmth of May in April. In ordinary soil, after the plant has taken root in the forest, its usual growth upward is from eighteen inches to two feet yearly.

Considered as an ornament, few trees claim our attention before the beech. Its stem is massive and powerful, its bark is

smooth and of a silvery cast, and when the heat of summer
unfolds its silken foliage, it displays a verdure of softness and
delicacy, and when viewed in the park, amidst the sunshine
and showers of summer time, it is a gem indeed.

In growing beech timber, the most approved mode is to plant
the tree in masses by itself, about six feet apart, except such
as are planted on the outside or more exposed parts of mixed
plantations, for shelter. When the species stands by itself,
the trees, by pressing gently on each other, readily assume
the best figure, and thus supersede the necessity of pruning.
It luxuriates in a dry calcareous soil, and even in poor sands,
where many kinds of hard-wood plants can scarcely exist.
When interspersed throughout a plantation, it readily adds
depth to the forest, but it is apt to prevail and subdue more
valuable kinds. It is found more destructive to plants under
its drip than almost any other tree ; so that even the holly
and yew are apt to suffer by it. It is seldom profitably
grown after the age of seventy years.

Beech timber is not generally valuable. It stands well
under water, and when it can be obtained of great length, it
is esteemed for keels to vessels. It is often used as piles,
flood-gates, and sluices, in the manufacture of chairs, bed-
steads, carpenters' tools, shovels for maltsters, and wooden
rollers. The turner, joiner, and toy-maker consume large
quantities of it. At fishing stations along the sea-side it is
employed as firewood in smoking haddocks, herrings, etc.,
and sometimes it is used as railway sleepers. It is very
superior as firewood, and its charcoal is esteemed in the
manufacture of gunpowder. The price of the timber varies
much in different localities, ranging from 6d. to 1s. 6d. per
cubic foot, and, when adapted for ship keels, at double that
price. The timber takes a high polish, and is often stained
in imitation of other substances. The beech cannot be
recommended as a suitable tree for coppice wood, as it
springs irregularly, and often fails altogether.

The mast, or nuts, are often yielded in great profusion by
old trees, and afford in autumn the choice food of the deer,

pheasant, partridge, etc. On the Continent, where the nuts attain to the greatest maturity, they are used in the manufacture of oil, which is said to be little inferior to that of the olive, and are esteemed both for food and light.

Although the vigour of the tree rarely extends beyond 150 years, yet it sometimes exists to a very great age. Of the recorded trees of England, the old beech in Windsor Park appears conspicuous; its girth is stated to be 36 feet, and it is said to have existed before the Norman Conquest. It is still in life, but has become a ruin. The celebrated beech at Woburn Abbey, commonly known as " Pontey's Beech," in 1837 was 100 feet high, with a clear trunk of fifty feet in height, and containing 317 cubical feet of timber, exclusive of that of the top. During the eight years preceding the above date this beech only produced in its trunk five cubical feet of timber, while a silver fir in the park at Woburn 114 feet high, and containing 350 cubical feet, exclusive of the head, added to its trunk eleven cubical feet of timber in the same period, thus showing a much greater rapidity of growth in the silver fir than in the beech at an advanced age. Castle Howard is celebrated for its beech trees, the highest is recorded at 110 feet, length of clear bole 70 feet, cubical contents 940 feet, diameter of head 96 feet. The finest trees I have seen in England stand at Studly Park, Yorkshire; they are about 100 feet high. Some of these trees are very remarkable : their trunks are short, but having ample space, they display a wonderful ramification of branches and bulk of top.

In Scotland, Morayshire is celebrated for fine specimens of the tree ; that at Earlsmill measures about sixty feet high, with a trunk sixteen feet in circumference at four feet from the ground. Dalvey, Grange Hall, and Burgie Castle are also noted for many trees of great magnitude.

F. s. purpurea.—The purple beech is a well-known ornamental tree, a native of Germany, where it was discovered in a wood between the middle and the end of last century. From this tree all the purple beeches in Europe have been produced, partly from seed, but chiefly from grafting. The growth of the purple is equal to that of the common tree, and the same treatment is adapted to the seeds of both sorts. The seeds of the purple, however, do not produce plants generally so dark in the foliage as the parent tree, although few of them are altogether destitute of a dark hue. The copper-coloured, and a few of different shades, have taken their rise from seed cultivation, and all the varieties are multiplied by grafting without deviation. The common beech is the stock employed, but as the ordinary mode of grafting is apt to fail to some extent, seldom 40 per cent. succeeding, on account, it is supposed, of the hardness of the wood, the surest method is to propagate by inarch-grafting. This is accomplished by selecting a tree of the purple variety having a bushy top, hanging down to the surface all round. Select a few hundred healthy fibrous-rooted common beech plants, about two feet high, more or less in proportion to the size of the tree, or the number of purple plants required. They should be planted around the purple plant, placing them so that a young twig of the parent tree may be appropriated to each stock or plant inserted. If the stocks are well rooted, and removed with balls of soil at their roots, the working or inarching may be performed in March, immediately after the stocks are planted ; but the most successful way is to let the stocks get established in the ground for a year before the time of inarching, which is the month of March. The handsomest plants are formed by making the junction not more than ten or twelve inches from the ground. The operation of inarching is very simple ; it consists in taking a thin slice off the side of the last year's shoot of the purple beech, in depth about one-third of its thickness, about an inch and a half in length, and nine or ten inches from its extremity, then at the exact point on the side of the stock where the scion can be brought in contact, and will lie close, make a similar incision on the stock, cutting only a little into the wood under the bark. The diameter of the stock being always much greater than that of the scion, the incision on the stock is commonly the broadest, therefore the scion should be placed on the one side, bark to bark, so that the circulation which flows immediately under the bark may meet in both. *Tongueing* is sometimes practised in the operation; that is performed by making a slit at the upper end of the incision nearly a fourth of an inch long, cutting up towards the point on the scion, and downward on the stock, and by fitting these little wedges into each other, the twig is thus held in position till tied up, and the future junction strengthened thereby. Strands of bast form the most suitable ligatures. After tying, no time should be lost in putting on the clay, which should have been mixed up a month previously with an equal bulk of droppings from the stable. In applying it on the graft it should be made of suitable consistency, not too thin, lest it should fall off, nor too thick, lest fissures should occur and admit air, which should be guarded against; after a day or two, the clay will be dry and may be coated over with moss (*sphagnum*), or the dried mowings of lawn, etc. Should excessive drought ensue, the grafted plants may be watered overhead occasionally of an evening. After the inarched twigs have grown for two months, the top of the stock should be cut off above the ball of clay, and the junction with the old plant severed when the summer's growth is nearly completed. In winter or spring the young plants should be lifted, dressed, and removed into nursery lines, each having a stake or support, tied below and above the graft; and the space around the purple tree may be filled up with stocks, for being worked after a summer's growth. By this means a tree the size of a gooseberry bush will yield a few dozen plants. And one several yards in diameter should produce several hundred plants every second year. When the branches of the purple tree are apt to get too high for this purpose they should be pegged down, and kept in a suitable position, or pruned off, so that the vigour of the lateral spray may not be weakened. The usual price for a strong plant of purple beech is 2s. 6d., and small plants two years transplanted are generally sold to the trade at 50s. a hundred. Although the tree forms a striking embellishment when planted at wide intervals throughout the demesne, yet closely inserted, or where a considerable number of them at once come under the eye, they have a sad and mournful appearance.

F. s. pendula.—This is a beautiful weeping tree, propagated by being grafted on a common beech. It proves an elegant object when placed on the top of a handsome and lofty stem. Its growth is direct downward, seldom ascending above the point of junction with the stock.

XXIX.

THE CHESTNUT TREE.

CHESTNUT TREE.—*Castanea* belongs to the natural order *Corylaceæ*. The genus derives its name from Kastanea, a city in Pontus in Asia, of which locality the tree is a native.

C. vesca (Gaertner) : The Sweet or Spanish Chestnut.—The term *sweet* is applied to it in reference to the fruit, and in contradistinction to that of the horse-chestnut, which is bitter ; it is named Spanish, because the best chestnuts for the table, sold in the London market, are imported from that country. The tree is believed to have been brought to Europe by the Greeks, from Sardis, in Asia Minor, about 500 years B.C. It was probably introduced into Britain by the Romans for the sake of its fruit, and being a tree of great duration, and ripening its fruit in favourable districts, it could hardly fail to become a permanent inhabitant. Linnæus united the genus *Castanea* with *Fagus*, the beech, which was not done by any botanist before his time, and which has not been adopted by any since. The distinctive characteristics of the two genera are—that the chestnut has male flowers on very long catkins, with the seeds farinaceous, while the beech, on the contrary, has male flowers on globular catkins, and the seeds oily.

The chestnut is one of our most ornamental large-growing trees. The diameter of the trunk is commonly large in proportion to the diameter of the head or height of the tree, and the trunk generally assumes a convoluted or twisted appearance ; its leaves are broad and long, strongly veined and serrated, of a dark green and glossy appearance, which change to a mellow yellow, or ripened hue in the end of autumn. It forms an important element in growing for picturesque effect.

As a park or lawn tree it is more tender than the oak, and seldom arrives at the height or diameter of head attained by that tree.

It prefers a deep sandy loam, or rich gravelly soil, such as the *débris* of rock, where the subsoil is open and dry ; on such its usual growth for eight or ten years after it takes root seldom falls short of three feet yearly. On favourable soil, in a close plantation and mild climate, it rises with a straight clean bole, and usually attains the height of from fifty to sixty feet. In exposed situations, and on retentive and wet subsoils, the tree ramifies near the surface of the ground, and seldom ripens its young shoots sufficiently to resist frost.

One of the oldest chestnut trees in the world stands on Mount Etna. M. Houel, in his *Voyage en Sicile*, states that he visited it, and found it in a state of decay. It had lost the greater part of its branches, and its trunk was quite hollow. A house was erected in the interior, with some country people living in it, with an oven, in which, according to the custom of the country, they dried chestnuts, filberts, and other fruits, which they wished to preserve for winter use ; using for fuel, when they could find no other, pieces cut with a hatchet from the interior of the tree. Kircher, about the year 1670, affirms that an entire flock of sheep might be enclosed in the Etna chestnut tree as in a fold. Brydon records his tour through Sicily in 1770, and states that the decayed trunk of this tree measured 204 feet in circumference. The oldest chestnut tree in England is supposed to be that at Tortworth, the property of Lord Ducie, in Gloucestershire. It stands on a soft loamy clay soil, on the north-west declivity of a hill. Evelyn states it to have been remarkable for its magnitude in the reign of King Stephen (1135) : it was then called the Great Chestnut of Tortworth ; from which it may reasonably be presumed to have existed before the Conquest. Strutt, in his *Sylva Britannica*, in 1820, states its measurement, at five feet from the ground, to be fifty-two feet in circumference, and its cubical contents, according to the customary method of measuring timber, to be 1965 feet. It is now about forty-five feet in circumference. The tree at that time ramified, at the height of ten feet, into three limbs; one of which, at the distance of forty feet from the main trunk, at the period already mentioned, was stated to be twenty-eight feet in girth. The largest chestnut tree that we know in Scotland stands on the lawn at Castle-Leod, in the Vale of Strathpeffer, Ross-shire. When we last saw it, at the height of three feet from the surface, it measured twenty feet in circumference. It was sixty feet high. The soil is rich and loamy, elevated about fifty feet above the level of the sea, and the tree is surrounded with high mountains. The timber of chestnut is less valuable than that of oak, but it bears so striking a resemblance to the timber of some species of that tree, that frequently the oak in old buildings throughout the country has been mistaken for chestnut. After the tree attains the age of fifty or sixty years, the timber generally begins to deteriorate, and the wood has the remarkable property of being more valuable when it is young than when it is old. The sap, or outer wood, soon changes into heart-wood; hence the great value of the tree for posts, fences, and all purposes where timber comes in contact with the ground, or is alternately wet and dry. The timber usually sells from 1s. to 2s. per cubical foot, according to circumstances; but it is for coppice-wood that the tree is chiefly esteemed, being possessed of the properties of growing as underwood, and of springing freely when lopped over. Its bark is only half the value of that of oak. Full-grown chestnut timber is generally brittle, and apt to become *shaky;* that is, the annual layers divide from one another, and fall into laths; but when sound it is much esteemed in the manufacture of liquor casks, and for this purpose it should be felled before the trunk exceeds a foot in diameter.

As a fruit tree, in Britain the chestnut has not a high reputation; but from its rapid growth and close foliage it is well adapted for a screen or shelter to the orchard. In Devonshire and some of the counties possessed of the best climates, the fruit is yielded in great maturity. In Spain it is grown chiefly for its fruit, which has become not only a common food for the peasantry, but an article of exportation. According to M'Culloch, chestnuts imported into this country from Spain and Italy, during the three years ending 1831, averaged 20,948 bushels; and the duty of 2s. per bushel produced, in 1842, a sum which proved that the consumption of that year must have amounted to 23,216 bushels.

As foreign seeds are often kiln-dried, to adapt them for travelling in packages, the tree is generally propagated from seeds of English growth, or from those produced in the earliest districts in Scotland, where they ripen during favourable seasons.

The seeds are frequently sown in October and November; when this is the case the young plants generally rise through the ground in April, when they are in danger of receiving injury by frost, unless protected. In some climates, where late spring frosts are apt to prevail, nurserymen preserve the seeds during winter, on a loft floor, and sow them early in spring. By this method the young plants do not rise until the middle or end of May, a season when, without protection, they are more likely to be exempt from injury. The seeds are sometimes sown in drills, and placed two or three inches asunder, and the drills sixteen or eighteen inches apart; but the more common method is to sow in beds, four feet wide. One bushel of fresh seed is sufficient for a bed thirty yards long. The cover on the seeds should be one inch deep. If the soil is very rich, it should be early and dry; otherwise the plants will grow to a late period in the season, and fail to ripen their wood sufficiently to resist the influence of frost; consequently they will lose their tops and become branchy.

The plants are sometimes removed from the seed-bed and transplanted into lines, at the age of one, but more frequently when two years old. In removing them they should be classed into two sizes, and have the extremities of their tap-roots pruned off. The plants in the lines should stand six inches asunder, and the lines should be about sixteen inches apart; where a greater space is allowed they are apt to become crooked and branchy, and in need of pruning. After remaining two years in the lines, the plants are commonly from two to three feet high, which is the common size used for forest planting. Such plants are worth 40s. per thousand. If required of a larger size, it is necessary to transplant them every second year, increasing the space between the plants and between the lines; and any time during open weather between October and the middle of March is suitable for their removal.

Many writers consider the chestnut a native of Britain ; we are not of this opinion ; and although it is not very subject to disease, all who are accustomed to cultivate the plant will readily admit that, even in the earliest and most favourable districts of our country, it is influenced by unfavourable seasons in a degree not felt by any tree indigenous to the climate of Britain.   There are many varieties of the chestnut cultivated as ornamental and as fruit-bearing trees ; the most distinct sorts are *C. Americana,* which has broader leaves than the common tree ; *C. variegata,* variegated with yellow and white streaks ; *C. glauca, C. glabra,* and *C. aspenifolia.* The varieties cultivated in France for the sake of their fruit are termed *les marrons;* they are large, farinaceous, and sweet ; and when roasted yield a rich aromatic odour.   The seeds of the varieties cannot be depended on to produce the same, therefore the true kinds are propagated by grafting.

XXX

THE HAZEL.

The Hazel, *Corylus Avellana* (L.), of *Monœcia Polyandria* in the Linnæan system, and of *Corylaceæ* in the natural order of plants, is a well-known deciduous shrub, or low tree, indigenous to the temperate climate of Europe. It is common in all quarters of Britain, and rises with numerous stems of rapid growth. Being hardy, it occupies elevated positions in the Highlands of Scotland, forming a jungle on waste and uncultivated spots, preferring situations partially shaded, and a dry soil incumbent on sandstone, chalk, etc. It abounds on slopes and precipitous places along the banks of rivers and streams, and it often springs from fissures of rock, no doubt from nuts deposited by the squirrel and other vermin. As an underwood it adds density to the forest of deciduous trees, but it does not endure the confinement of fir woods.

It is readily propagated from nuts, which it yields abundantly, and which ripen throughout the autumn and beginning of winter, at times varying with the climate and altitude of the situation. As the size and vigour of the seedling plants generally correspond to the size of the nuts, it is of importance, when they are very unequal, to submit them to the ordeal of a wire-riddle adapted to class them into two sizes. As the weak seeds produce a similar progeny they may be rejected, or, if sown, they will have a better chance of success by themselves than if interspersed with those of more robust growth. During winter, or early in spring, is the time for sowing them, and the soil should be light and sandy. A bushel of seed is sufficient for thirty lineal yards of a bed four feet in width.

The covering should be one inch in depth. The young plants break through the ground towards the end of May, and only require to be kept clear of weeds during the summer. A strong crop is usually removed into nursery lines when one year old, but weak plants are allowed to remain two years in the seed-bed.

In nursery lines the plants should be placed about ten or twelve inches apart, and the lines two feet distant. After remaining two years in lines, the plants are commonly removed to their final situations.

The hazel is profitably cultivated in coppice, and as under-wood. When lopped, it stools, and shoots up vigorously, producing growths from established roots five or six feet long in one season ; and in favourable soil it is not apt to become enfeebled by the treatment. In Staffordshire, and many other quarters, it is cultivated in the vicinity of potteries and manufactories, where it is formed into crates of various sorts, adapted for the transmission of goods, for which purpose it is well adapted, being extremely tough and elastic. It is also used in forming hurdles, hoops, walking-sticks, garden-seats, and all descriptions of rustic work. The roots yield a valuable timber for veneering, and for the manufacture of toys and fancy wares. The wood makes a charcoal esteemed in the manufacture of gunpowder and black paint. The hazel never produces wood of a sufficient size for building purposes ; but a good specimen of the tree is sometimes found to stand thirty feet in height, with a· trunk three feet in circumference. When trained as a standard with one stem, it forms a very handsome ornamental tree, and often yields a rich display of catkins, which commonly continue in bloom during winter and spring, before the leaves expand, the gay appearance of which are generally appreciated at that season. It is also one of our best deciduous plants for underwood ; as such it adds richness, closeness, and seclusion to narrow beltings and clumps. Few plants retain their leaves longer after they are affected by frost, the influence of which imparts to the foliage a rich yellow colour, which continues ornamental for months before the leaves are shed. The effect of the plant as an underwood is greatly enhanced, by the display of contrast, when it is interspersed with the box, the holly, or the yew.

The common hazel comprehends an endless number of varieties, many of which are cultivated exclusively for their fruit, among which are the following :—Nuts or filberts, *C. A. tubulosa, C. A. crispa, C. A. tenuis;* also varieties of the Northampton and Barcelona nuts, great cob nut, etc.; and to insure the continuance of the respective qualities for which these sorts are distinguished as fruit-bearing trees, their propagation must either be by layers or by grafts. Great quantities of nuts, both of the wild and cultivated kinds, are sold in the English markets. Besides those of British growth, large importations are made from France, Portugal, and Spain. The latter are in the highest estimation, and are generally sold under the name of Barcelona nuts. The following quotation from the *Dictionary of Commerce,* published in 1834, furnishes an idea of the consumption of this article :— " The entries of nuts for home consumption amount to from 100,000 to 125,000 bushels a year; the duty of 2s. a bushel producing from £10,000 to £12,000 clear." Mr. M'Culloch adds, " The kernels have a mild, farinaceous, oily taste, agreeable to most palates. A kind of chocolate has been prepared from them ; and they have sometimes been made into bread. The expressed oil of hazel nuts is little inferior to that of almonds."

C. Colurna (L.)—The Constantinople Hazel is the only species of the genus which attains to the dimensions of a timber tree. It has been introduced into England, and cultivated, but not extensively, in the neighbourhood of London, for 200 years. It grows to the average height of our timber trees. It has a white bark, horizontal branches, and is altogether handsome, and quite hardy. Although it is readily propagated by nuts, which it yields in favourable seasons, by layers, and by grafting on the common hazel, yet it has never become common throughout England. In Scotland it is still more rare. One of the best specimens of the tree near London is at Sion, where it stands upwards of sixty feet in height. This species is well worthy of general cultivation. There are a few other kinds of the tree, natives of America, of dwarf growth, rarely met with except in collections kept for the sake of variety. They possess no qualities to induce their extended propagation.

XXXI

THE HORNBEAM.

THE HORNBEAM (*Carpinus*).—This tree belongs to the natural order *Corylaceæ*. The genus comprehends only three or four species, all of which are deciduous trees. They yield flowers unisexual ; those of the two sexes are in distinct catkins upon the same plant.

C. Betulus (L.)—The Common Hornbeam, which is the principal tree of the genus, is a native of England, of Ireland, of the south of Scotland, and of many parts of central Europe, avoiding the extremes of heat and cold. It bears a striking resemblance in its leaves to those of the common beech, but it is destitute of their glossy varnish or polish, and it presents the appearance of a tree intermediate between the beech and the elm. It is perhaps cultivated to a less extent than any other hardy timber tree adapted to the climate of Britain. It attains to a middle size, but is not valuable as an ornamental tree, nor as timber. Although instances have occurred of the tree attaining the height of eighty or ninety feet, with a trunk three feet in diameter, yet trees two-thirds of these dimensions are generally reckoned large specimens. In magnitude, therefore, it stands between the beech and the birch, and, like the latter, it often produces trunks of a flat and irregular figure.

As a hedge plant the reputation of the hornbeam is considerable. Compared with the beech, it endures pruning better, and it has the advantage of being much less subject to disease, and atmospheric influence, when grown for a long time in a confined form, complaints which of late years have prevailed to an alarming extent in several districts. It is also superior to beech in being less injurious as a hedge plant to crops in its vicinity; no doubt from its rooting and deriving support at a greater depth from the surface. It grows close and twiggy, and retains its leaves to a late period in the season, but not throughout the winter so universally as does the beech; and although it generally attains a greater height during the first six or eight years than that tree, yet it seldom forms a hedge so compact, bushy, and stout as the beech at that age. As an agricultural plant it is valuable for screen fences to shelter exposed fields. It endures a rough and windy situation, thrives well on the common sorts of soil, and on cold, clayey ground it grows quickly, particularly when planted closely, where it forms a thicket or a compact mass of spray and foliage, which admit of being confined by pruning to prescribed limits. It readily springs when lopped over at the surface, or at any height from the root.

The seeds are formed in a small nut, and are usually ripe in the end of autumn. When sown immediately on becoming ripe they spring irregularly; a few vegetate the first spring, but the principal crop the second year. The usual mode of treatment is to sow the seeds in spring, at the rate of one bushel of clean seeds to fifty yards of a bed four feet wide; the covering should be about half an inch thick. The seeds, thus treated, remain dormant the first year after sowing, and vegetate in the spring thereafter; therefore where the ground is valuable, a light annual crop may be grown above the seeds for the first year, such as that of onion, lettuce, radish, cabbage-plants, etc. When the crop of hornbeam is a very close one, the young plants may be removed, or thinned out, and transplanted when one year old; but if they are thin, and have sufficient space for becoming two-year seedlings, it is usual to allow them to remain in the seedbed until the plants are of that age. When lifted, the extremities of their roots should be pruned off, the plants should then be inserted into nursery lines, about sixteen inches asunder, and the plants a few inches apart in the lines. After two years in lines the plants are usually fit for hedges; and if they remain longer without being removed, and a greater space allowed, they are

apt to become tall and bare near the surface, which renders
them unfit for hedge plants, until they are made bushy by
being cut down near to the ground.   The plant being hardy,
and almost exempt from disease, its nursery treatment is very
simple, it being required only to allow the plants sufficient
room to keep them shapely, and to keep them free of weeds.

The wood of hornbeam is white, tough, and durable, adapted
for handles and stocks for tools, yokes for cattle, milk vessels,
wheelwright work, and other kinds of rural carpentry.   Evelyn
states that for milk vessels it excels either yew or crab; and
Linnæus observes that it is harder than hawthorn, and capable
of supporting great weights.   Loudon records a piece of it,
two inches square and seven feet eight inches long, having
supported 228 lbs., while a similar beam of ash broke under
200 lbs.; one of birch, under 190 lbs.; of oak, under 185 lbs.;
of beech, under 165 lbs.; and of other woods at a less weight;
and yet, notwithstanding its powers of resistance, the hornbeam has very little flexibility, it having, before it broke, bent
only 10°, while ash bent 21°, the birch 19°, and the oak 12°.
The timber is not in ordinary demand; and where it is to
be met with it usually sells at the same rate as the beech.
As firewood, it is placed in the highest rank for heat, bright-
ness, and durability; its charcoal is esteemed for its use in
cooking, in forges, and in the manufacture of gunpowder.   Of
the common species, the incised-leaved and the variegated are
the chief varieties.   The other species, namely, *C. Americana*
and *C. Orientalis*, being of small growth, are not cultivated as
timber trees.

XXXII.

THE ALDER.

ALDER TREE—*Alnus glutinosa* (Gaertner).—The glutinous or Common Alder belongs to the natural order *Betulaceæ*, and is considered the most aquatic tree growing in the climate of Britain. It is indigenous throughout Europe, and is met with commonly on the banks of streams and in swampy situations. In its wild state it seldom attains to a greater height than thirty or forty feet; but cultivated in good soil, in various places throughout Britain and Ireland, it has attained to the height of from sixty to eighty feet, with a trunk three to four feet in diameter. It is generally at full maturity at fifty or at most sixty years of age, and, when timber is an object, it should then be cut down. The finest trees of the species in North Britain stand at Gordon Castle and Huntly Lodge. There existed, during the early part of the present century, in the parish of Abernethy, in Strathspey, a clump of alders, adjoining part of the native pine forest, containing trees of very remarkable dimensions. When I last saw them they were in a state of decay, without a sound trunk among them. They had lost their tops, and ranged only from twenty to thirty feet in height; but six or eight of them had trunks ranging from ten to seventeen feet in circumference, dimensions greater than usual for trees of this species. The largest recorded tree in England grows on the right-hand side of the road upon entering the village of Haverland, Norfolk. It stands in a damp, favourable situation, near to a rivulet, and reaches the height of sixty-five feet. The trunk, at one foot from the ground, is twelve feet in circumference. The next largest recorded trees are in the Bishop of Durham's park at Bishop-Auckland.

As an ornamental object, the alder is often deficient, more particularly when young; and the praises which have sometimes been bestowed on its appearance arise from its association with rivers and rills, which are pleasant and enduring beauties in nature; the alder, being a marginal accompaniment of these, comes in for a share of respect and esteem. In a cultivated state, however, and in good soil, when the tree attains to a considerable size, and is about to complete its upward growth, it often assumes a change of character, becomes more picturesque in figure, and displays a ramification little inferior to that of the oak.

The timber of the alder is not valuable. When used for a post, it soon perishes when it comes in contact with the ground; but it is well adapted for piles in laying the foundation of bridges or embankments beneath water. It has also been successfully employed in a young state as brushwood, when cut in winter, for filling drains in the reclaiming of marshy land. It is found suitable for the staves of fishcasks and for packing-boxes, and it is sometimes employed by last-makers, turners, patten and clog makers, along with other soft wood, such as the willow and poplar. As the wood is subject to the attack of insects, the finer pieces, immediately after felling, are sometimes in Highland districts immersed for a few months in a pit of water dug in a peat bog; and the effect of this preparation is reckoned more complete if the water is impregnated with a quantity of lime. The wood prepared in this way is well adapted for cabinets, tables, etc., and has some resemblance to mahogany. The charcoal of alder wood is esteemed in the manufacture of gunpowder.

The propagation of the tree is speedily effected from seed, which is found in small cones, which ripen in October and November. These should be collected when dry. The seeds are usually extracted by spreading out the cones on a loft floor, to the depth of eight or ten inches; and, if situated

over an apartment where a strong fire is kept, the operation will be greatly facilitated. The cones are turned over
every day or two, until they become quite dry and open;
after that they are threshed, or trodden with the feet, and
then the seeds are sifted out. The season of sowing the seed
is March, or early in April. Any rich friable soil will suit.
After being dug over smoothly, the beds should be marked
out four feet wide, with an alley of one foot between each.
A pound of seeds of ordinary quality is sufficient for ten
yards of a four-feet bed; but as the seeds often differ in
quality in different seasons, the sower should exercise his
judgment in trying the seeds by bruising a few of them, when
the fresh white appearance of the kernel readily indicates
their vitality. If the seeds are not sown until April, and the
weather be dry, they had better be moistened for a day before
sowing. Sow regularly on the surface of the ground; after
which, rake the surface slightly and evenly, which is all that
is necessary for covering the seeds, unless the weather is dry,
when it is an advantage to roll the ground after sowing and
raking. The seeds being generally of little value, it is well
to sow thick, and thin out the young plants where they are
too close, in June or July. At the end of the first year's
growth, the seedlings commonly range from four to eight
inches, according to the quality of the soil and the season.
These plants should be transplanted during the following
spring into lines a foot apart, and the plants three or four
inches asunder in the lines. They are of rapid growth
when young, and often attain the height of eighteen inches
at the end of the second year. They then require more
room, and should either be removed to their final destination, or thinned out, allowing them space in the nursery in
proportion to their height. Generally speaking, plants of
whatever age take most readily to the ground when finally
transplanted, if they have undergone that nursery operation
the year previously.

The alder is seldom a profitable plant to the nurseryman.
This arises partly from its being commonly sold at a lower

price than other forest trees, on account of being easily grown; partly from the want of a regular demand; but mainly on account of its very rapid growth after being two or three years old. The plants must then either be sold, or allowed a space so large that it is unprofitable to keep them, and their destination is, not unfrequently, the fuel-heap. When dried, they become brittle, and rot rapidly, which renders them unsuitable even for stakes or props to support other plants.

Although the tree is often seen in an aquatic state, or in very marshy ground, yet it luxuriates and becomes a large tree in land that is well drained and has no superabundance of moisture. When a plant, it is tenacious of life, and from its rapid growth it is well adapted for being interspersed as nurses in plantations, in bleak and exposed situations, where it readily makes an appearance, and serves as a protection in establishing more valuable trees. It may afterwards be removed in the early thinning of such plantations, but it is for the most part planted alone in marsh ground and along the margins of streams. There it is particularly valuable for binding and consolidating the banks by its close spreading roots, which are rendered the more effective when the tree is kept as coppice-wood, or cut at intervals of eight or ten years, so that the trees do not get so tall that the roots would become disturbed during rough weather.

Its uses, in a growing state, consist in its supplying fagot-wood and hurdle-wood, and in reclaiming low meadow land, either continually or partially flooded. This is best done by ridging up part of the soil in summer time, and then planting the young trees on those ridges in spring. In a few years, by the fall of the leaves and the operation of the roots, the soil becomes dry and firm. That it has the opposite tendency, viz., to create moisture, is a vulgar error, current in many books of Arboriculture. It is also highly to be valued as a free grower for several years in exposed districts, *if the soil has been trenched;* and it is especially serviceable in nursing more valuable trees planted by the sea-side. It is particularly useful in some districts, such as the fens, as a screen-plant, because it rises in particular spots, where other trees refuse to grow. In addition to all these, its services, in a growing state, are by no means inconsiderable among dyers in country districts, and especially throughout the Highlands of Scotland. If cut in spring, the shoots dye a cinnamon colour, and the catkins produce a green.

XXXIII.

THE BIRCH.

THE BIRCH.—*Betula* belongs to *Monœcia Polyandria* in the Linnæan system, and to *Betulaceæ* in the natural order of plants. The tree is indigenous throughout the north, and on high situations in the south of Europe. It is extremely hardy, and only one or two other species of trees approach so near to the North Pole. There are two tree varieties natives of Britain, *Betula alba* (L.), and *B. a. pendula* (Smith); the latter is by far the more valuable and ornamental. When a plant, it may readily be distinguished by the touch; its bark being covered over with rough exudations, while that of the common tree is soft and velvety. Each variety is found exclusively in some districts, but frequently they are interspersed. Throughout the most remote parts of the Highlands of Scotland, the birch is often found in extensive coppice on rocky elevations, where no other ligneous plant is to be met with. It also stands in glens and ravines, adorning the margins of lakes and rivers, where the silvery whiteness of its trunk, and the light and airy habit of its spray, form scenes beautiful and interesting, even in the absence of every other tree.

Though it is often found associated with the alder on swampy ground, yet few trees more successfully resist drought. Adapting itself to various soils and situations, it possesses a wider range than any other plant. It is well suited to form a cover on ground from which Scotch pine timber has been recently removed; the exuviæ which always overspread such places, though hostile to plants in general, are favourable to the birch, which commonly springs up and becomes the successor of the pine. The common tree, where it grows wild,

U

attains a height of about thirty feet, and the weeping variety about forty feet; but both sorts rise to a much greater height when formed into plantations, particularly when interspersed with other trees. Some of the finest weeping birches in Britain stand on the banks of the river Findhorn, near Forres, in Morayshire; these are sixty feet high, with trunks upwards of two feet in diameter, and display pendant masses of spray ten feet in length, adding a graceful variety of verdure to scenes in themselves of great beauty. The soil is a surface stratum of sandy peat earth, with gravel, incumbent on sandstone. In nurseries the weeping variety is the kind chiefly cultivated, and notwithstanding the disposition of the plant to grow wild, it is generally the most uncertain nursery crop of any hardy tree.

The seeds are usually ripe in September, when they are collected, and to prevent fermentation, should be thinly exposed in an airy situation, until they become quite dry, when they may be packed up until March; they should then be sown. Respecting the nursery treatment of this tree, it is of importance to notice that a crop of seedlings will seldom or never succeed on ground which has been much used in raising nursery crops. All the best crops of the tree I have ever seen have been produced on new ground. Failure is so certain to follow the sowing of the seed on ordinary nursery ground that of late I do not practise it unless I can obtain an inch or two of fresh surface soil, clean and rich from some adjoining field, such as that after a crop of potatoes or turnips, or any other well-manured green crop. A friable sandy peat soil is generally preferred for their growth, which should be smoothly dug over, and beds formed four feet broad, with an alley of one foot between each. The seeds are then spread regularly on the surface, at the rate of one bushel of clean seed to each bed of thirty lineal yards. In dry weather they are closely pressed into the ground with the feet, and no other covering is necessary; if protracted drought ensue in May or June, it is of advantage to shade the beds, which is usually done with spruce or fir branches. Plants a year old, if well

grown, are about six inches high, and usually sell at from
3s. to 4s. per 1000. They are transplanted, when one year
old, into lines one foot apart, and the plants three or four inches
asunder; when they remain two years in lines they are
generally from two to three feet high, and fit for the forest,
and then they usually sell at from 20s. to 25s. per 1000.
The tree is frequently grown as a nurse for other trees; such
as the oak, the chestnut, etc. It is also used as underwood,
and for coppice-wood; but of the two varieties the common
sort is best adapted for being planted for coppice-wood; it
springs more readily from the root, and is not so apt to die off
as the weeping variety. The pruning and lopping of the
birch should be performed in autumn or early winter, to avoid
bleeding. For hoops it is commonly cut every five or six
years, and for bark every eighteen or twenty years.

The Weeping Birch differs from our other pendulous trees,
which are commonly grafted on the tops of tall stocks. It
commonly attains the age of fifteen or twenty years, and in a
cultivated state twenty or thirty feet in height, before it
assumes a pendulous form. The extremities of the more
vigorous lateral branches then begin to droop, while the top
shoots generally hold on in an upward direction until the
vigour of the tree subsides. The beauty of the tree, where
embellishment is the object, may generally be much enhanced
by a careful thinning out of the branches when they appear
in a thicket in the top. Immediately after the lateral branches
first begin to take a downward direction is the time for thinning them, at same time giving the surface of the ground, for
a few yards in diameter around the trunk, a coating of two or
three inches deep of thoroughly dissolved byre manure and
decayed leaves, removing all surface vegetation on the ground
before applying it. This adds vigour to the descending spray,
the beauty of which, on well-grown specimens, is surpassed by
no other kind of pendulous vegetation.

Not long ago, in many parts of the Highlands, the birch
may be said to have been the universal wood, and used by
the Highlanders for every purpose. They made their beds,

chairs, tables, dishes, and spoons of it, and even manufactured ropes and horse harness by heating and twisting the spray. The brushwood is used in forming wicker fences, to prevent the inroads of cattle and sheep, in thatching cottages, and in forming brooms or besoms. The tree attains maturity at various ages, according to the description of soil and situation which it occupies, but it seldom increases after seventy years old.

The wood of both varieties is of a light colour, shaded with red, and its grain is intermediate between coarse and fine, and not of great durability, except when grown in a stunted form at a great elevation. It is chiefly employed in the manufacture of fish-casks, and commonly sells at from 6d. to 8d. per cubical foot. When formed into stave-wood and delivered at the shore, the common price is £3, 15s. per 1000 superficial feet. As fuel, it occupies the twelfth place among twenty-one different kinds. It is much esteemed in smoking hams and herrings, in consequence of the fine flavour which it imparts. The tree occasionally produces knots of timber, beautifully marbled and veined, which are susceptible of a high polish, and are valuable in veneering. In remote parts throughout the Highlands of Scotland manufactories are in operation for converting birch wood into bobbins, and in the absence of roads these are carried to the nearest rail or water communication with manufacturing towns.

The bark of the birch is among the most incorruptible of vegetable substances. . It is in demand for tanning, and it is preferred by fishermen to the bark of any other tree for preserving nets and cordage ; it yields a softness and elasticity to nets, which cause fish to enter the meshes more readily than when they are preserved by any other material.

It commonly sells at from £5 to £6 per ton, according to its quality, or the state in which it has been prepared.

Birch wine is produced by the tree being tapped, by boring a hole in the trunk during warm weather, in the end of spring or beginning of summer, when the sap runs most copiously. It is recorded that during the siege of Hamburgh in 1814, many birch trees in that vicinity were destroyed in this manner by the Russian soldiers.

Several kinds of the tree, with large leaves, and likely to attain a greater size in this country than any of the native sorts, have been introduced into Britain ; among them *Betula lenta*, or the cherry birch of Pennsylvania, appears to yield the largest and most beautiful timber.

The cut-leaved weeping birch is one of the most elegant pendulous trees in cultivation. It is propagated by being engrafted on the common sorts. Although it often produces shoots of great vigour when placed on strong stocks, sometimes three feet in one season, yet it only becomes a small tree, seldom in the north of Scotland exceeding twenty feet in height. It has the advantage of at once assuming the pendulous habit of growth ; but the tree, though of great elegance in foliage and form, is not so robust and enduring as the native weeping tree. It is cultivated purely for ornament.

XXXIV

THE WILLOW.

WILLOW.—The genus *Salix* belongs to *Diœcia Diandria* of Linnæus, and it is the type of the natural order *Salicaceæ*. It comprehends many species and varieties, natives of all quarters of the world, very diversified in appearance, and ranging in stature from a few inches to about eighty feet in height.

There is no genus of plants in general cultivation whose species are more confused than that of the willow. This is accounted for partly from the more prominent kinds having hybridized and yielded intermediate varieties without number, and partly from each species containing male and female plants, and the same species differing to some extent in appearance at certain seasons of the year; add to this the circumstance of old trees assuming a very different appearance from young ones, and that no tree is more apt to change its appearance from a change of soil and climate, and it will not be surprising that some confusion should exist in the genus, and that the more prominent species only should be readily recognised.

From twenty to thirty very distinct species are natives of Britain, but these have been divided into a much greater number.

His Grace the Duke of Bedford, in 1829, published his *Salictum Woburnense*, in which 150 species of the tree were figured and described. The interest of the publication was greatly enhanced by the circumstance, that at the time all the kinds existed in the Salictum at Woburn, where every opportunity was afforded for dispersing the most esteemed varieties throughout the country. The kinds of British and foreign willows enumerated in more recent publications amount to

several hundred species or varieties. For our purpose it is
only necessary in this article to notice a few of the more prominent kinds, best adapted for the purposes for which the
tree is useful.

For the produce of timber there are three species, which,
together with their allied varieties, are generally recognised,
and deserve notice. They attain the ordinary size of timber
trees, and being indigenous in this country, they are extremely
hardy, and grow to some size in soil of almost any description, viz., *Salix caprea* (L.), *S. alba* (L.), and *S. Russelliana*
(Smith).

S. caprea, the Goat Willow, or Sallow, is a common tree, found
indigenous in waste ground, particularly in cold and marshy
situations. In rich wet ground, a two-year-old seedling plant
will sometimes produce several shoots three or four feet high;
and under the most inhospitable circumstances it generally
ripens its growth to the uppermost bud.

This tree and its varieties are among the broadest-leaved of
the willow tribe. The vigorous young shoots have a dark-
brown glossy bark, which in spring time forms a fine contrast
with the buds, which are white and prominent; this, together
with the profusion of catkins displayed by the male plant
during the opening up of the season, give the tree a gay
appearance, and render it one of the most ornamental of the
genus. The goat willow is a valuable tree for coppice, where
there is a demand for hoops, poles, rods for crates, sheep
fences, or for similar purposes, and where it may be cut down
every three or four years; during a short period no other
tree will produce so great a bulk of fagot-wood. On a congenial soil, a healthy stock will sometimes in one season
produce a sheaf of straight clean shoots eight to twelve feet
long, and many of them an inch in diameter, at a yard from
the ground. This species frequently attains the height of
forty or fifty feet, with a trunk one and a half to two feet
in diameter. It affords a valuable shelter in maritime situations, withstanding the influence of the sea better than
most plants, and its timber is reckoned the best of any of the

willows. Like almost all the willows it is propagated by
cuttings; strong one-year shoots, formed fourteen to sixteen
inches long, and sunk ten or twelve inches in prepared ground,
are often attended with as much success as rooted plants; but
where the ground is not of the best description and well prepared, rooted plants should be used.

S. alba, the White or Huntingdon Willow, is a well-known
tree, and in favourable soil and situations it frequently ranges
from fifty to eighty feet in height, with a trunk two or three
feet in diameter. If we except the grey poplar, *P. canescens*,
perhaps there is no other tree, native or foreign, which, in this
country, attains to so great size during the first twenty or
thirty years of its growth. About that age it generally, in
favourable situations, reaches the height of from sixty to
seventy feet, when its trunk often yields one cubical foot for
every year of its growth. It is more frequently planted as
a timber tree than any other willow. It is a good coppice
willow, and often grown as a pollard tree. Its year-old shoots
are remarkably strong and tough, but being twiggy, or full
of laterals, it is rendered unsuitable for the finer purposes of
basket-making. At Hopetoun House, near Edinburgh, this
tree is recorded to have attained the height of seventy feet,
with a trunk four feet and a half in diameter.

S. Russelliana, Russell's, or the Bedford Willow, is one of
the best tree-willows in cultivation. It was brought into
notice about sixty years ago by His Grace Francis Duke of
Bedford, and ever since it has been honoured by the family
name. Johnson's willow, at Lichfield, is of this species; its
height is recorded to be forty-nine feet, with a trunk twelve
feet in circumference. One of the best English specimens of
this tree stands at Sion, eighty-nine feet high, with a trunk
upwards of four feet in diameter. In Banffshire, the banks of
the Deveron, at Duff House, are embellished with some fine
specimens of this tree, upwards of fifty feet high, with trunks
from eight to nine feet in circumference.

The bark and leaves of all the willows are astringent, and
the bark of most sorts is used for tanning leather.

A deep rich soil, in the vicinity of water, is that which de-velops the tree-willows in great perfection, but being extremely hardy and tenacious of life when young, they will grow to some extent in ground of almost any quality. In the driest exposures, and at the highest altitudes, the willow is often found among the few plants of ligneous form. In places where a close cover of fir, larch, or other timber, has been felled, which does not spring from the root, it is often found difficult to form a cover of any kind of vegetation; for this purpose willows are valuable, and are often indigenous to such situations, where they are the means of sheltering and estab-lishing more valuable trees. As a nurse, however, the tree-willows require the attention and care of the forester, as they are apt to confine and destroy the trees with which they are associated. As a shade and nurse to the silver fir, the holly, and the yew, the willow is of much advantage.

Willow timber is white, soft, and light, adapted to various agricultural purposes, such as rake and scythe handles, sheep flakes or hurdles; these are easily formed, and, from their extreme lightness, are readily removed from place to place, and not apt to be damaged by their own weight. When sawn into boards, the timber is well suited for lining for carts and barrows; lightness for such purposes is a recommendation; but the chief advantage in this timber is that it is not apt to splinter or receive damage from the fall or friction of hard materials. It is much sought after of late, not only for the construction of railway carriages, but for timber for brakes, as it does not readily ignite when employed for that purpose.

The branches of the different species of the tree-willow are useful, when cut between October and April, for being formed into frames for embankments, to prevent the encroachments of rivers and streams. These frames should extend from the channel of the water to the top of the flow bank with a gentle slope. The larger timbers being interwoven with the smaller spray, and the whole being covered a few inches with sand, gravel, or the common soil of the banks, the branches readily push out numerous fibres, and create a surface vegetation in the proper form for resisting the influence of water. By this means materials of a shifting character are consolidated, and by being yearly lopped the willows establish a permanent embankment.

Of the dwarf willows or osiers, adapted for basket-making, the variety is very great. The most approved kinds are *S. viminalis* (L.) *S. rubra* (Hudson), *S. Forbyana* (Smith), and the numerous varieties related to these species, among which the pack-thread, whip-cord, and red Dutch are known in some districts. The kinds are always propagated by cuttings, which are planted on the ground where they are to remain. A cutting is made of the strongest portion of a one-year shoot, about fourteen inches long. About two-thirds of its length is placed into the ground. They are formed into lines eighteen inches to three feet asunder, and the plants from one to two feet apart, in proportion to the strength of the species. They reach their full strength the third year, and endure yearly lopping variously, according to the nature of the plant, the soil, and climate. Some varieties are more tender and subject to disease than others. Of late years an atmospheric blight has been severely felt in some quarters. It has usually occurred in the end of July or beginning of August; its influence is known at the time by a faded curl on the foliage near the top of the shoot, where a black spot is formed on the twig, and marks the height of the growth at the time of the casualty. Each spot covers a small portion of dead wood, which occasions the rod to snap when bent at that point. Some of the best varieties of *viminalis* have suffered by this influence, particularly such as were old, and had outlived their most vigorous period, while the young and vigorous plants of the same species were almost exempt from injury.

Among ornamental willows the *S. Babylonica* forms a very graceful and interesting tree. It is a native of Asia and of the north of Africa. Being somewhat tender, it is only in the best of soils and seasons, in this country, that it ripens its young twigs at their extremities. Unlike most willows, it does not grow freely from cuttings, and should therefore be propagated by layers. The male plant of the purple, the black, the yellow, and many other osiers, form very ornamental small trees, which become particularly interesting from the gaiety and richness of their early blossoms, forming the harbingers of summer and the first food of the bee.

The new American weeping willow, recently introduced, is spreading rapidly throughout Britain. Being a plant of very feeble growth, to exhibit it in its most attractive form it requires to be grafted on the top of a strong stem, such as that of *S. caprea*. Its long slender twigs then droop down with much elegance, and become agitated by the slightest impulse of wind, like the spray of a playing fountain. The value of this species is greatly enhanced by its being extremely hardy. A drooping variety of the *S. caprea* has lately been discovered in the west of Scotland, and is now cultivated in nurseries under the name of the Kilmarnock Weeping Willow.

XXXV

THE POPLAR.

POPLAR (*Populus*).—The poplar tree belongs to *Diœcia Octandria* in the Linnæan system, and to *Salicaceœ* (Loudon) in the natural system of botany, where the willow and poplar form the order. All the poplars, like the willows, are unisexual—either male or female plants. Any deviation from this rule is very rare among poplars; and as all the species abound in varieties, differing to some extent according to sex, age, influence of soil and climate, some of the species, as might be expected, are not easily recognised; and botanists differ in some instances as to what are species and what varieties. The tree, in some species or other, is a native of all quarters of the world.

All the poplars are remarkable for rapidity of growth, therefore no tree is more frequently employed to furnish immediate effect in a bare locality. Although many of the species are short-lived, and destitute of that elegance of ramification so desirable in trees, yet, interspersed with other kinds, they furnish a shelter and embellishment until those of a slower growth, possessed of a more elegant form, have attained to a size adapted for the purpose for which they were intended.

I may confine my remarks to the following species, with their more prominent varieties, being the kinds best adapted to our climate :—

P. alba (Linnæus), the white poplar or abele tree; *P. a. canescens* (Smith), the grey poplar ; *P. tremula* (L.), the trembling, or aspen ; *P. fastigiata* (Desfont), the fastigiate, or Lombardy ; *P. monilifera* (Aiton), the necklace-bearing or black

Italian; *P. balsamifera* (L.), the balsam ; *P. candicans* (Aiton), the Ontario poplar.

The three species first-named are propagated by layers, as they do not successfully strike root by slips or cuttings. The last-named four species strike readily from cuttings.

Plants grown from layers are necessarily three or four times the price of the kinds that grow from cuttings, and plants in general raised from cuttings are dearer than those produced from seed.

·· An idea is common that all plants that are produced from layers or slips are apt to become dwarfed or stunted by this mode of continuous propagation. However correct this theory may be with regard to some plants that are apt to form tap-roots, and derive their support from far underneath the surface, I believe that respecting such trees as poplar, willow, lime, etc., their vigour is reduced in no perceptible degree by their propagation from layers or cuttings for any length of time.

The mode of propagation by layers is very simple, and is detailed under LIME-TREE. The shoots on a young stool commonly double in number every year till about four years of age, when a vigorous stool should yield three or four dozen plants yearly. After the young layers are removed, it is the usual practice to prune and equalize their roots, and place them in nursery lines two feet apart, and the plants one foot distant, for one year at least, before inserting them into the forest. As poplar plants are very apt to overgrow, they are un-profitable for nurserymen when there is not a constant demand.

Poplar Cuttings.—The kinds adapted for this mode of propagation strike very freely in any ordinary soil, but a sharp sandy soil is that which yields the most fibrous roots. Strong one-year shoots are commonly employed for cuttings when they can be obtained, but two-year-old wood will suit. Cuttings are formed ten to twelve inches long, and are inserted in winter or early in spring, about two-thirds into the ground, in lines eighteen inches apart, and at the distance of six or eight inches in the lines. In summer, when they have sprung a few inches, they should be dressed by rubbing off all the young growths except the strongest on each plant, which should be chosen for the stem of the future tree. One-year-old plants are often about two feet high, and are sometimes planted out at that age. After two years, in ground of ordinary quality, plants range from three to five feet. This is the ordinary age for transplanting into their final destination. If required larger for hedgerows, or special purposes, they should be replanted into greater space in the nursery.

P. alba : The White Poplar (synonymes: The abele tree ; the downy poplar; the Dutch beech tree).—This tree is a native of different countries in Europe. Although it grows wild in many parts of Britain, yet my experience of it makes me believe that this form of the tree is not a native, but an imported species. It is indigenous to Palestine, and the name *Abele* is supposed to be derived from the name of a place where the tree abounds in the plains of Nineveh on the banks of rivers; and the timber of this species is said to be the shittim-wood of Scripture. This tree has several varieties, all remarkable for the striking contrast in colour between the upper and under surface of their leaves; the upper being of a dark-green, while the colour underneath is pure white and very downy. The distinguishing points of the varieties are size, vigour of growth, figure of foliage, distinctness in colour of foliage, and hardiness of constitution, and generally the finer and the more beautiful the contrast of colour in the leaves, the more weak and tender the variety. That known under the name "white Egyptian," is particularly fine in the colour of its foliage, but it is the most tender; while the common white, and that known as *acerifolia*, with a deeply lobed leaf like a maple, are the strongest growing of the species *alba*. Propagation by cuttings of the branches is not successfully practised in this species ; but as it is apt to produce suckers from the root, they are sometimes resorted to, as well as cuttings of the root, as a means of increasing the tree. Seeds of this species are seldom to be obtained, and layers are found to make the best plants. It is important that the plants to be used as stools be of the best varieties. From the remarkable contrast in the colours of the leaves of this species, the tree has a very ornamental effect when agitated by the wind; standing on the margins of lakes, or on the outskirts of large plantations, it appears to the greatest advantage. Though it is less hardy and less vigorous, yet it is more ornamental than the kind next noticed, which is considered by many a distinct species.

P. a. canescens; The Grey, or Hoary Poplar.—This tree is a native of Scotland and of many parts throughout Britain. It is often found on low and uncultivated soil having a tendency to moisture, where it spreads from the root, yielding suckers profusely. It blossoms in April, when the male catkins appear very conspicuous on the tree, of a purple colour, just before the expansion of the leaves. As the tree does not readily strike by cuttings, the usual mode of propagation is by layers, as already described. In soft alluvial soil, somewhat moist, no tree will attain to a greater size in the space of ten, twenty, or forty years. After the age of sixty it commonly takes heart-rot, and is often at maturity at forty or fifty years of age. It is found to luxuriate in very exposed situations at a great elevation, and does not object to a considerable mixture of bog or peat-earth in the absence of stagnant moisture. Its rapid growth when young is not confined to any particular description of soil, provided it is soft, rich, and moist. The following is a statement of the comparative growth of trees formed for a screen or shelter plantation, twenty-eight years of age, on ground far too wet for growing large timber of the ordinary kinds, as the surface is frequently flooded for want of an exit at high tides, and where the surface soil is seldom more than eighteen inches above the rise of water :—

Average girth at three feet above the surface.

Poplar, Grey,	. . . .	5 feet 2 inches.
Do. Black Italian,	. . .	4 „ 4 „
Do. Lombardy,	. . .	2 „ 5 „
Lime, ash, beech, and birch,	. .	2 „ 3 „

Another small plantation adjoining the above, and similarly situated, is thirty-two years planted :—

Girth at three feet from the ground.

Poplar, Grey, ...	6 feet 8 inches.
Do. Italian, ...	5 „ 2 „
Birch, ...	3 „ 2 „
Alder, ...	2 „ 8 „

Excess of moisture has here produced dead wood on every kind of tree except the grey poplar, which is in a healthy and vigorous state. Both species of poplar are about one-third taller than the other trees, which range from twenty-eight to thirty-two feet.

One of the finest *P. canescens* ever produced in Britain grew on the lawn at Bicton Gardens, in Devonshire, and was blown down by the memorable hurricane of the morning of 11th January 1866. Its dimensions were—girth, 14 feet 6 inches ; length of stock, 49 feet ; height of the tree, 114 feet ; and its age was between eighty and ninety years. It grew on good soil, that formerly had been a kitchen garden, and for many years turfed, and with a number of well-made American-plant beds dotted about it. A beautifully clear stream, about six feet wide, and four or five inches deep at all seasons, with the bottom paved with pebbles, ran through this ground. Under this stream the roots of the tree spread in all directions, as well as under the turf into all the American beds, and often choked up drains in the vicinity. The timber for the most part was sound, and beautifully white and clean, but heart-rot had begun in the bottom of the trunk. Seven years ago, in forming a pond to propel machinery at a farm-steading situated in a very bleak exposure, 650 feet above the level of the sea, for the sake of economy and ornament, I formed the pond with a few islands. On these, and on the margins, was deposited the soil excavated during the progress of formation. These islands and hills in the vicinity were then planted with a variety of the common firs and broad-leaved plants usually

grown in the country, to the extent of about fourteen or sixteen sorts ; and, after six years' growth, this tree, the grey
poplar, is very far ahead of every other sort, both in height
and girth, and already assumes a timber-tree appearance ;
showing that a disturbed or loose surface soil in the vicinity
of water is an element favourable to the most rapid growth of
this tree, as it is to every other species of poplar.

P. Tremula, the Trembling-leaved Poplar, or Aspen, is
indigenous to Britain, and to mountainous situations throughout Europe and Asia. Throughout the Highlands of Scotland
it is frequently associated with the natural birch. It is a
beautiful tree, of a stately and elegant appearance, round-
headed, and tall in proportion to its girth, of rapid growth, and
extremely hardy. It attains a considerable height in almost
any description of soil, growing luxuriantly when young in a
soil dry and sandy, as well as in that which is wet and strong.
The roots of the aspen spread over the surface of the ground,
and when the vigour of the young tree subsides, the greatest
objection to its cultivation in many situations is its propensity
to produce suckers from the roots, which, in neglected grounds,
form a jungle around the tree.

In the Highlands, the young shoots are greedily eaten by
sheep and cattle; and for this purpose, in Germany and
Sweden, the spray is frequently used both in a green and
dried state ; and, according to one writer, this is the most
valuable purpose to which it can be applied. The seeds of
the tree ripen in summer, and may be immediately sown ; but
propagation by layers, or suckers, is the more speedy method.
This species grows freely from cuttings of the roots, but not
from cuttings of the branches. Plants one or two years transplanted from layers are commonly from five to six feet high,
when they should be removed from the nursery. In soil of
ordinary quality, the average growth of the aspen for the first
ten years is not under three feet annually. One of the loftiest
trees of the species in Britain is reported to stand at Castle
Howard, in Yorkshire, 130 feet high, with a trunk three and
a half feet in diameter. Few trees are more ornamental or

X

interesting in the landscape ; standing alone as a lawn tree, it
assumes a pendulous form. On the outskirts of plantations,
or when the species is grouped by itself, and seen either on
the hill top or on the sloping sides of the forest, it forms an
object of great interest. Its foliage forms a contrast with
most other trees throughout the summer, being of a beautiful
glaucous green, which, after the first frosts of autumn, changes
into a fine mellow or ripened hue, and ultimately into a bright
yellow. The trunk of advanced growth is of an ash colour.
With tasteful arrangement, this tree, being perfectly hardy,
forms a valuable element in embellishment, at any altitude
adapted for arborescent vegetation. The trembling of the
leaves is far more conspicuous in this than in any other species
of the poplar, arising from the construction of the leaf and
petiole, or leaf-stalk. The leaves are round and smooth, and
stand on long slender foot-stalks, which cause them to become
agitated by the slightest impulse of the air, so that their
quivering is quite perceptible, and well heard, during weather
comparatively calm—

> " When zephyrs wake,
> The aspen's trembling leaves must shake."

In the *English Flora*, Sir J. E. Smith says,—" The colour
of the upper surface of the leaves is a fine dark glaucous
shining green, and that of the under surface of a paler shade.
The disk of the leaf has a small point, and three ribs. It is
somewhat wavy, and often shorter than the foot-stalk, which,
being vertically compressed in its upper part in relation to the
plane of the leaf, counteracts the ordinary waving motion of
the leaf in the wind, and causes it to quiver with the slightest
breeze."

It is worthy of notice that no tree so interesting, ornamental,
and hardy, is so sparingly planted throughout Britain, so it is
not often to be seen in the nurseryman's catalogue. The wood
is the strongest, hardest, and heaviest of the poplars ; and the
bark is sometimes used in tanning leather.

P. Fastigiata, the Fastigiate or Lombardy Poplar, is a well-
known tall tree, readily distinguished from every other species

by its upright growth; and having its lateral branches closely
gathered around the stem, it forms a taper shape; possessing
that habit of growth among deciduous trees which the upright
cypress has among evergreens.   It is a native of Italy, particularly of the banks of the Po in Lombardy.   It was introduced into Britain about the middle of the eighteenth century,
and is propagated by cuttings as speedily as any description
of willow.   Being easily increased to any · extent, it soon
became common throughout Britain.   Some other species of
the genus yield a greater bulk of timber in twenty years, but
no other tree will attain to a greater height in that period.
When young it grows vigorously in any soil, but to become of
a great size the ground should be rich and deep, with water
within reach of its roots.   On the Continent it is usual to
form fences of the tree, by inserting two or three year old
plants, which commonly stand six or seven feet high, in
straight lines.   The plants, about six inches apart, are connected by a horizontal rod, placed at the height of three feet
from the ground, thus forming a fence in one season.   The
plants are commonly lopped in course of a few years, and
sometimes are thinned out, which enables those remaining to
become of considerable girth, and useful as timber.   Throughout a great part of the Continent this is the only tree in the
divisions and along the margins of fields; but the uniformity
of its growth, the straight lines, and the flatness of many
parts of the country, form scenes monotonous and wearisome.
In this country the principal use of the tree is embellishment
and shelter.   It speedily rises to a great height, and forms a
screen of verdure during summer.   In landscape-gardening it
forms a very striking object; judiciously introduced into the
masses of round-headed forest trees, it changes and improves
the features of the scene.   It has a commanding effect situated
at the end of buildings, or arising from the background,
diversifying the regularity of the sky-mark usually formed by
the tops of ordinary plantations.   As an adjunct to an old
ruin, a group of this species often adds greatness and dignity to
the scene, when skilfully placed and viewed from a distance.

As a town tree it has no rival. It rises in a narrow space, and endures the influence of smoke better than almost any other ; and its value is greatly enhanced by its springing rapidly, and appearing in all its characteristic form in a very few years. Dr. Walker records a tree, on the banks of a canal near Brussels, which in fifteen years grew to the height of eighty feet, with a trunk from seven to eight feet in circumference. In France, on the banks of the Seine at Rouen, trees of this species, bordering some of the public promenades, stand 150 feet high. In England, some of the finest specimens produced were planted by a cottager in the village of Great . Tew, in Oxfordshire, who lived to see them, at the age of fifty years, 125 feet high. The finest tree in Scotland lately was supposed to be one that was blown down on the lawn at Darnaway Castle, in Morayshire, in a gale in 1860. At Taymouth, in Perthshire, this tree stands upwards of 100 feet in height.

The trunk, in an advanced state, is deeply furrowed, and the tree is not apt to produce undergrowths. Unlike trees in general, its whole top waves with the wind in one gentle sweep, and its shade is very harmless to the crops in its vicinity. The extensive use of this poplar in the division of fields on the Continent suggests the practicability of employing the living tree as a support for wire-fence along fields. Of all others it grows most upright, and fresh posts of it inserted into good soil take root and vegetate. In some parts of the east of England it is grown on marshy ground in thickets, two and three feet apart ; and no tree admits of being grown more closely. In five or six years the timber is serviceable for poles in fences, and for hurdles for enclosing sheep, and similar purposes. Although when exposed to the atmosphere it proves less durable than many sorts of timber, it should be remembered that its propagation is simple and cheap, its timber is soon obtained, very easily worked, and handy from its lightness, and the ease with which it can be removed from place to place. Although the tree cannot be recommended as coppice-wood, yet when cut while it is young, and not very near to the ground, and trained to one stem, it furnishes a good second or third crop of poles.

P. monilifera : The Necklace-bearing or Black Italian Poplar. —This poplar is generally believed to be the same species as *P. nigra Canadensis* of Micheux, or a variety of that tree. Its name "Necklace-bearing" is derived from the structure of the female catkins, resembling strings of beads. This tree is remarkable for its rapidity of growth. It comes into leaf in Scotland about the 1st of May—rather later than the Lombardy poplar. The bark of its young shoots is of a darker hue than that of that tree. The yearly lateral shoots take a more wide and horizontal range, and instead of being round, as in the Lombardy poplar, they are fluted or ribbed, particularly towards the extremities. Its upward growth under similar circumstances is equal to that of the Lombardy poplar; while, on account of the strong side branches and larger leaves of the black Italian poplar, its girth is greater, at once showing a decided superiority even during the second year of growth from cuttings, from which both grow readily.

The tree is a native of the continent of America, and some are of opinion that it was introduced into Italy, and from thence into Britain. Its first introduction into this country was in 1772, from Canada; but its second introduction appears to have made it better known, and from which it is supposed to have taken its popular name. Respecting this tree Loudon says,—" The rate of growth in the climate of London, in good soil, is between thirty and forty feet in seven years; and even in Scotland it has attained the height of seventy feet in sixteen years." From strong cuttings, I have often seen plants six feet high in two years, and the second year lateral shoots are nearly equal in strength to the leader. Nevertheless pruning is not necessary; the vigour of the side-branches increases the diameter of the bole, and the strong flow of sap naturally gives the top an ascendency.

I am not aware of any distinguishing mark by which the sex of the tree can be ascertained before it comes into flower ; but in many situations the female tree of this species is objectionable on account of the profusion of seeds which it sheds. It blossoms in spring, before the leaves are expanded, and the seeds fall commonly early in June. I have seen the produce of two or three trees in a clump alongside a turnpike road drifted into the side-drains like wreaths of snow, and after a few weeks of moist, warm weather, it sometimes becomes a mass of vegetation obstructing the sewerage. Near well-kept lawns and walks, etc., the male plants should therefore be preferred. Notwithstanding the ease with which the tree might be grown by the thousand, growth from seed is not practised in nurseries, on account of the tree being seldom required by the thousand, and a cutting being equal in strength to a two-year seedling, while a two-year-old plant from a cutting will furnish on an average upwards of a dozen of slips or cuttings without injuring it. Although the sexes are seldom kept distinct or separate in nurseries, propagation by the branch is the sure method by which the sex of the young plant can be known.

When the tree is to be grown from seed, the treatment should be that afterwards described for all the poplars. In ordinary arable land of a medium quality, the tree often yields from twenty to thirty cubical feet of wood in mixed plantations in little more than twenty years.

The following is a note of trees planted twenty-four years since, on rather poor, sandy soil, about twenty feet above the rise of water, and in a good climate, about forty feet above the tide-mark, in a mixed plantation—the girths taken three feet above the surface of the ground :—

	Girth.	Height.
Black Italian Poplar, . .	3 feet 6 in.	60 feet.
Larch, ...	3 „ 4 „	48 „
Elm, Scotch, ...	3 „ 0 „	33 „
Beech, ...	2 „ 6 „	40 „
Sycamore, ...	2 „ 1 „	34 „

P. Balsamifera, the Balsam Poplar, is a native of North America, where it rises to the height of eighty feet. In

Britain it is of much more humble stature, and grows vigorously
only for a few years when young. It is purely ornamental.
The bark of the young wood is of a rich chestnut colour, and
the buds are large, and encased in a glutinous balsam. The
trunk has an ash-coloured bark. In the opening of the spring
it early discloses its leaves, of a pale yellow, possessed of a
rich balsamic fragrance, which it diffuses throughout the
atmosphere. The leaves eventually become of a rich dark
green. There are many varieties of this tree, which differ in
the size, shape, and the shade of the leaves, in the vigour of
growth, and in the time in which they expand their foliage.
The tree and its varieties are all readily propagated by cuttings,
or from suckers, which it is in the habit of producing. It
will grow when young in almost any description of soil with
considerable vigour, and sometimes it produces in one season
a shoot four or five feet long; but its early vigour resembles
that of several species of willow, the laburnum, and locust
trees; it is of short duration, and in this country it only
attains the stature of a dwarf timber tree, seldom exceeding
thirty or forty feet in height.

P. Candicans : The Whitish-leaved Balsam-bearing or
Ontario Poplar; syn. *P. macrophylla* (Lindley).—This tree is
a native of America, and is generally known in this country
as the Ontario Poplar. In appearance and fragrance it is
very like the balsam poplar, but it is a week or two later
in coming into leaf. Its leaves are larger and rounder, and
altogether the tree is of more robust growth; but in this
country it fails to attain great dimensions, usually getting
stunted or set in growth on reaching forty feet in height,
which it generally attains in twenty years; then its branches
soon become so brittle that they can scarcely support a crow.
Its vigorous growth, broad leaves, and pleasant fragrance
when young make it an interesting object in newly-formed
plantations by the road-side; and as it is readily grown from
cuttings, it is often employed for thickening up ornamental
plantations till slower-growing and more permanent trees get
established.

There are other species of the poplar introduced, and sometimes cultivated in this country ; but they are less important
than those described, on account of their being either of small
stature or too tender for our climate.

The growth of the poplar from seed.—All the poplars blossom
early in spring, and the female plant usually ripens and sheds
the seed in the end of May or early in June. On account of
the facility with which young plants can be produced from
cuttings and layers, the growth of the tree by seed is seldom
practised. The following, however, is the mode of raising
the tree by that means :—

The seeds appear a white, cottony substance, and where the
ground is smooth, such as on short meadow grass, or along a
road-side, they sometimes can be readily swept up. On being
collected, they should be immediately sown on the surface of
smoothly-dug, friable ground, then raked in, and rolled with
a heavy roller, or smoothly tramped in with the feet. This
bed should then be shaded by being overspread with straight-
drawn straw, or with broom, or heath, or some such substance,
that will furnish shade and admit air. If the weather is not
very moist, watering all over the covering should be carefully
performed with rain or pond water, or that which has been
for some days kept stagnant in the sunshine. The plants will
spring in a week or two, according to the heat of the weather,
and the cover may then be removed. They will be a few
inches long before winter, and in spring should be transplanted
into nursery lines about fifteen inches apart, and the plants at
the distance of six or eight inches.

The soil most suitable for all the species is that which is rich,
loose, and moist, or in the vicinity of water. The cost of
trenching the ground is always well repaid by rapidity of
growth. When a poplar becomes stunted it rarely regains its
vigour. If it ceases to grow freely from any other cause than
unsuitable soil, it will generally show a tendency to spring
from the surface of the ground ; and if young, on being lopped
over at the surface, and confined to one shoot, it will become
vigorous.

Pruning is not to be recommended for the poplar. The species which ramify, such as *P. canescens* and *P. monilifera,* can be grown with clean trunks by being inserted so that they will gently press on one another, which will have the effect of diminishing the vigour of the side-branches, and of producing clean timber. In amputating a large branch, unless skilfully performed so as that the wound will not lodge water, the tree is apt to suffer. The wood being extremely soft is apt to crack and admit moisture into the tree, which becomes ruinous. For this reason none of the species can be relied on as coppice-wood for more than a crop or two.

Disease.—The poplar furnishes food for the larvæ of several species of moth; but I have never found the tree assailed to any serious extent by insects, or any disease except the atmospheric blight known as the potato disease, and its influence was confined purely to the varieties of *P. alba.* The nearly-related species, *P. canescens,* on the most trying visitations of this influence, I have found always exempt from injury when growing in the same quarter. It was noticed that the finer the varieties of the tree, or those that produced the greatest contrast in the colours of the leaf, were the sorts on which the disease fell with the greatest severity. My experience of this disease was not confined to one locality. I have seen the influence most distinctly in various places, some of them 100 miles apart. About twenty years ago my attention was first directed to it early in August, one year when the potatoes in the fields had flagged and fallen by the disease. I observed in the stool ground, or propagating ground, in the nursery, the drooping leaves of the white poplar to have fallen simultaneously with those of the potato. I found afterwards that during a succession of bad years of the potato disease the stools of this species of poplar died along with those of the *Solanum variegatum,* or shrubby potato, situated in the same propagating ground. At the same time the trees of the white poplar throughout the country became much enfeebled, and showed quantities of dead branches, and in many districts the tree is now seldom met with where it formerly abounded.

The deadly influence of the same malady was quite perceptible on some of the more tender exotic willows, and on some other kinds of vegetation.

Recorded opinions of the tree.—Hitherto the genus *Populus* has been esteemed chiefly for its services in a living state, and writers on this tree have differed greatly respecting its appearance and worth. Gilpin says,—" When the gentle breeze pressing upon the quivering poplar bends it only in easy motion, while a serene sky indicates the heavens to be at peace, there is nothing to act in concert with the tree. It seems to have taken its form from the influence of the sea-air, or some other malign impression, and, contracting an unnatural appearance, disgusts." Muskau says, " It is too fluttering." Cobbett represents it as " a great, ugly tree." Evelyn seems fully to apprehend its value when he says, " It puts a guise of such antiquity upon any new enclosure, that whilst a man is on a voyage of no long continuance, his home and lands may be so covered as to be hardly known by him at his return." And although some associate its appearance with the idea of an upstart, and despise it for want of the elegance of ramification, and the ancestral dignity of the British oak, yet in a good specimen of two or three of the species, the massive bole, and the magnitude of the branches, frequently command the respect of the beholder ; and we have the high authority of one of the first Greek writers in calling the tree beautiful—Homer—who compares the fall of Simoisius by the hand of Ajax to a poplar tree just cut down :—

> " So falls a poplar that in watery ground,
>   Raised high its head, with stately branches crowned ;
> So down it lies, tall, smooth, and largely spread,
>   With all its beauteous honours on its head."

Poplar Timber.—There is a great similarity in the timber of all the different species and varieties of the tree ; indeed, the difference in soil sometimes changes the appearance of the timber of the same species as much as the difference between one species and another. The appearance of the wood of all the species is white, with a yellowish tinge sometimes near the centre when it becomes old. The centre timber very soon becomes short, and decays in most of the kinds. The timber is remarkable for its lightness and softness ; that of *tremula* is the hardest and heaviest of any species when grown in firm soil. The softest and lightest wood is that which is fast grown, on rich, soft, mossy ground. In remote districts throughout the country, where its uses were unknown till lately, I have known it to be sometimes sold at sixpence per cubical foot, and cheaper than any other wood ; but since its uses are become better known, it often fetches the usual prices of the different kinds of hardwood. The wood of all the species contracts very much on being seasoned. It suits all the purposes of willow timber. It is esteemed for cladding for carts, waggons, barrows, railway-trucks, and the like. Employed as such it does not splinter like most kinds of deal, nor crack in the driving of nails. When struck, the blow only indents the soft wood, without piercing or cracking it. Its lightness recommends it for many purposes, such as doors and gates, more especially such as are of great dimensions and in frequent use, where heavier wood is apt to strain the hinges. It is also esteemed for kitchen furniture, dishes, rollers for silk, carver and cooper work, butchers' and bakers' trays, and all purposes where a clean appearance and lightness are a recommendation. On the Continent it is the common wood for clogs, and various small wares ; of late it is much sought after, on account of its softness, for railway brakes, it being difficult to ignite by friction.

XXXVI.

THE ELM.

The Elm (*Ulmus*).—This genus is the type of the natural order *Ulmaceæ*, and occupies a very important place among our British timber trees. Two species deserve particular notice, *U. campestris* and *U. montana*.

U. campestris (L.), or English elm, is a tall, elegant tree, of rapid growth ; and although it is not famous for its ramifications, yet from the density of its foliage and its clustering habit of growth, in bright weather it displays a variety of light and shade such as painters appreciate in such objects. It is of erect growth, and yields a tall bole, remarkable for the uniformity of its diameter throughout ; and few trees produce the same quantity of valuable timber in so short a period. If we except the oak, it is more common than any other timber tree in England, where it is said that upwards of forty places take their names from that of elm, such as Nine Elms, Barn Elms, etc. Around the palaces, castles, cathedrals, old halls, etc., it is commonly met with of large dimensions and venerable character ; but as it rarely produces seed in England, it is questionable whether it is a native of that country. If not truly indigenous, it must have been introduced at a very early period, and propagated by art. It is a native of the south and middle of Europe, and the west of Asia, and there it yields seed abundantly. In France, plants are raised from seed in immense numbers. The tree is remarkable for its propensity to produce seedling varieties ; many of these imported, of very feeble growth, have spread throughout this country, and are of no value as timber trees. The aptitude of the tree to vary from seed has rendered the genus very confused. About twenty sorts of *U. campestris* are cultivated in nurseries, and botanists are unable to determine which are species and which are varieties. It is therefore of importance that the planter of timber should know that he is possessed of a kind of the tree adapted to the purposes intended.

In England, the tree is chiefly propagated by layers from stools, or from suckers from old trees. The former is the better mode. A stool is formed by lopping over a plant which has become established in the ground. During the ensuing summer, the root or stool produces a number of young shoots; when these have completed their growth for the season, they are bent down to the depth of five or six inches into the ground, and fixed with the earth, having their extremities clear above ground, and in an erect position. During the succeeding summer these layers become rooted, and another crop of young shoots are produced by the stool. The layers should be removed any time in open weather during winter or early in spring, which will make way for the next crop of young shoots being laid down as already described. Thus, a stool yields a crop of plants yearly; and that they may become well rooted, the soil should be rich, friable, and sandy.

Unless the ground into which the young plants (layers) are to be planted is clean and of the best quality, they should be inserted into nursery lines for two years previously to their being finally planted out. The plants should be a foot apart, and the lines two feet distant. If they are intended to remain for a longer period, a greater space should be allowed to them in the nursery.

The tree grows freely in soils of very opposite qualities; it may be seen in dry sandy soil, and in the most unctuous clay, and in both it grows better than most trees; but that of an intermediate description, rich, soft, and open underneath, is best adapted to its development. The young tree forms a root more fibrous and bushy than most plants, which adapts it for being transplanted when of a size and age beyond that which is common. In ordinary soils, the tree usually attains the height of twenty-five feet in ten years. It naturally rises into a straight form, and seldom requires pruning. Its fibrous roots, upright and rapid growth, recommend it for a hedgerow tree, and as such it is capable of being transplanted of a size almost sufficient to resist injury from cattle.

Of the recorded trees of the species in England, one of the most remarkable is the Crawley Elm, which stands on the high road from London to Brighton. It measures sixty-one feet in circumference at the ground, and is seventy feet high. Its trunk is perforated to the top, and it measures thirty-five feet round the inside at two feet from the base. At Hatfield, in Hertfordshire, there is one which measures forty-eight feet in girth, and contains 493 cubical feet of timber. In Warwickshire, at Croombe Abbey, a tree 200 years old measured in the diameter of its trunk nine feet six inches; diameter of its head, seventy-four feet; and is said to have stood 150 feet high. This is perhaps the loftiest tree of the species produced in England. One at Milbury Park, Dorsetshire, 200 years old, is 125 feet high. One at Strathfieldsaye is 130 feet high.

Of this species of elm, Scotland possesses comparatively few specimens, and these do not seem remarkable for old age; but generally, they rise to a greater height than the native tree of the genus. The best specimens are found at Bothwell Castle, Barnton House, and Cullen House, where they range from eighty to a hundred feet in height.

The wood is brown, hard, and of a fine grain; and as it is not apt to crack, it is adapted for the manufacture of articles that require lateral adhesion, such as the wooden furniture in the rigging of ships; and its quality, as well as its figure of growth, render it very suitable for ship keels. It is also employed by the block and pump maker, the cart-wright, and cabinet-maker; and in London it is the common wood used in making coffins. Its value varies much in different districts, and is regulated by its quality and adaptation for particular purposes; generally ranging from 1s. to 1s. 8d. per cubical foot. The tree generally attains maturity in seventy or eighty years; after which it has a tendency to become hollow in the centre, a casualty from which young trees are not exempt.

The elm, particularly the English elm, is subject to the ravages of several insects. The most fatal of these is the *scolytus destructor*, a small beetle, only about a fourth of an inch in length, which perforates the bark of large trees during the heat of summer, commonly about the month of July. It forms a channel, generally upwards, between the inner bark and the wood, into which the female insect deposits her eggs to the extent of from twenty to fifty. About the month of September the larvæ are hatched ; these ramify, forming horizontal channels on each side of the original perforation, which is commonly an inch or two in length. The caterpillars feed on the inner bark, rendered congenial by the descending sap in autumn. When the insect finishes its course of feeding, it stops in its progress and turns to a pupa, then to a beetle. It then gnaws its way straight through the bark, and emerges early in summer. The bark of the tree presents the appearance of being pierced by shot. The circulation of the sap is interrupted ; the foliage becomes scanty, and of a sickly hue, which is followed by numerous dead branches, and sometimes by the death of the whole tree. No practical cure has been discovered. In Scotland, the effects of the scolytus are seldom observed, but in England and on the Continent, and most frequently where trees ornament populous towns and frequented promenades, the ravages of the insect are most fatal, and very likely owing to the absence of the numerous birds which in forests and rural districts are sustained by such insects.

U. montana (Bauhin) : The Mountain, Scotch, or Wych Elm. —This species is a native of Scotland, and very valuable as a timber tree. It grows to a less height than the English elm. Compared with the best varieties of that tree, it is of slower growth, yields a much shorter bole, but is far more spreading in its habit, more bold in its ramification, and more picturesque in figure, more hardy in constitution, and better adapted to adorn the glens and mountain sides which it is destined to occupy. It usually attains to the height of about fifty feet, but it is often found much higher when interspersed with taller-growing trees. It does not produce suckers like the English elm, but it yields seed freely. Its propagation from seed has given rise to several fine kinds, which are propagated for their various peculiarities. In Britain none of this species have, as in the case of *U. campestris*, degenerated so far as to be unfit for cultivation as timber trees.

The blossoms of the *U. montana* appear in April, just before the leaves expand, and the seeds are usually ripe about the middle of June, when they should be gathered and immediately sown in rich, clean, friable soil. The beds should be four feet broad, and a bushel of seeds is commonly sufficient for twelve lineal yards of a bed ; but as the seeds are very often so unequal in quality that one half are not possessed of the power of germinating, the quantity on a given space must be regulated by the judgment of the individual who conducts the work, who should try what proportion of seeds are well filled, and aim at raising the plants about two inches apart. The cover on the seeds should be about half an inch deep, and in dry weather the beds should be watered and shaded. The plants often appear in a week after sowing, and then the shading is discontinued ; they require no further care but to be kept clear of weeds during the summer.

In the following winter, or early in spring, they may be removed into nursery lines ; but when they do not stand too close in the seed-bed, they are frequently allowed to remain for two summers before being removed. The distance between the plants in the lines should be six inches, and that between the lines eighteen inches. Two years is the usual time that the plants remain in the lines ; if allowed to stand much longer without being disturbed, the roots of this species are apt to get bare, after which the plant becomes stunted on being transplanted. The tree will grow rapidly, and yield heavy timber in a rich deep soil only, and it prefers an open subsoil. Where water stagnates near the surface, its growth becomes feeble, and lichen overspreads its bark.

The period at which pruning is most frequently of advantage is that when the tree is from eight to fourteen years of age. It is then apt to ramify near to the ground, and to form a very short trunk. By shortening the competing shoots and diminishing the strongest of the lateral branches with the pruning-knife, the bole may thus be lengthened into a more useful figure than that to which the tree is naturally inclined, particularly in open situations.

Few trees are more difficult to uproot than the Scotch elm: it is a rare occurrence indeed to see a tree of this species thrown over by the influence of the wind; it also resists the influence of water better than most trees. During the great flood of August 1829 in Morayshire, the river Findhorn formed a new channel through a mixed plantation, and changed its course; the only Scotch elm known of in the space continued to stand for a few years in the river, and hold its own, forming a small island, till a winter flood accompanied with ice removed it.

The timber is much in request for agricultural purposes, such as naves, shafts, rails, and frames for carts and barrows, handles for spades, forks, and other implements where strength and elasticity are required. It is also extensively used in ship and boat building. For naves, it is generally preferred in Scotland to any other wood; and when the tree is of the size adapted to that purpose, namely, from twelve to fifteen inches in diameter, it usually sells at about 2s. 6d. to 3s. per cubic foot, which is about one-fourth higher than its usual price for general purposes, and occasions the tree to be felled more frequently at the diameter stated than at any other size. The tree often yields large protuberances of gnarled wood, finely knotted by an accumulation of growth, and richly veined,—these are much esteemed for veneering, and are often very valuable.

U. m. glabra (Miller): The Smooth-leaved Wych Elm.—This is one of the fastest growing trees of the genus, and valuable for timber. It appeared about the middle of last century, and is known in some districts under the name of the Huntingdon elm. The plant is propagated by layers, but more frequently

Y

by grafting, using the Scotch elm as the stock. There are
several varieties of cork-barked elm (*U. suberosa*), also of American elm, variegated elm, and curled-leaved elm.

The most ornamental and picturesque tree of the genus is *U.
pendula* (Loudon), weeping elm, of which there are also several
varieties. This is a tree of very peculiar character. It seeds
freely, and from that circumstance, as well as from its large
leaves and habit of growth, it is believed to have sprung from
U. montana, the Scotch elm. It began to be cultivated in
nurseries about the close of the last century. Plants raised
from seeds are apt to lose the peculiarities of the species. It
is therefore propagated by being grafted on the tops of the
stems of any of the common elms. In this way it grows freely,
and soon forms a head of considerable magnitude, and of the
most wild, diversified, and rugged form. In the vigour of
youth it shoots forth in a frond-like manner, often directing
its branches horizontally, some downwards, some upwards,
and some obliquely, displaying a majesty and grandeur in its
ramifications which are never seen in any other young tree.
As an object for the lawn, the park, or the pleasure-ground,
it has no equal among fast growing plants. Its picturesque
effect resembles that of the cedar. Some of the finest weeping elm trees in England are at the Fulham and Hammersmith
nurseries, near London. The best specimen of the tree in
Scotland stands in Blackfriar's Haugh, Morayshire, about
thirty feet in height, with a trunk four feet six inches in girth
at a foot above the surface, and the spread of its branches
measures 108 feet in diameter.

XXXVII

THE WALNUT.

WALNUT.—The *Juglans* or walnut tree belongs to *Monœcia polyandria* in the Linnæan system, and to *Juglandaceæ* in the natural order of plants. The flowers of the genus are unisexual, and both sexes are produced by one plant. Several species are cultivated in the British nurseries. They are deciduous trees, of large growth, yielding valuable timber, and the common species is much esteemed on account of its fruit.

J. Regia (L.), the Royal or Common Walnut, is a native of Persia, and one of our earliest introduced fruit trees, having been in cultivation in England since the middle of the sixteenth century. It forms a large spreading tree, which blossoms in May, and ripens its fruit during the following autumn.

It is propagated from the nuts or seed, which usually part with their outer husks on becoming ripe and falling from the tree. They vegetate during the first season, and may be sown in winter or early in spring; they should be planted a few inches apart in drills, with a cover about two inches deep.

As the crop is very apt to be injured by any degree of frost after the plants appear above the surface, the drills should be protected by twigs of spruce or silver fir, or some other cover. It is necessary to raise the seedling plants in a dry early soil, otherwise they will not mature their shoots sufficiently to endure frost. Dry sandy soil, although poor, is preferable for young plants to soil of a more fertile description. In the latter the largest plants are produced, but unless the climate or the season is very superior the first winter deprives them of their tops. The seedling plants form strong tap-roots; therefore, to adapt them for removal it is necessary to transplant them when one, or at most when two years old, pruning off the extremities of their tap-roots, in order to promote that bushiness of fibre so necessary to the safe removal of plants. The same mode of transplanting should be practised every second or third year, allowing the plants additional space, according to their size, till they are placed into their final situations.

In a deep, dry soil, with a good climate, the tree in early life grows rapidly. Where it is not retarded in its progress by being removed, it usually attains the height of twenty feet in twelve years; at that age it generally begins to yield fruit, when its upward progress becomes more slow and the figure of the tree more ramified.

The vigorous tap-root which the tree naturally strikes into the ground renders it particularly suitable for some situations; wherever the surface is bad and the subsoil of good quality, the walnut is more likely to suit than almost any other tree. Its propensity to form a vigorous tap-root exceeds even that of the oak. It sometimes occurs, through the influence of wind or the sudden overflow of rivers, that spots of rich ground are rendered barren by a deposit of sand-drift or of débris. These may be turned to good account by the walnut raised from seed on the ground, and by such other trees as draw their support from a great depth. In the roughest situation the common walnut maintains a well-balanced head, and when it is raised from seed, and not transplanted, it is less likely to be uprooted by wind than almost any other tree. Striking deep into the ground, its roots do not obstruct the cultivation of the fields, and it is perhaps the most common hedgerow fruit-tree throughout the Continent. Cultivated, however, for the sake of its fruit, in most situations it is all the better of being transplanted. Trees that have never been removed seldom ripen their fruit so early in the season as those that have at some period undergone the check of transplantation. Frequent removal has also the effect not only of making the trees bear sooner, but it improves the quantity and the quality of the fruit. This is on account of the roots of transplanted trees ranging nearer the surface of the ground, and more under the influence of sunshine. Trees that are of a great age are also found, in a late climate, to ripen their fruit better than young trees in the same locality, unless the latter have been frequently transplanted, which has the effect of occasioning their sap to flow less copiously, and imparts a ripening tendency.

Large quantities of fruit are yearly imported into Britain from France and Spain, to an extent varying according to the crops in the respective countries. It is recorded in the history of this tree that the winter of 1709 was very fatal to the walnut throughout Europe, and occasioned so great a scarcity of its timber that in 1720 an Act was passed in France to prevent its exportation, and the extensive propagation of the tree was encouraged by the French Government.

Throughout Britain the walnut attains to the size of a large timber tree. It is of great duration, and, when grown in sufficient space, in old age it presents a picturesque and elegant form, with ramifications resembling that of the oak. Many fine specimens of the tree are to be found throughout Britain. In the north of Scotland it attains a greater size than any other fruit-tree. At Gordon Castle, Banffshire, it stands sixty-six feet high, with a trunk eleven feet in circumference at three feet from the ground, and yields fruit abundantly. At Altyre House, Morayshire, a tree has attained the height of sixty-two feet, with a trunk four feet in diameter, and contains about 200 cubical feet of timber. The soil is a deep sandy loam incumbent on gravel. It yields large crops of fruit, which ripen almost every year. There are other trees in Morayshire of nearly similar dimensions, but on account of the soil and situation which they occupy being somewhat later, it is only in very favourable seasons that their fruit becomes fit for dessert.

The timber of the common walnut, when young, is white and soft, but as it advances in age it gets more dark and solid, and ultimately it becomes shaded and veined, of a light brown and black; and generally the pieces most ornamental are found in the vicinity of the root. It soon becomes seasoned, when it neither cracks nor warps, and it is reckoned the most ornamental of European timber, and much esteemed for cabinet-work, gun-stocks, etc.

The fruit yields a valuable oil, which, in the south of Europe, is much used in culinary preparations and at table, in medicine, and for burning in lamps. It is also used to impart a polish and lustre to timber, and by artists in fixing white and delicate colours. The mass of husks which remain after the oil is expressed is used as food for swine, sheep, and poultry. The fruit in a green state is frequently pickled and preserved.

The roots of the tree, by boiling, yield a valuable dark-brown dye, which becomes fixed in wood, wool, or hair, without the aid of alum. There are several varieties of this species of walnut, which are cultivated chiefly on account of the size, early maturity, thinness of shell, or some other estimable property of their fruit. They are increased by grafting, budding, or inarching on the common tree.

J. nigra (L.), the Black Walnut of America, was introduced into Britain about the middle of the seventeenth century. Its leaves are about twice as long as those of the common walnut, and are composed of six or eight pairs of opposite leaflets, with a single or terminal leaflet; and, as in the common species, the leaves emit a strong aromatic odour. In its native country this tree attains a great size, being sometimes found 100 feet high, and of proportionate diameter. It is readily raised from seed, which is commonly imported. The mode of cultivation is the same as that detailed for the common species. In England it advances with all the vigour which characterizes the tree in its native country; but as a fruit-tree it is inferior to the common species, and being later in ripening, it is only suitable for being grown as a timber tree, or in an isolated situation on the lawn, where, in good soil, it becomes a large spreading tree of great beauty. The timber is of a dark colour, finely grained, susceptible of a high polish, and in great reputation in the manufacture of the finer articles of furniture.

J. cinerea (L.)—The grey walnut tree is a native of America. It is propagated like the other species, and in appearance bears a striking resemblance to *J. nigra.* It is less common than that tree.

All the species of American hickory are related to the walnut, and comprehended in the same natural order. They are of no value in the climate of Britain.

XXXVIII.

THE MAPLE.

THE MAPLE (*Acer*).—This genus composes almost entirely the natural order *Aceraceæ*, and comprehends upwards of twenty hardy species which are cultivated in Britain (natives of Europe, America, and of India), besides a great number of kinds which are too tender to endure the climate of Britain. It is believed that there are many valuable hardy kinds of the maple, which flourish on the lofty mountains of India, in Japan, and China, which have not yet been introduced into this country. Some of our cultivated species consist of distinct and interesting varieties. Two or three of the species attain the full size of timber trees, and the growth of the others ranges from that of trees of a medium size down to the stature of shrubs. The whole genus is remarkably handsome. Some of the species grow rapidly at an early age in almost any soil, yielding fine green smooth shoots. Many kinds are interesting on account of their flowering early at the time of the expansion of the foliage ; and from their elegantly-lobed leaves, of the finest texture, which in autumn furnish the most exquisite tints of every shade of yellow and scarlet, they are much esteemed in ornamental plantations.

A. pseudo-Platanus (L.): The Mock Plane-tree, or Sycamore. —This tree is the largest and most common, although certainly not the most ornamental, species. It is a native of Switzerland, Germany, Austria, and Italy, where it is found associated with other trees in hilly situations. It was introduced into Britain about 300 years ago, and it appears to have been one of our first cultivated timber trees. In Scotland, it generally marks the first spots of reclaimed land throughout the country. Up to the present time, few deciduous trees are found better adapted for standing singly in a rough and exposed situation. Regardless of the prevailing winds, it generally rises with a large and well-balanced head, valuable in affording shelter for cattle; and, from its yielding a deep shade, it has been recommended as suitable for being planted on the south side of the dairy, in order to equalize the temperature during the heat of summer. It is also well adapted to withstand the injurious effects of the sea spray.

This species is easily propagated by seeds. It blossoms in spring, and the seeds become ripe early in the following autumn, when they should be collected and mixed up in a pit of dry earth or sand, and sown during the succeeding spring. When sown in autumn, immediately on being gathered, the young plants generally appear too early in the season to escape the influence of frost, and are consequently destroyed. The ground for seed should be dry and well pulverized, and only moderately rich. Very fertile, moist soil creates an excess of growth, so that the seedling plants do not mature their wood to resist the effects of frost,—a circumstance which is only likely to occur to plants of one year's growth. One bushel of seed is sufficient to sow a bed four feet broad and twenty-four yards in length. The covering of the seeds should be half an inch deep.

Seedlings of one year's growth should be transplanted into nursery lines two feet apart, and the plants six or eight inches asunder in the lines, where, in two years, they are generally from three to four feet high, and suitable for forest planting. Although the sycamore will grow in soil of very opposite qualities, yet a deep, soft, dry soil is that most congenial to its development, and in such it is generally twenty feet high at the age of ten years; but instances are known of the tree attaining the height of forty feet in less than twenty years.

It comes into leaf early in the season, when its foliage presents a lively green, particularly attractive in the month of May; but from the circumstance of the leaves exuding a glutinous or clammy substance, which retains dust and every impurity of the atmosphere, the foliage soon becomes dingy, and the tree loses half its charms.   One of the finest trees of this species stands at Burgie Castle, Morayshire, with a trunk upwards of thirteen feet in circumference, and the diameter of its top, or spread of its boughs, measures ninety feet.

Not only was the sycamore well suited to the soil and climate of Scotland, but its timber, which is white and closely grained, was well adapted for the manufacture of household vessels, such as bowls, cups, and other articles of turnery in general use during a bygone age, but now superseded by earthenware.   The timber is susceptible of a fine polish, is not apt to warp, and is employed as moulds and pattern-blocks in manufactories, stair rails, and like purposes, and commonly sells at very various prices in different districts.

Norway Maple—A. Platanoides (L.)—This species is quite hardy, and when young its rapidity of growth exceeds that of the sycamore, although ultimately it does not attain the dimensions of that tree.   Its foliage is of a fine form and texture, of a glossy polish, which retains a lively green throughout the summer.   In autumn the leaves assume various tints, in which yellow prevails.   It luxuriates in a deep, well-drained soil, and its mode of propagation and sub-sequent treatment should be the same as that recommended for the sycamore.   This species includes several distinct varieties, conspicuous among which is the cut-leaved or eagle's-claw maple, a very ornamental tree, which is readily propagated by grafts or buds, using the common sycamore as the stock.

The Striped-Barked—A. striatum (L.)—or snake-barked maple is very ornamental at all seasons.   The bark is longitudinally marked with black and white stripes.   It is indigenous to North America, and generally attains a height of from twenty to thirty feet.   It is sometimes propagated from imported seeds, but more frequently by being grafted on the common sycamore.   Its wood is white, and esteemed in cabinet-work.

The Sugar Maple—A. saccharinum (L.)—This tree forms extensive native forests in North America, throughout Canada, New Brunswick, and Nova Scotia. The tree has been known in Britain for upwards of 100 years, where it seldom attains a height beyond forty feet. In its native districts it frequently reaches the height of from sixty to seventy feet; but the diameter of its trunk is commonly small, generally ranging from twelve to eighteen inches. In appearance the tree is very ornamental; it resembles the Norway maple, but its leaves are more glaucous, or white, underneath, and in autumn they assume a rosy tint, which adds richness to the variety of foliage peculiar to that season. In Britain the tree is propagated from imported seeds, which are treated like those of the sycamore, but as the sugar maple is more tender, it requires a dry and sheltered situation.

In its native country, during the opening of the season, the tree, when pierced, yields a copious flow of sap, which is readily converted into sugar. The flow of sap is said to be more abundant during a hot sunshine which has been preceded by a frosty night, and particularly if the surface of the ground is covered with snow. The sap richest in saccharine matter is that extracted from trees which occupy isolated situations, or are somewhat detached from the density of the forest, where they enjoy the full influence of sunshine and air.

In Britain, sugar has been extracted from this and several other species of the maple; but it is not believed that in this country it can be cultivated profitably. Numerous samples of maple sugar were produced at the Great Exhibition of 1851, and several prizes were awarded for this article; but valuable as it is to the settlers in a district where the tree is indigenous, it is not expected that, as an article of commerce, it will extend far beyond the locality in which it is manufactured.

The wood of this maple is esteemed in cabinet-making; it assumes a rosy tinge, takes a fine polish, and has a silken lustre, but it is not durable if exposed to moisture. In old trees, it sometimes displays rich undulations of fibre, and yields gnarled and spotted timber, much esteemed as veneers, etc.; and known, together with the knotted timber of the common field maple, *A. campestris*, under the popular name of Bird's-eye maple. Its timber is esteemed as fuel. Its charcoal has the highest reputation; and its ashes, abounding in alkaline principles, are said to furnish more of the potash of commerce than any other American tree.

The large-leaved maple—A. macrophyllum (Pursh)—is a native of North America; hardy, and of rapid growth, it attains to a great size. It is very ornamental, and yields timber beautifully veined and valuable. It was introduced into Britain in 1812, and is not yet in general cultivation; it is readily propagated by layers.

The red or scarlet maple—A. rubrum (L.),—a native of North America, is a low-growing tree, ornamental in its development of red blossom, late in spring or early in summer. It luxuriates in a rich soil, and endures moisture better than any other species. When old, it sometimes produces timber very valuable, owing to its curled and undulating fibre yielding a display of light and shade rarely surpassed by any other wood. It is difficult to raise this species from seed, therefore it is commonly propagated by layers.

A. circinatum (Pursh).—This is a recently introduced hardy tree of the Oregon, where it forms impenetrable thickets, rising to the height of from twenty to forty feet, with pendulous branches, yielding leaves which, in autumn, surpass the brilliancy of the finest scarlet oaks.

A. villosum (Wallich).—This is a hardy maple, recently introduced by Osborne and Co., of the Fulham nursery, from the Himalayas, where it is said to attain a great size, and to possess an appearance much superior to the sycamore. Speaking of it in its native country, Dr. Royle says, it is only seen with pines and birches on the loftiest mountains, which are for many months covered with snow.

The other species of this genus, although very ornamental, are of smaller growth.

A. Negundo (L.), or Ash-leaved Negundo, is a native of the United States and of Canada. Although it has been introduced and cultivated in Britain for nearly two centuries, it is a rare tree in Scotland. It grows very fast when young, but it seldom exceeds thirty or forty feet in height, and in late seasons its vigorous shoots are apt to be killed back by frost. It has lately given rise to a new variety, *A. N. variegata*, which on account of its bright white foliage is very attractive, and is much sought after. It is readily propagated by grafting on any common maple. It is purely an ornamental tree.

XXXIX

THE LIME TREE.

LIME TREE.—The *Linden*, or *Teil-tree*, is the *Tilia* of botanists. It belongs to *Polyandria monogynia* of the Linnæan system, and it is the type of the natural order *Tiliaceæ*. Some botanists have divided the genus into two species, which embrace a number of distinct varieties, while others consider these possessed of characteristics sufficiently distinct to constitute species.

Tilia Europæa (L.), or common Lime, is the principal tree belonging to the genus. It is a native of the north of Germany, Russia, and Sweden. It is also found wild on the Alps of Switzerland, and in the north of Italy, in Spain, in Portugal, and in Greece—varying considerably according to the soil and climate which it occupies. It is stated to be indigenous to England, and found wild in some parts of the counties of Kent and Essex. But whether a native or foreign tree, it does not shed its seeds and spring up in uncultivated ground as indigenous plants generally do, which has occasioned some doubt as to its being really a native of Britain.

It affects a good climate and a rich alluvial or loamy soil, and the extended cultivation of the districts congenial to its support may counteract the natural growth or spread of the plant. It is found, however, that seeds are only ripened in the best seasons, and on trees most favourably situated. It is unsuitable for bleak situations or dry poor soils. Its chief use is to form an embowering shade along the avenue, and as a park tree or lawn ornament; and in towns throughout the Continent of Europe the tree is planted in lines along the streets and public promenades. Its blossoms expand in July, and are fragrant, particularly in hot weather. Where heat is reflected by pavements and buildings, its odour is strengthened and its shade is rendered exceedingly desirable. It attains to a great size in a short period.

Though destitute of the wild ramifications and picturesque character of the oak or Scotch elm, it forms a vigorous, pliant, well-balanced tree, with a great number of lateral branches of an easy and graceful habit, and generally accords well with the meadows and cultivated grounds with which it is frequently associated.

In favourable seasons the seeds are ripe in autumn, and may be sown in winter or early in spring. The plants come up in the ensuing summer. Where several kinds of lime trees stand together, and blossom at the same time, the seeds readily become hybridized, and produce various sorts, although they are gathered from one tree. The growth of plants from seeds, however, is seldom practised; even when the seeds can be had fully ripe, their progress is slow compared with that of layers, which always perpetuate the exact variety of the parent tree.

The method of raising plants from layers is therefore generally practised by nurserymen. When topped over at the surface of the ground, the plant readily produces a number of young shoots, and is easily formed into a stool. The young shoots are bent down into the earth to the depth of three or four inches, with their extremities placed in an upright position, which form the young plant. The laying of these shoots may be performed in winter, or early in spring, and the plants will become rooted and ready for removal by November following, when the young shoots, the produce of the preceding summer, should be inserted into the ground into the same position as described, to form another crop of young plants, and so on, removing a crop and laying down another yearly. It is often necessary to manure the stools by adding a few inches in depth of rich compost or vegetable mould, and in some soils, destitute of a considerable propor-

tion of silex, it is of advantage to mix a quantity of sharp
sand with the soil.   After the third year's produce a healthy
stool will furnish from sixty to seventy plants yearly.   The
plants on their removal are generally about two feet high ;
they should be transplanted into nursery lines about two and
a half feet apart, and the plants about fifteen inches asunder.
Having stood two years in lines, they are usually six feet
high, and fit for being finally planted out.   But the lime
admits of being grown in the nursery to a much greater size,
and transplanted in safety, provided it is removed every
second year, which has the effect of preserving its roots in a
fibrous or bushy state ; as it advances in size the space in the
nursery lines should be increased, in order to preserve the
proportion of the plant, and when such are finally removed
they command an immediate effect.

Not only is the tree adapted for the avenue, the lawn, or
park scenery, but it forms a thicket in the belt or screen fence,
where it is frequently serviceable to an agriculturist.   It admits
of being frequently pruned, and although destitute of foliage
in winter, it becomes close and twiggy.   The ordinary progress
of the tree in rich, sheltered soil is about two feet yearly in
height for the first fifteen or twenty years, after which its
progress is greatest, where it has space, in adding diameter to
the trunk, and expanse to the lateral branches.   The British
specimens of the tree seldom exceed the height of eighty feet,
and in open situations it maintains nearly the same diameter
in the spread of its branches, from the surface upwards to a
great height.

Of the recorded trees of this species, perhaps the largest is
that which has been beautifully portrayed by Mr. Strutt.
It stands at Moorpark, in Hertfordshire.   Nineteen large
branches, six or eight feet in girth, strike out horizontally
from sixty-five to seventy feet in length, and these support
three or four upright limbs.   The tree is in full vigour, and
its branches droop down and root on the ground.   The trunk
girths upwards of twenty-three feet ; the head is 122 feet in
diameter, and nearly 100 feet high ; and the tree contains

875 cubical feet of timber. At Cobham Hall in Kent a tree stands ninety-seven feet high, and nine feet in diameter.

One of the largest and handsomest trees of the species to be met with in Scotland stands on the lawn at Gordon Castle, with a head of nearly 100 feet in diameter. Happily situated in a congenial soil, it has become a lofty tree, with branches depending to the surface of the ground. It displays a mountain of foliage of the finest form and texture, with a trunk upwards of sixteen feet in circumference ; and the magnitude and beauty of the tree harmonize with the scenes by which it is surrounded, and recall to memory the passage in Landor's *Conversations,*—" Old trees in their living state are the only things that money cannot command. Rivers leave their beds, run into cities, and traverse mountains for it ; obelisks and arches, palaces and temples, amphitheatres and pyramids, rise up like exhalations at its bidding ; even the free spirit of man, the only thing great on earth, crouches and cowers in its presence. It passes away and vanishes before venerable trees. What a sweet odour is there ! Whence comes it ?—sweeter it appears to me, and stronger than the pine itself." " I imagine," said he, "from the linden." " Yes, certainly." "Oh, Don Pepino," cried I ; " the French, who abhor whatever is old, and whatever is great, have spared it. The Austrians, who sell their fortresses and their armies, nay sometimes their daughters, have not sold it. Must it fall ? Oh, who upon earth could ever cut down a linden ? "

The wood of the lime tree is very soft, of a pale yellow or white colour, very light, and not apt to be attacked by insects. It is esteemed in preference to any other kind for carving, and shoemakers, glovers, and saddlers prefer it to every other kind for cutting leather upon ; and for this purpose it is frequently retailed in planks by leather-merchants throughout the country. It is used at foundries in forming moulds, and is appreciated on account of being easily worked, and for producing a fine surface, and it is not subject to warp. In the manufacture of gunpowder its charcoal is said to be inferior only to that of the hazel. It is also used by architects in forming models of buildings. The demand for these purposes being irregular throughout the country, the price of the timber varies exceedingly.

The inner bark of the lime furnishes material for the manufacture of mats, which forms an important article of commerce, and is extensively used in gardening, and in the covering of packages in general. The manufacture is chiefly confined to Russia and Sweden. The tree is cut immediately after the ascent of the sap, when the bark is readily removed. It is then macerated in water till it divides freely into layers or strands. These are then formed into mats, ropes, nets, and coarse cloth; and iu Russia the outside bark of the tree is frequently employed as tiles for covering the roofs of houses.

Loudon states that the honey produced by the lime-tree blossoms is considered superior to all other kinds for its delicacy. It sells at three or four times the price of common honey, and is used exclusively for medicine, and in the manufacture of liqueurs. It is procurable only at the little town of Kowno, on the river Niemen, in Lithuania, which is surrounded by extensive forests of the tree.

T. E. microphylla, the Small-leaved European Lime, is very distinct from the common tree, and easily recognised when in leaf. Its foliage is much smaller, and at first sight it has a striking resemblance to that of the *Populus tremula*, or trembling poplar. Its blossoms, although smaller, are more abundant, and have a stronger perfume than those of the common species. The tree, though very rare in this country, abounds in the Duchy of Nassau in Germany, and is perfectly hardy. It forms the principal tree embellishment in the avenues at Schwalbach. It is of dwarf habit, and does not appear to attain a height beyond that of thirty feet. Its profusion of blossom when I saw it obscured its foliage, and presented one mass of cream-coloured inflorescence, which continued unfading for a month, accompanied with a constant hum of bees. In the Continental nurseries it is trained with clean stems to the height of six, eight, or ten feet, and the tree naturally forms a round, bushy head. It is of little value as a timber tree, but seldom equalled for embellishment.

T. Americana (L.), the American Lime tree, is of a more robust habit than the European tree. Its leaves are larger, of a dark-green colour, cordate, acutely-pointed, and generally smooth and shining. In Britain it is a month later than the common tree in expanding its blossoms. Its twigs are of a dark-brown colour, and the branches of the young trees generally take a wide range. In its native country the tree commonly attains the height of eighty feet. It has never become very common throughout Britain. Like the European species, it affects a rich, loose, deep soil ; on such it flourishes on the borders of Lake Erie and Lake Ontario. The species contains several varieties, and their mode of propagation and treatment is the same as that stated for the common tree. The timber of both species is very similar; and the images affixed to the prows of American vessels are often carved on the timber of the *Tilia Americana*.

XL.

THE ASH TREE.

ASH TREE: *Fraxinus excelsior* (L.)—The tall or common ash belongs to the natural order *Oleaceæ*, and some species of the tree is to be found in all quarters of the world. The common ash is a native of Britain, and is one of the most important timber trees adapted to the climate of this country. It attains to a great size, reaching in fine specimens to about 100 feet, with a trunk four to five feet in diameter of sound timber. It luxuriates in rich deep soil of various qualities, inclining to moisture rather than drought. The best specimens are commonly met with in glens and valleys, but sometimes at considerable elevations where the soil is rich. The tree is very hardy, breaking into leaf late in the season, and shedding its leaves earlier than most other trees. Although it often possesses a very elegant figure, and forms during summer a desirable object in lawn or park scenery, yet for the greater part of the year its want of foliage renders it less ornamental than many other trees. The ash-blossoms appear in April, a few weeks before the development of the leaves; and the seeds have often a prominent but not ornamental appearance for months after the leaves are shed. In growing the tree from seed it is apt to sport, that is, to produce varieties; sometimes the tree yields male blossoms only, other trees produce only female blossoms, but for the most part both sexes or hermaphrodite blossoms are produced. The male trees are the handsomest, and are generally recognised in autumn by the absence of seed, the profusion of which gives the other variety a faded appearance at that season of the year. The foliage of the male tree is commonly most abundant and glossy, and is often retained to a later period in the autumn. These qualities render the male tree the most suitable for embellishment as a park or lawn tree, but the variety can only be produced with certainty by engrafting.

Viewed as a landscape decoration in summer, the ash is nearly perfect. Its whole outline is easy, and, in a good specimen, we look in vain for anything lumpish or rigid. It may sometimes be deficient in grandeur, but this is amply counterbalanced by its gracefulness. We are wont to dwell with pride on the stubborn massiveness of some trees ; but an attentive observer of nature will have noticed that there are times when this unbending framework has its disadvantages. When the wind is up, the ash acquits itself much better than the oak. The one frets and flutters, the other sways itself easily about, and appears to have really a rejoicing tone with it.

In Scotland perhaps no tree is more frequently employed as hedgerow timber ; this may be attributed to the important purposes for which the timber is adapted in agriculture ; but it certainly is employed at a great sacrifice : its roots spread near the surface and impoverish the soil to a great extent, and its top ramifies where it has space into an unprofitable shape, unless early and frequently attended to by pruning, which is very seldom the case ; for although the tree is often seen of a great size, it is rarely met with to a great extent possessed of a profitable figure in any other than close plantations.

The tree is propagated by seed, which becomes ripe in October or November, and is generally yielded in great abundance. Care should be taken to make choice of seed from trees free from canker, with a clean bark, vigorous and freely grown. When gathered they should be laid in a pit made in a light friable or sandy soil; with this soil they should be intimately mixed up, at least equal to the bulk of the seed, and allowed to remain in this state, like haws, for fifteen or eighteen months, with no further care than that of turning them over every three or four months; this brings them forward to the end of February, when they should be sown in open dry weather, then, or early in March.   If sown a month or two earlier the tender plants would appear too soon, and be apt to suffer from frosts ; if sown a little later, the seeds are apt to vegetate in the pit, which occasions the production of crooked-rooted plants.[1]

In the second February or March, then (about fifteen months after the seeds are gathered), the seeds should be taken from the pit and sown in good loamy friable soil, into beds four feet wide.   It will suit if they are made to follow turnips, potatoes, carrots, or any crop which had a fair quantity of manure applied to it.   The beds should be opened up by a deep cuffing, and the seeds should be spread with a spade, and afterwards regulated with a rake, so that on the average each seed may occupy about two square inches of surface space, thus allowing from eight to nine hundred plants to occupy the lineal yard of the bed.   After the seeds are sown the ground should be either rolled or beat with the back of a spade, after which the cuffing should be drawn on covering the seed fully half an inch deep.

Another mode of laying down an ash crop sometimes practised, which may be most convenient to those who grow it on a small scale, is to sow the seeds at the time they are collected, or the spring thereafter, allowing the beds to remain the first summer either vacant, or growing any light vegetable crop, such as garden-turnips, radishes, cabbage-plants, etc. In this case the young.ash plants are apt to appear before the

---

[1] It may be proper to notice here a great mistake, very common in books giving directions on the growth of the ash from seed, which is calculated to mislead the inexperienced cultivator.   Thus London (*Arboretum Britannicum*, page 1224) directs that the seeds, which are ripe in October, should be taken to the rotting heap, where they should be turned over several times in course of the winter, and in February they may be removed, freed from the sand by sifting, and sown in beds of any middling soil.   "The plants," he continues, "may be taken up at the end of the year, and planted in nursery lines."   This mistake appears to have been copied by subsequent writers on the ash tree. Now to an experienced grower this error is harmless, because he knows that at the end of the year the plants will not be in existence ; but any other, expecting a crop as stated, concludes that the seeds are lost, and crops the ground with something else.

frosts of the succeeding spring are over, when they require to be protected by any light covering, such as that of straw, fern, or the twigs of firs or other evergreens. When the seeds are rotted in the pit for fifteen months, and sown about the last week of February or early in March, they are later in appearing above ground, and are generally exempt from injury by frost. If the ground is rich the plants will be six or seven inches high the first year, and may then be transplanted ; when of a less size, and thin enough, they had better be allowed two years in the seed-bed, after which they should be transplanted into lines eighteen inches apart, and the plants six inches distant. When they have been two years in this situation they generally stand about three feet high, and are fit for ordinary forest planting, but when required of greater strength they should be again transplanted in the nursery, with space allowed in proportion to the height desired.

The most approved method of growing ash timber is to plant the tree by itself in a congenial soil. It being naturally a loose-headed open object, and inclined to ramify, it is then apt to form straight clean timber, and to require less pruning than when it is interspersed among other trees. The soil it delights in most is a sweet hazelly loam of considerable depth, and the situation it prefers is the bottom of a hill or slope near to a river, where, without being saturated with water, its roots may yet have access to moisture in the heat of summer. In the case of some trees we look for an advantage after slow growth in having hard and durable timber. Not so with the ash ; its growth should be hastened by good soil and a somewhat sheltered situation ; for it is always found that when stopped midway in its career, either by its roots reaching a sour and wet subsoil, or by the poorness of the land, the wood becomes brittle and shaky. In the case of large plantations, therefore, where a variety of soils exists, it is always well to remember that the properties of the ash in perfection—strength, toughness, and elasticity—are the result of a free and unimpeded growth.

The final situation of the ash should therefore be carefully chosen; and nothing gives the plant a more rapid start than trenched soil. In firm or hard soil, however, large pits are made; the roots on reaching the undisturbed ground make little progress compared to their vigour in land trenched two feet deep. In cases where the soil is wet it must be drained, for although the tree is partial to moisture, no plant is more readily injured by stagnant water. In mossy soil it does not attain to a great size; but on such it grows and forms valuable coppice-wood. The time for planting is from October to the middle of April. In ordinary sheltered ground the ash should be inserted in holes in the trenched land, about five or six feet apart. Early thinnings are more valuable than that of most other trees. Where the ground is high and exposed, nurses may be introduced, either as a mixture, or along the margins of the most exposed parts. The nurses may be the pines, larch, or alder, etc.—these to be thinned out as circumstances require, giving the trees uninterrupted possession of the ground after they have become established, which is usually ten or twelve years after being planted. At this period, pruning, so far as to prevent a plurality of leading shoots, is of much advantage, and is generally but little required in a healthy plantation of this tree at any other time. The ash was planted extensively by the late Earl of Leicester at Holkham. It there occupies various soils and situations, and the result illustrates in a striking manner how much judgment is necessary in fixing for it a permanent site. Some of the trees when fifty years old contained only thirty cubic feet, whilst others are said to have contained seventy-six.

Important uses belong to ash timber. The manufacturer of agricultural implements employs it in the construction of carts, waggons, carriages of all sorts, ploughs, harrows; and it is specially adapted for the handles of rakes, forks, spades, shovels, mattocks, and the construction of dairy utensils. It is used for important purposes by the millwright, and it is the chief wood of the coachmaker and wheelwright. The cooper makes his hoops of it, and the fisherman his oars, and it is unrivalled in all purposes where strength and elasticity are required. As it springs readily from the root, it is an excellent tree for yielding coppice-wood, which is adapted for various important purposes at various ages, such as hoops, crate-work, hop-poles, and tool-handles.

The ash rarely attains to the height of 100 feet.[1] Some of the best specimens of the tree in this country stand on the western extremity of the county of Moray, at Earlsmill, near Darnaway Castle, at Brodie Castle, and at other places, ranging from sixty to eighty feet high, and at three feet above the surface in girth from fourteen to seventeen feet. One of the largest ash trees in the county of Moray was lately blown down in the parish of Duthil. It measured upwards of twenty feet in circumference—a great girth for a hardwood tree in the Highlands. Throughout Scotland ash has been planted only to a very limited extent of late years, compared with what it was in the early part of the present century. In some districts ash timber is becoming scarce, and for many purposes larch is employed as a substitute. The price of ash timber varies much in different localities, according to the quality and distance from market. It arrives at maturity at various ages in different soils, and seldom improves after seventy or eighty years, when it generally becomes short in the grain, or brittle. The usual price ranges from 1s. 6d. to 2s. 6d. per cubic foot. That which is rapidly grown in close woods, clean, and free of knots, is always the most valuable.

[1] The lately published statistics of old and remarkable trees throughout the country give a startling account of the incredible height attained by ash trees in several counties : some are stated from 100 to 110, and even up to 140 and 160 feet high ! It is to be regretted that these measurements were not taken with greater exactness, as it renders of no value the report, which would otherwise have been useful and interesting.

XLI.

THE PLANE TREE.

THE PLANE TREE.—*Platanus* belongs to the natural order *Platanaceæ*, and to *Monœcia polyandria* of the Linnæan system. The genus comprehends only two species, the Eastern and Western Plane, which are cultivated in Britain chiefly as ornamental trees. This tree is quite different from the *Acer pseudo-platanus*, the great maple or sycamore, which in Scotland is popularly termed plane tree (for a description of which see MAPLE).

Of all the broad-leaved deciduous trees which grow in the climate of Britain, there is perhaps none which yields foliage more beautiful than the plane.

The seeds are formed in round balls, and are suspended from the branches by slender thread-like stalks, which form a singular but graceful feature in the tree at all seasons of the year. Both species attain to a very great height in their native countries; and some of the more favoured spots throughout Britain contain samples of both kinds, which in magnitude and beauty are not surpassed by the best specimens of our timber trees.

The great objection to their general cultivation is the change of weather, which, frequently occurring at the opening of the season, is apt to destroy the leaves immediately after the expansion of the buds; and unless the soil is early and the situation warm our summers are hardly sufficient to mature the young wood so as to endure the frosts of winter.

P. Orientalis (L.)—This species is a native of the east of Europe and the west of Asia. It abounds on the banks of the Grecian rivulets, on the coast of Asia Minor, and, according to Royle, it extends southward as far as Cashmere. It was introduced into Britain about the middle of the sixteenth century. The branches of the tree are wide-spreading. The leaves are five-lobed, palmate, with the divisions lanceolate. On young, vigorous shoots they are frequently upwards of a foot broad, and ten inches long; but in old trees of less vigour they are not much beyond half of these dimensions. The tree blossoms in May, and in favourable seasons ripens its seeds in October. The beauty of the tree is celebrated in the earliest records of Grecian history. Herodotus and Elian tell us that when Xerxes invaded Greece, he was so enchanted with a beautiful plane tree in Lycia, that he encircled it with a collar of gold, adorned it with jewels, necklaces, scarfs, and infinite riches; confided the charge of it to one of the ten thousand, caused a figure of it to be stamped on a medal of gold; and, by compelling his whole army to encamp in its neighourhood for days, occasioned a delay which was the cause of his defeat. The embowering shades which were formed around the schools of Athens, the groves of Epicurus, the shady walks planted near the Gymnasium and other public buildings, the groves of Academus, in which Plato delivered his celebrated discourses, were all formed of the plane tree; and the enthusiasm of the Greeks and Romans is said to have been more extravagant in the cultivation of this than of any other tree. All travellers in the East have been enchanted by its grandeur and gracefulness. Lamartine says that near Constantinople three-decked vessels are constructed and launched into the sea under the very shadow of the plane. It is the most gigantic vegetable production on the banks of the Bosporus, embellishing the beautiful meadow of Buyukdere with trees of immense size, "one of which would overshadow a regiment." A distinguished tree at this place is recorded by many writers, the trunk of which presents the appearance of seven or eight trees, having a common origin, like that of shoots from a stool. Some of the trunks proceed from the surface, others as high as seven or eight feet; the circumference at the base is 141 feet, and its branches extend over a space of 130 feet in diameter, which, if it can be considered a single tree, is probably the largest in the world; and it possesses additional interest, from the circumstance that, as De Candolle conjectures, it must be upwards of two thousand years old.

In Britain the tree is sometimes grown from seeds, that are produced in round balls, which should be broken and the seeds sifted, to separate them from the cottony substance with which they are mixed. They should be sown in March. They require scarcely any cover of soil, but should be pressed into the surface of the ground, and kept moist by being covered with leaves or spray of trees. But the most speedy method of propagation is to grow the plants by layers, in a similar manner to that recommended for the LIME TREE.

The plane is of rapid growth, and young plants are frequently four feet high when only one year transplanted from layers, when they may either be planted out or nursed another year or two, and then removed.

It requires a deep, rich, soft soil; and it is generally found that in all places where the tree has become of a remarkable size, its roots have had access to water. The tree also requires shelter, without being crushed or confined. It will not luxuriate at a great altitude; but in alluvial soil, along the valleys, and near the banks of streams, the beauty and stateliness of the tree are worthy of every care.

As a tree for bordering and shading public walks, streets, or promenades, where the soil and climate are favourable, it has no equal. Its branches range horizontally, and contain imbricated masses of large foliage, which, from their construction, are favourable to the admission of air, and exclude both sunshine and rain; its canopy of foliage is of a finer green than the foliage of any other large tree; and from the motion of its large leaves, which are easily moved by the passing breeze, it often displays during sunshine what are called "flickering lights" throughout its excursive top. In Britain, the tree attains the height of from seventy to ninety feet, and generally is of rapid growth, according to the soil and climate.

At Mount Grove, Hampstead, a plane about 100 years old

is upwards of seventy feet high, with a trunk from four feet
four inches, and a head upwards of ninety feet in diameter.
Its timber, when young, is of a yellowish white ; in an old
tree it assumes a brownish gloss, of a fine grain ; it takes a
high polish, and is esteemed in cabinet-making.

Several specimens of this tree are found in Morayshire
associated with other trees in hedgerows standing on soil of
ordinary quality ; compared with the Scotch elm, ash, and
sycamore, they are about one-fourth less in height and in
cubical contents, which appears to arise from the plane failing
in late seasons to ripen the extremities of its young shoots,
which, to some extent, are killed back in the succeeding winter.
At Gordon Castle the tree has attained the height of sixty-
eight feet, with a trunk three feet in diameter.

P. Occidentalis (L.), the Western Plane, is a native of North
America, abounding in the fertile valleys on the banks of the
Ohio and its tributaries, and along the great rivers of Pennsylvania and Virginia. It was introduced into Britain about
1630. Its appearance is very like that of the Oriental species;
the leaves are large, thin, angled, lobed, soft to the touch, and
somewhat downy underneath, with the fruit-balls much
smoother than in the other species. The bark scales off in
longer pieces than in the Oriental tree. In common with the
other species, its young shoots proceed in a zig-zag direction,
and in cold weather in May it assumes a scorched appearance ;
but as soon as steady summer weather comes on, it gradually
furnishes itself with the rich garb of green for which it is so
celebrated as a park ornament or embowering shade. It grows
from cuttings, but it is more frequently propagated by layers.
The treatment suitable for this is similar to that recommended
for the other species. This is a more rapidly growing, but a
less hardy tree ; its vigorous shoots seldom become matured
to the extremity, and consequently they die back to some extent from the effects of frost. Yet the best specimens of plane
throughout Britain are of this species. At Croome, in Warwickshire, it attained in seventy-five years to the height of
100 feet, with a trunk three and a half feet in diameter in

that time; and it frequently attains the height of thirty feet in fifteen or eighteen years. A tree of this species in the palace garden at Lambeth is recorded to have grown, near a pond, to the height of eighty feet in twenty years. Perhaps the tallest tree of the species in Britain is one at Chelsea Hospital Gardens, where it extends its roots towards the Thames; it is 115 feet high, with a trunk five feet in diameter at a foot above the ground.

XLII.

THE HORSE-CHESTNUT.

THE HORSE-CHESTNUT (*Æsculus*) is the type of the natural order *Æsculaceæ*, and it belongs to *Heptandria monogynia* in the Linnæan system.

Æ. Hippocastanum (L.): The Common Horse-Chestnut.—The native habitat of this tree remains very uncertain, but it is believed to have been introduced into this country from the Levant, about the middle of the sixteenth century. It is most esteemed in a living state, and chiefly for embellishment. We have no other ornamental tree which attains to such dimensions, with blossoms so rich and beautiful. It is therefore much employed in avenues and hedgerows, and frequently in conspicuous parts along the margins of plantations. Happily situated in a congenial soil, it forms a beautiful object as a lawn tree, particularly when standing full grown on grounds of a corresponding magnitude. It rises with a straight trunk, and produces a large umbrageous head of a pyramidal form, closely clad with opposite digitated leaves of a deep green. After the foliage begins to expand, the tree is remarkable for its rapidity in forming the whole season's growth, which is usually completed in three, or at most four weeks. The flower-stalks emerge above the leaves, and the blossoms are generally expanded by the end of May. The association of flowers of such beauty and delicacy with one of the loftiest of timber trees have caused the sportive imagination of some authors to compare the tree to many extravagant figures. It has been likened to an immense lustre or chandelier, from the structure of its blossoms tapering above its foliage like lights. Others have termed it the giant's nosegay, a gigantic hyacinth, the lupine tree, etc. ; and, alluding to the gaiety of its blossoms, and the manner in which it scatters them on the grass, with the comparative worthlessness of its fruit and timber, it has been considered an excellent emblem of ostentation. The young wood being matured early in the season renders the tree adapted to endure cold and unfavourable situations. It is only, however, in warm and sheltered places that it yields blossoms abundantly, and a high degree of temperature is necessary to expand them, and exhibit the tree in its most attractive form.

In favourable seasons the tree generally yields a heavy crop of nuts, from which young plants are readily propagated. The nuts become ripe about the end of October, and may be sown any time during the winter. The seedlings will grow in almost any description of soil, but a rich, deep, free loam produces them in greatest vigour. The size of the seedlings, however, is regulated considerably by the size and soundness of the nuts. A bushel of nuts is sufficient for a bed four feet wide and twenty yards long. They should be either rolled or beaten down with the back of a spade to fix them in their places previously to the spreading of the covering, which should be from one to one and a half inch deep.

The seedlings may be transplanted into nursery lines either at the age of one or two years, allowing a space of from eight to twelve inches between the plants, and about two feet between the lines. The plants should be removed every third or thereafter, increasing the space as they advance until they are finally situated. The fibrous nature of the roots of this tree, particularly after being frequently transplanted, prepares it for being removed in safety at a size greater than that at which trees in general admit of being transplanted. This valuable property fits it, when properly nursed, for immediate effect in forming lines and avenues of verdure, and in decorating park or lawn scenery, when of a size sufficient to withstand the bad effects of cattle or sheep.

The most celebrated avenue of this tree in England is that at Bushypark, near London, and one of the loftiest trees of the species stands at Enfield, where it measures 100 feet high. But the largest in Britain is supposed to be one at Nocton in Lincolnshire. Its height is about seventy feet, and the space occupied by the spread of its foliage measures upwards of 100 yards in circumference, its immense branches being supported by props give it the appearance of the banyan tree of the East. The horse-chestnut is never planted for the sake of its timber, but sometimes for its nuts, which are eaten by deer. The timber is suitable for the purpose of common deal; when dry it only weighs from 35 to 37 lbs. per cubical foot; being lighter than many kinds of wood, it is commonly cut into boards for packing-cases, for lining to carts and barrows, and being easily worked it is sometimes used by carvers and pattern-makers.

Æ. H. rubicunda (Lois).—The red flowering horse-chestnut is a very showy dwarf tree, and flowers at an earlier age than the common tree; also all the yellow flowering and smooth-fruited kinds known as the genus PAVIA are of dwarf growth, and are readily propagated by being engrafted or budded on the common horse-chestnut.

XLIII.

THE COMMON OR WILD PEAR TREE.

THE Common Pear Tree : *Pyrus communis* (L.)—This tree belongs to *Icosandria Di-Pentagynia* of the Linnæan system, and to the natural order *Rosaceæ*. It is a native of many parts throughout Britain. Nurserymen grow not only the seeds of the wild tree for stocks on which to engraft the various kinds of valuable pears, but the seeds of the cultivated garden varieties are also employed for the same purpose. This has given rise to an endless number of varieties, none of which ultimately attain to the size of very large timber trees ; but their rapid growth when young is seldom surpassed by any other forest tree, and the varieties are all quite hardy.

Mode of Propagation.—The seeds of the common or wild tree, and that of all cultivated pears, if picked out of ripe fruit and immediately sown, will come up during the first spring ; but if the seeds are allowed to become quite dry, and are kept till spring and then sown, a great proportion of the crop frequently lies dormant, and only appears above ground in the following spring.

The soil should be a deep rich loam. Manure has a powerful influence in advancing the growth of the plants. A pound weight of fresh clean seed is sufficient for fifteen lineal yards of a bed four feet wide, and the cover should be about half an inch deep. One-year seedlings are often upwards of a foot high, and proportionally stout.

At this age the plants should be inserted into nursery lines, fifteen inches apart, and the plants six inches asunder. After being two years transplanted they are fit for being put into lines for stocks, or for being planted into good clean prepared ground in the forest; but if they are required larger, they should be again transplanted into nursery lines and allowed additional space. At the age of five, plants established in the ground often yield shoots three feet long in one season in ordinary rich soil.

As a fence plant it grows far more vigorously than hawthorn, but none of the varieties make so close a hedge, or present so even a surface as that plant. Many of the varieties raised from the seeds of grafted fruit have few or no spines, yet as a rough fence and shelter forming a thicket on the outside of an exposed plantation, or belting on dry ground, such as that along a sunk fence, it readily springs up, and is a useful tree.

The figure of the wild tree is generally pyramidal, and to that shape the varieties commonly incline, although some are very dissimilar in figure. All yield white blossoms, and in good soil the ordinary rate of growth is about twenty feet in ten years, or about forty-five to fifty feet in thirty years, with a trunk from one to one and a half feet in diameter, which is the ordinary size of the tree in this country.

It is of great duration, not subject to disease, and has been known to live for several centuries.

The wood is much valued by artisans. When well dried it is not apt to crack or warp; it takes a fine polish, and is esteemed as the most suitable timber for being died black, in imitation of ebony, as is the wood of all the allied species.

P. Malus (L.), the Wild or Crab-Apple Tree. Syn. *Malus communis* (De Candolle.)—This tree is a native of Britain, and is also found wild throughout Europe. It is the parent of the innumerable varieties of fine apples in cultivation in our gardens and orchards. The tree is grown extensively in nurseries for the supply of stocks, on which the cultivated sorts are engrafted ; and supplies of seeds are readily obtained from the cider manufactories throughout England.

In every respect the mode of treatment detailed for the raising of the pear is adapted to the apple. The seeds of the numerous kinds of cultivated apples yield a great variety of the plant, and generally, for timber trees, they are fully as vigorous in growth, and attain to greater dimensions than the common crab apple; but at best, in point of size they are inferior to the pear tree. The apple is quite hardy, and endures a cold and bleak exposure, although it only attains to a timber size in good soil, well drained and moderately sheltered.

It forms a gaudy ornament while in flower in April and May, and presents a great variety in the colour of its blossoms, from a delicate pink to a deep red. It is to be met with in the Highlands of Scotland in an indigenous state, associated with the sallow, the birch, and hazel, and some of the varieties retain the leaves and fruit far into winter.

The tree is subject to disease, the chief of which is the insect known as the "American bug" or "apple aphis," which appears like a woolly substance on the branches, and sometimes under the surface, on the roots. It eats into the bark, impedes the circulation of the sap, disfigures the branches, and retards their growth. With some trouble it can be cured on a garden tree by train-oil, alkaline, or urinal wash of any kind. The best mode of applying the liquid is by heating and brushing it into the crevices of the bark on every branch and twig, with a painter's brush. In the orchard and forest this is impracticable, and there is no cure.

The wood of the apple varies in quality, according to the soil and the variety of the tree. All are hard, fine grained, of a brown colour, and sought after for articles of turnery.

The Mountain Ash, or *Rowan Tree*—*P. aucuparia*—(Gaertner). —This tree is indigenous to the mountainous situations throughout Europe. It is a well-known, beautiful, deciduous tree, with smooth branches, and the leaves pinnate, with uniform, serrate, smooth leaflets. In the months of May and June it produces numerous panicles of white blossoms, highly fragrant. They are generally followed by a profusion of scarlet berries, which become ripe in October.

In Britain the tree attains to its greatest perfection in rich soil throughout the Highlands of Scotland, where it rises in a wild state, frequently associated with the birch, the alder, and the poplar (*P. tremula*); and in the rocky recesses of some upland districts, away from cultivation, it may be seen in autumn with its terminal shoots bending under the weight of scarlet berries, which sparkle amongst the varied tints of foliage peculiar to that season. But although the beauty of its foliage and fruit is hardly surpassed by any other deciduous tree adapted to our climate, yet it is possessed of other more valuable, though less conspicuous qualities.

From its being extremely hardy it luxuriates in a cool soil at a great altitude, where many other trees perish from exposure. Although it never attains to a great size, yet it advances rapidly during the first eight or ten years of its growth. It is therefore suitable for rearing plantations of trees of slower growth, and in establishing a shelter round buildings and reclaimed land, in bleak and inhospitable situations. Several instances could be referred to where, on exposed ground in the north of Scotland, of an altitude about 1200 feet, the mountain ash and other hardy trees of this genus have readily established a shelter, where, without protection, many sorts cannot live. The tree admits of being planted at the height of about four feet even in a bare situation; it generally readily takes to the ground; and although deciduous, it soon affords shelter, from the number and closeness of its branches. As a hedgerow tree in the worst exposures, it has no superior. It rises in an upright form when young; afterwards it forms a shapely head, which stands unaffected by the prevailing winds; and whether in the wild acclivity of the Highlands, or in the low and cultivated avenue, its appearance and habits of growth remain more fixed and unchangeable than that of most other trees.

The mountain ash is raised from the berries, which ripen in autumn, when they should be collected and mixed up with sand or light sandy earth in a pit, where they should be turned over every second or third month, in order that the berries may become regularly decomposed. They should be sown during the second winter, or early in the second spring, after they are collected. The soil should be rich and friable, and the seeds should be spread on the beds and regulated with a rake, with the view of the plants arising about two or three inches asunder. At the age of two years the plants should be removed from the beds into nursery lines, after pruning off the extremities of their straggling roots ; the lines should be two feet distant, and the plants six inches apart. After remaining two years in nursery lines, the plants are commonly from four to five feet high, when they may either be planted out permanently or transplanted into greater space, to adapt them to situations where, on being planted, they will form an immediate effect.

Some of the finest trees of the species in the north are produced in Badenoch, Inverness-shire. The best of these, and perhaps the finest tree of the species in Britain, stands at the west entrance gate on the estate of Belleville. It is fully forty feet high, with a very large well-balanced head, and a handsome trunk from eight to nine feet in circumference. The tree is indigenous on the estate, but very likely it had been placed in its present situation by Macpherson, celebrated for his translation of the "Ossian" and other Gaelic poems, who purchased and improved the estate towards the end of last century. The tree affects a cool soil and a moist atmosphere. It attains to the age of many centuries, and in open situations it is seldom attacked by disease. Few trees are superior to it in an ornamental point of view ; and its berries are the choice food of our singing birds, particularly the thrush and the blackbird.

The tree forms an excellent coppice, and as such furnishes

valuable shoots for hoops, crates, and similar purposes, and the bark is possessed of a valuable tanning principle. The timber is strong and elastic; and, like that of all the species of the genus, it is closely grained, susceptible of a fine polish, and being well adapted for being stained of any colour, it is valued by the wheelwright, and for articles of turnery, tool-handles, etc.

The species consists of many cultivated varieties, all of which are very ornamental, particularly the weeping mountain ash, which, when grafted on the top of the common mountain ash, or on that of the service tree, forms one of the most elegant pendant trees in cultivation.

The yellow-berried variety is also very conspicuous. The others differ from the common tree chiefly in yielding both larger foliage and fruit, but none of the kinds attain the ordinary size of timber trees. All of the sorts yield seeds, but their peculiarities are only accurately multiplied by layers or by budding or grafting, and the common tree is generally used as the stock.

All the kinds of pears in cultivation will grow on the mountain ash, but grafted on this stock, though they readily bear fruit for a few years, yet they become very stunted; the stock seldom swells in growth equal to the graft, and the union is neither perfect nor permanent.

Service Tree—Pyrus Aria (Ehrhart), or White Bean Tree. —This species, which comprehends a great many varieties, is found indigenous on the hilly districts throughout Europe. Its stature is very diversified, according to the soil and situation which it occupies, and, cultivated under the most favourable circumstances, it seldom attains to the height of forty feet. It grows with great rapidity for the first six or eight years of its age, frequently attaining the height of twelve or sixteen feet in that time, after which it becomes bushy, when the progress of its upward growth becomes very slow. It is a hardy tree, of great duration, and in every respect resembles the mountain ash, except that its leaves are entire, light green above, and downy underneath, which causes it to present a striking and varied appearance when ruffled by the wind. It is late in coming into leaf; it grows on soil of very opposite qualities; it forms a shapely round-headed tree, unaffected by the wind, and establishes a shelter where many kinds of trees fail to exist. The mode of propagation and treatment suitable for the mountain ash is adapted to this tree, and the timber of both species is very similar. The varieties are propagated by being engrafted, using the common tree or the mountain ash as the stock.

XLIV.

THE CHERRY TREE.

THE CHERRY (*Cerasus*) belongs to the section *Amygdaleæ* of the natural order *Rosaceæ* (Loudon), and to *Icosandria monogynia* in the Linnæan system. The cherry tree comprehends all the wild or common varieties of the tree which in Scotland are commonly termed geans, all the engrafted kinds cultivated for their fruit, the various kinds of bird cherry commonly named after the different countries to which they belong. The Portugal and common laurel belong to the same family.

The numerous kinds of which this genus is composed are all very ornamental, but none of them can be classed among our timber trees except the wild cherry or gean—*C. sylvestris* (Ray). It is indigenous throughout Europe, and consists of very numerous varieties, which differ considerably in the size and shape of their leaves, the size and quality of their fruit, in their rapidity of growth and ultimate bulk. These last-mentioned differences are of great importance in an arboricultural point of view ; indeed, as much so as usually exists in a difference of species among other tribes. In the Highland glens of Scotland the best timber varieties usually attain the height of from forty to fifty feet, with trunks two feet in diameter ; and under favourable circumstances they seldom fail to attain these dimensions. The tree will grow in any description of soil, provided it is dry, and not a pure clay ; that most congenial to its development is a sandy loam on an open subsoil. Very few trees will thrive so well on chalky or rocky situations where there is only a scanty layer of soil. It is very hardy, and grows at considerable elevations, but it attains a large size only on low sheltered ground. It springs wild in the north of Scotland, along the cliffs of the Findhorn, and other romantic rivers and streams of the same character, and often shoots from the crevices of rocks in a picturesque figure. Nothing can be more beautiful than the purity and richness of its blossoms in spring, or more brilliant than the gorgeous hue of its foliage in autumn.

Some fine specimens of this tree are to be seen in Ross-shire. On the approach to Rosehaugh House, the seat of James Fletcher, Esq., in a mixed plantation of magnificent timber, stand some of the largest cherry trees of various species that are to be found in the north of Scotland. Here the common wild cherry ranges from fifty to sixty feet in height, and some are upwards of eight feet in girth, and quite vigorous. One of these trees forms a very conspicuous object, by extending a comparatively small branch, suspending over the roadway a large bunch or cluster apparently four or five feet long, and a yard in diameter, of close twiggy vegetation, of quite a different character from that of the common tree, the twigs and leaves being less than half the size of the species, and all of spontaneous growth. Instances of this sort are frequently observable in the growth of the birch, the purple laburnum, etc., but such rarely occur with the cherry.

The tree is raised from the stones of the fruit, which are the seeds; the stones also of all the cultivated or garden cherries produce the wild tree. When collected, they should be mixed up with double their bulk of sand or dry earth, and may be sown any time from November to February. The cover should be one inch deep. Some of the plants commonly appear in the end of the ensuing spring; but it not unfrequently happens, that owing to the stones not becoming sufficiently decomposed, many of the seeds lie dormant until the second spring, when the principal crop appears. The plants should, at the end of their first year's growth, be lifted from the seed-bed and transplanted in nursery lines two feet apart, and the plants six or eight inches asunder in the lines. When two years transplanted, they are commonly about four or five feet high, and are fit for being transplanted into the forest.

When young, the plants are remarkable for their rapid growth, and are not particular as to the quality of the soil, provided it is dry.

In pruning this tree, the operation should be performed early in autumn; at any other season the wound is apt to gum, and continue open, while in the end of August or beginning of September the descending sap forms a cicatrix, and the wound immediately closes. The tree generally attains its full size in sixty years.

The wood is valuable; it is of a reddish colour, close-grained, and takes a fine polish. When steeped in lime-water it assumes a deeper and prettier shade; the process prevents the colour from fading when exposed to the influence of sunshine. It bears a strong resemblance to the common kinds of mahogany, and is much sought after by cabinet-makers, turners, etc., and is commonly used in the manufacture of tables, bed-posts, and other articles of furniture. It sells at from 2s. 6d. to 3s. 6d. per cubical foot, and is profitably grown as a timber tree. The plant is the most common stock on which nurserymen bud and engraft the varieties of double-blossomed cherries, and all the kinds cultivated for their fruit.

C. Mahaleb (Miller), or Perfumed Cherry, is indigenous to France and the south of Germany. In Britain it rises to the height of about twenty feet, and is grown as an ornamental shrub or low tree. Its wood is hard, brown, susceptible of a high polish, and emits a pleasant fragrance. It is sought after by cabinet-makers. The kernel of the fruit is employed by perfumers to scent soap; and in France, the branches, both in a green and dried state, are prized as fuel for their fragrance while burning. The plant is quite hardy, and grows either in a bleak exposure or as an underwood in any soil, however poor, if dry. Of the tree there are many seedling varieties.

C. Padus (Dec.) : The Bird Cherry; *C. Virginiana* (Michaux), the Virginian Bird Cherry ; and *C. V. serotina* (Lois), the late flowering, or American Bird Cherry, are kinds which grow from twenty to thirty feet in height; they are adapted to form an underwood, and yield a good shelter for game; they thrive best on dry ground. The wood of all these kinds possesses a richness of colour and veining that recommends it for the purposes of the cabinet-maker and turner.

As plants, they are all very ornamental in the season both of their flower and fruit; and no tribe of plants yields a more abundant supply of food for singing birds, particularly for the blackbird and thrush. The plants grow freely either from seed or layers. The seeds should be sown, when thoroughly ripe, a few inches apart; they all spring the first season, and should be placed in nursery rows when one year old.

The Laurel Cherry—*Cerasus Laurocerasus* (Lois),—or Common Laurel, is a native of the west of Asia, and grows wild in woody and sub-Alpine regions in Caucasus and in the mountains of Persia. It was introduced into Britain in the beginning of the seventeenth century. Parkinson in his *Paradisus*, published in 1629, is the first who records it, under the name of the Bay Cherry, growing at Highgate, at the country house of Mr. Cole, a merchant of London, where it had "flowered divers times, and yielded ripe fruit also." The plant, in a healthy state, is one of our finest evergreens; its leaves are large and massive, with a shining polish. It not unfrequently produces shoots from two to three feet in length yearly, and when young is of far more rapid growth than most evergreens in the climate of Britain; but as the plant fails to yield a thickness proportional with its length of growth, it spreads out to a large bush, and assumes the character of a shrub rather than that of a timber tree. The chief use of the plant is embellishment, but it is only in a soil thoroughly suitable that it is adapted for that purpose; in a wet, hard, or retentive soil it is affected by frost, when it becomes unsightly, yielding only a scanty foliage of a yellowish green, and a number of dead twigs. It should therefore be planted in a rich, deep, free soil, and in a sheltered situation. It affects the shade, and forms a highly ornamental underwood. It is propagated by seed, by cuttings, and by layers.

The berries become ripe in autumn, when they should be

washed, to relieve them of the pulp, and immediately sown;
but it is more readily increased by cuttings, which should
be planted in September, in sandy soil partially shaded;
the cuttings should be twigs twelve inches long, of the
same season's growth, with an inch of wood of the previous
year's formation, from which the roots will spring to form
the plant; cuttings should be inserted half their length
into the ground. During the following summer the cuttings
will form roots, and shoot out to some extent. When the
plants become close on one another, they should be transplanted into nursery lines, at distances sufficient to enable
them to become bushy, which in a year or two will adapt
them for the shrubbery, or underwood in beltings, or along
the drives throughout the forest, where it is an excellent cover
for game, and its berries form a favourite food for pheasants,
etc. In sheltered situations it forms a very ornamental hedge
or screen-fence. Its green leaves, in consequence of their
flavour, are frequently used in small quantities for culinary
and confectionary purposes, to impart the flavour peculiar to
bitter almonds, and to the kernels of the other amygdaleæ.

The distilled water from the leaves of the plant is a deadly
poison. The principal varieties of this plant are the variegated
and the narrow-leaved.

The Portugal Laurel Cherry—*C. Lusitanica* (Lois),—or common Portugal Laurel, is a native of Portugal and Madeira. It
was introduced into Britain about the year 1648, where it attains the size of a dwarf tree. It is one of the best evergreens
adapted to our climate. It does not grow so rapidly as the
common laurel, but is more hardy and luxuriant on soils of
very opposite qualities. Its leaves stand so close on the plant
as to display only the recently formed wood, of a dark purple
colour, emerging from a dense mass of lucid green foliage.
The flower spikes appear early in June; and the blossoms are
followed by oval-shaped berries, which change from a green to
a deep purple by the end of autumn.

The berries should be collected and dried in a mixture of
sawdust or dry sand, and kept spread on a floor till March,

when they should be sown. If the seeds are committed to the ground at the time of their becoming ripe, they generally spring during March, when the young plants require protection, as they are apt to suffer by late frosts. If the crop of seedlings is close, they should be transplanted at the end of the first season's growth ; but if they have sufficient room in the seed-bed, they may be safely left till they complete their second year's shoots, and then transplanted into nursery lines, where they should stand so far apart as to prevent the foliage of one plant pressing on another. After remaining two years in lines, they are fit to be permanently placed.

Of all the seasons recommended for transplanting laurels, and evergreen shrubs in general, September is the best. The Portugal laurel is frequently grown from cuttings, treated as recommended for the common laurel, but those produced from seeds form the handsomest plants. The Portugal laurel forms a very compact and ornamental evergreen hedge, well adapted for breaking the influence of wind. It admits of being pruned into shape, and forms a suitable boundary to the garden or parterre.

Early in the present century the north of Scotland possessed many fine specimens of the Portugal laurel. One of the largest stood on the lawn of Gordon Castle. It had a trunk eleven feet in circumference ; it was nearly thirty feet high, and the diameter of its head was fifty-four feet. In favourable seasons these trees ripened their seeds abundantly ; and it is worthy of remark that the plants raised from them have proved far hardier than imported seedling plants, or those plants grown from imported seed : thus trees grown from Scotch seed continue to give the clearest evidence, during our severest winters, of the great importance of acclimatation.

XLV.

THE THORN.

THE THORN (*Cratægus*).—This genus belongs to *Icosandria Di-Pentagynia* in the Linnæan system, and to *Rosaceæ* in the natural order of plants.

The species are all small-growing ornamental trees, quite hardy, and remarkable for their number, profusion of blossom, fragrance, and display of fruit, of the various shades of scarlet, red, purple, yellow, and green, ranging from the size of a pea up to that of a small apple.

The species are chiefly natives of America and Europe, but some are to be found in all quarters of the world. Respecting this genus Loudon says, "If a man were to be exiled to an estate without a single tree or shrub on it, with permission to choose only one genus of ligneous plants to form all his plantations, shrubberies, orchards, and flower gardens, where would he find a genus that would afford him so many resources as that of the *Cratægus* ?" This genus of dwarf ornamental trees adapted for the decoration of gardens and pleasure-grounds, even of those of limited extent, became popular by the numerous rare and interesting specimens brought under the notice of the British public, at an early period in the present century, in the arboretum of the Horticultural Society at Chiswick, and in that at Hackney, where the numerous species of the rarer sorts were brought to a fruit-bearing state.

Cratægus Oxyacantha (L.) : The Common Hawthorn. (Syn. Hedge-thorn, Maybush, White Thorn, Quick, Quickset.)—This is the most common and useful species of the genus. It is met with in many parts throughout Europe, and appreciated everywhere in Great Britain as the best plant for forming hedges.

The many hundred thousands of this useful plant which are yearly formed into hedges throughout the country, are all produced from seeds, which are well known under the name of *haws*. When these become thoroughly ripe, which is generally the case in the end of October or November, they should be collected. As the seeds do not vegetate the succeeding season, they should be placed in a pit, and mixed up with as much sand or light sandy soil as will prevent fermentation, where they should be turned over every two months or so, to induce decomposition equally throughout the mass. They should be removed and sown after undergoing this treatment in the pit or rot heap for twelve or fourteen months. Dry open weather in winter is the best sowing time.

Another mode of treatment is to sow the seeds immediately after they are collected, covering them about two inches, and sowing any light annual crop on the surface of the same ground, such as onions, radishes, or cabbage plants, which should be carefully removed, after which the surface of the beds should be regularly dressed and raked, leaving half an inch cover on the seeds, with the edges of the beds and alleys well defined.

I have practised both these modes successfully with the seeds of hawthorn, holly, and the like, which do not vegetate the first year, and while I have found the latter method—that of immediate sowing when the seeds are quite ripe—to be the most convenient and sûre when the quantity is small and apt to be overlooked, I adopt the process of pitting and rotting for a year in all cases where they are to be grown on a large scale.

The soil most suitable for seedlings is that which is free or easily pulverized, rather dry than otherwise,—such as a sandy loam, and very rich. The thorn does not object to any quantity of manure, and if the soil is not very rich, well-made manure may be applied in digging for the seed-beds.

A bushel of haws newly gathered, or that quantity dissolved with the mixture of earth employed, is generally sufficient for about twenty lineal yards of a bed four feet broad. After being sown the seeds should be rolled in or beat down with the back of a spade. The cover on seeds that are kept a year should be about half an inch deep, and in March, just before the young plants break through, the surface of the beds should undergo a slight polish with a rake to remove all clods, and adapt the ground for the rising crop.

Hard frost during spring is often fatal to the plants on their appearance; it is therefore well to take the precaution of giving a slight covering of straw, fern, dry leaves, twigs of fir, or other evergreens, till the end of May.

When the plants have completed their first year's growth, they generally stand very unequal, ranging from four to nine inches, and sometimes a foot in height, according to the richness of the soil and the nature of the season. At this age they should be loosened in the ground with a nursery fork and lifted, except the very small ones, which should be left for another year, when it is usual for a second crop to arise from a proportion of seeds that remain dormant in the ground during the first summer and appear during the second spring, when they are sheltered to some extent by the plants that are left in the beds of the first year's growth.

The transplanting should be performed in winter or early in spring, for few plants appear earlier in leaf, or suffer more if their roots are exposed to the droughts of spring. In transplanting the plants for the first time, the lines should be ten or twelve inches asunder, and the plants three or four inches apart in the lines, in ground clean and in rich condition, or made so by manure.

After the vigorous growth of two years in lines the plants should be removed. Hedges are frequently formed of plants of this description, but they are far inferior for that purpose to strong plants which have undergone the process of transplantation two or three times. At their first removal from the lines, ' plants that are intended for trees, or for stocks for the propagation of rare and ornamental sorts by grafting and budding, should be selected, choosing the tallest and handsomest. These should have their straggling roots pruned, and be transplanted into ample space—say, eighteen inches between the lines, and six or eight inches plant from plant in the lines.

Those intended for hedges should have their roots pruned where they are long, and their tops beyond ten or twelve inches high cut off; also, in order to induce a stout and bushy habit, they should be replanted into greater space than formerly, according to their size.

Plants twice or thrice transplanted, stout in the stem, and well furnished with bushy roots, are of many times the value of those that have been only once transplanted, whatever their height may be. In forming a hedge with strong well-nursed plants, fewer are required for a given space; they are more easily kept clean, less subject to casualties, and they become a fence sooner by a few years than younger plants. The hawthorn cultivated for hedges forms a leading article in all the small country nurseries throughout Britain; a great advantage it possesses over many other plants is, that it never overgrows; by topping and replanting it is increased in value, and can always be kept in a fit state for fences, for which it is generally preferred to any other plant.

Dr. Walker tells us that the first hawthorn hedges planted in Scotland were on the road along Inch Buckling Brae, in East-Lothian, and at Finlarig at the head of the Tay in Perthshire. They were planted at both places by Cromwell's soldiers.—*Essays*, p. 53.

The hawthorn affects a rich loamy soil or clayey gravel, rather dry than otherwise. As an ornamental dwarf tree it has no superior among deciduous plants. Trained to a single stem it possesses a neat and elegant appearance in figure and foliage. In a wild state it becomes a large spreading bush. The profusion and beauty of its flowers, and the sweetness of their fragrance, enlist the favour of the beholder, and few trees hold a more prominent place in poetical or legendary lore. Nevertheless, Gilpin views this tree in a different aspect. He says, "Its shape is bad; it does not taper and point like the holly, but is rather a matted, round, heavy bush. Its fragrance indeed is great; but its bloom, which is the source of that fragrance, is spread over it in too much profusion : it becomes a mere white sheet, a bright spot, which is seldom found in harmony with the objects around it. In autumn the hawthorn makes its best appearance. The glowing berries produce a rich tint, which often adds great beauty to the corner of a wood or the side of some crowded clump." On this passage Sir Thomas Dick Lauder observes : " We think Mr. Gilpin is peculiarly hard on the hawthorn. Even in a picturesque point of view, which is the point of view in which he always looks at nature, the hawthorn is not only an interesting object by itself, but produces a most interesting combination or contrast, as things may be, when grouped with other trees. We have seen it hanging over rocks with deep shadows under its foliage, or shooting from their sides in the most fantastic forms, as if to gaze at its image in the deep pool below. We have seen it contrasting its tender green and its delicate leaves with the brighter and deeper masses of the holly and the alder. We have seen it growing under the shelter, though not under the shade, of some stately oak; embodying the idea of beauty protected by strength. Our eyes have often caught the motion of the busy mill-wheel over which its blossoms were clustering. We have seen it growing grandly on the green of the village school, the great object of general attraction to the young urchins who played in idle groups about its roots, and perhaps the only thing remaining to be recognised when the schoolboy returns as the man. We have seen its aged boughs overshadowing one half of some peaceful woodland cottage, its foliage half concealing the window, whence the sounds of happy content and cheerful mirth came forth. We know that lively season—

'When the milkmaid singeth blythe,
And the mower whets his scythe,
And every shepherd tells his tale
Under the hawthorn in the dale;'

and with these, and a thousand such associations as these, we cannot but feel emotions of no ordinary nature when we behold this beautiful tree."—Lauder's *Gilpin*, vol. i. p. 195.

Of the British trees of this species recorded by Loudon

when he wrote thirty years ago, the two largest belong to
Scotland. At that time he records one 110 years old at
Tyningham in Haddingtonshire, forty-six feet high, with a
trunk three feet in diameter, and a head forty-seven feet,
standing on light loam or clay. Another is recorded to stand
at Duddingstone near Edinburgh, of similar dimensions.

The tree also attains an unusual height farther north. At
Gordon Castle in Banffshire, in Morayshire, and on the sheltered slopes of Badenoch in Inverness-shire, instances are found
of its approaching the height of forty feet, and yet in full vigour.

The hawthorn is a tree of great duration, and has been
known to exist for several centuries.

The timber of the tree is very hard, and of a creamy-white
colour. It is not much used in the arts, owing to the difficulty
of getting it of a large size. It takes a very fine polish, and
is esteemed in turnery for knobs and handles to small implements, for mallets, teeth to rakes, and all purposes where
hardness and durability are required.

The following are the leading varieties of the hawthorn,
and are propagated by grafting or budding, using the common
tree as the stock :—Double blossom white, a well-known tree,
the fading flowers of which change to a delicate pink. Double
red. Single and double scarlet—all the blossoms of these are
of great beauty.

C. O. aurea is a variety which yields yellow haws, not common, but very ornamental.

C. O. præcox, the early flowering or Glastonbury thorn, is
remarkable for coming into leaf in winter, and in favourable
weather instances are recorded of its having blossomed freely
in England at Christmas. In the north of Scotland I have
seen it covered with foliage in the end of January and early
in February, but the severe frosts every year rendered the
tree very unhealthy. I have not known it to be tried on the
west coast of Scotland, but I believe that under the influence
of the Gulf-stream, where the winters are mild, the tree would
find a congenial atmosphere.

Of the genus there are upwards of thirty species found in
the collections of this very ornamental dwarf tree.

XLVI.

THE ELDER TREE.

THE ELDER TREE (*Sambucus*) belongs to the natural order *Caprifoliaceæ*, and to *Pentandria trigynia* of the Linnæan system. It comprehends several species, intermediate between shrubs and low trees, deciduous and hardy, some of which are found indigenous to every quarter of the world. All the kinds are readily grown from slips or cuttings, and generally they are remarkable for their rapidity of growth previously to their yielding berries.

S. nigra (L.), the common black-berried elder, or bourtree, is a native of Europe, the north of Africa, and the colder parts of Asia. In Britain it usually attains the height of about twenty feet; but in rich sheltered situations, when associated with other trees, it frequently rises much higher. In open sunny situations it yields a profusion of creamy-white blossoms, which in June have a conspicuous appearance; they are produced on widespread terminal cymes, and are generally succeeded by a heavy crop of black berries, which become ripe by the first of winter. The tree is frequently met with in the neighbourhood of houses and gardens, and is seldom observable in more remote and unfrequented spots; yet it is by no means a desirable object. Its blossoms and foliage when grown extensively emit a faintish sickening odour, which is believed to be unwholesome, and during hot weather is said to produce narcotic stupor in those who sleep under its shade. Evelyn mentions a tradition of a certain family in Spain who became enervated and died in consequence of their house being seated among elder trees, and he adds, that when at last the trees were rooted out it became a healthy abode. Although there is a species of aphis that feeds on the elder, yet an infusion of the leaves proves fatal to various insects which infest blighted or delicate plants.

Medicinally, the plant has been known throughout Europe from the earliest periods of our medicinal history. In this respect it occupies a conspicuous place in the works of Theophrastus. The inner bark of the tree is an active cathartic. The flowers serve for fomentations and cooling ointments; when dried they make a fragrant but debilitating tea, useful in acute inflammation from the copious perspiration which it is sure to excite. A wine is also made from the flowers, which, in scent and flavour, resembles that produced by the Frontignac grapes. Elder-flower water is also employed to impart a flavour to some articles of confectionary, and it is esteemed as a cooling and refreshing lotion for the skin. The berries yield a very large quantity of spirit, and it is said their juice is employed to adulterate port wine. Loudon states that in different parts of the county of Kent there are fields or orchards planted with the elder entirely for the sake of its berries, which are brought regularly to market, and sold in immense quantities at from 4s. to 6s. per bushel, for the purpose of making wine, which is much drunk in cold weather in London in the houses of the lower classes, mulled as a cordial.

As a screen-fence in bleak exposures and maritime situations the tree is of great value; when the young plant has just become established in a deep fertile soil it will often produce shoots five and six feet long in one summer, but in inhospitable exposures the points of the shoots become weather-beaten; the tree readily makes new efforts of growth, and thus becomes twiggy at the extremities, affording shelter to its entire height. It generally rises, under the most adverse circumstances, with a vigour hardly equalled by any other plant. In wet situations the success of the tree, for shelter, is much assisted by thorough drainage, without which the late-formed growths fail to become matured, and consequently suffer from the influence of frost. As a nurse for other trees, particularly when interspersed throughout the outskirts of plantations in rough exposures, it is very valuable, but only as a shelter. As a fence, apart from shelter, it is objectionable in many situations : not only is it apt to get bare near the surface of the ground, but its roots take a wide range, and have a powerful influence in subduing the crops in their vicinity. For this purpose it is far inferior to the hawthorn, beech, and many other hedge plants. But as a screen-fence or shelter on poor dry soil where there is difficulty in growing the ordinary sorts of hedge plants, it is very useful, provided it is frequently pruned, say twice or three times a year, which renders it compact and good-looking.

The elder is readily propagated from young shoots ; the buds or joints of these are commonly from six to ten inches apart. Each cutting or slip should be formed with a joint close to its lower extremity, from which the roots will spring; another joint should exist near to the top of the slip, the buds of which will form the branches of the plant. The slips should be from eight to twelve inches long, and inserted about half their length into the ground, any time from the beginning of November till the first of March. They readily take root, and frequently grow several feet in height during the first summer. Their distance should be a foot apart in the lines, and the lines nearly two feet asunder. The plants are usually fit for removal at the age of one year, and never improve in nursery lines after being two years old.

The wood of the young tree is brittle, and contains a large pith. When old, it is hard, of a glossy yellow, and susceptible of a high polish. It is employed in forming articles of turnery, in the manufacture of mathematical instruments, combs, shoe-makers' pegs, and generally for the same purposes as box-wood.

S. n. virescens (Dec.) differs from *S. nigra* only in its bark being whiter, and in its berries being of a green colour. Its growth is equally vigorous. The other leading sorts connected with the common elder are the gold and silver blotched, and the parsley-leaved, all of which are of comparatively stunted growth, and only cultivated as ornamental varieties.

S. racemosa (L.)—The racemose-flowered Elder is pretty generally known as the scarlet-berried elder, and is by far the most ornamental plant of the genus. It is a native of the south and middle of Europe, and of the mountains of Siberia, where it forms a low tree from twelve to fifteen feet high. In Britain it attains a greater height, and it possesses all the vigour in growth of the common species, and is equally hardy. In its deciduous state, the chestnut-coloured bark and the prominent buds of its young shoots are very ornamental. The young wood yields racemes of flowers, which open with the expanding leaf in spring. Its foliage is of a bright green, pinnate, deeply serrated, and extremely handsome. Its panicles of fruit resemble miniature clusters of grapes of a bright scarlet, which attain their brilliancy early in autumn, and long before the leaves are shed. When in full fruit, in point of beauty it has no rival among deciduous plants. But it is shy in yielding fruit ; even in situations possessed of a good climate, the plant, though constant in blossom, unlike the common elder, often fails to produce fruit. I have seldom seen this species yield fruit abundantly in any part of Scotland, except in late districts, in the Highlands of Moray, Inverness, and Ross. In earlier situations the plant generally blossoms about the first of April, while frosts prevail ; and its success in the Highlands is attributed to the influence of a late climate, which retards the expansion of the blossoms until a period when they are exempt from injury.

XLVII.

THE LABURNUM.

THIS tree is the largest species of the very ornamental genus *Cytisus* of the natural order *Leguminaceæ*. It belongs to *Monadelphia Decandria* in the Linnæan system. Two kinds are common throughout Britain, *Cytisus Laburnum* (L.), and *C. L. alpinus* (Miller), the Alpine or Scotch Laburnum. The latter yields the finest timber; in other respects the kinds are so similar that they may be treated as one.

The laburnum is a native of the mountains of France, Germany, Switzerland, and Italy, and is not indigenous to any part of Britain. It was introduced into this country at the close of the sixteenth century. It is a low deciduous tree, with trifoliate leaves, and seldom exceeds the height of twenty feet, except when situated on rich soil, and sheltered and nursed among trees of taller growth, where it often attains a height of from thirty to forty feet. Like many of our low trees, it grows vigorously during the first few years of its age. It is perfectly hardy, and luxuriates for some years in almost any description of ground, and in a moderately dry soil it is certainly one of the best ornamental trees adapted for bleak and inhospitable situations, where its value is enhanced by being rarely subject to disease. The early rapidity of its growth renders it suitable for being interspersed as a shelter to many kinds of young ornamental trees, which ultimately take its place and attain a greater size. It will be always highly esteemed as an embellishment along the margin of plantations, roads, and avenues, and it is used as the stock on which the beautiful pendulous kinds of the genus are engrafted.

The laburnum blossoms in May and June, with all the richness and profusion peculiar to its tribe. The seeds are poisonous, and are yielded in pods, which become ripe in the beginning of winter.

In spring the seeds should be sown on light, friable ground, and placed two or three inches apart, and covered nearly an inch deep. One-year-old plants are generally about a foot high, when they should be transplanted into lines two feet asunder, and the plants one foot distant in the lines. During their second year's growth in the lines, the plants often make an average advance of three feet. After having been two years in the lines, or at most three, they should be removed either into their permanent situations or into additional space in the nursery lines. By frequent removal plants may be grown to a large size and still kept in a suitable state for being safely transplanted into situations where their ornamental effect is immediately required. As the bark and young shoots of laburnum are preferred by hares and rabbits to those of any other tree, young plants, which can be raised in great numbers at a few shillings per 1000, are frequently interspersed throughout young plantations as food for these animals, and as a means of preventing their ravages on other plants. Notwithstanding the rapid growth of the young tree, it soon forms a beautiful heartwood, varying from a brown to a dark-green, and sometimes a black colour, which contrasts remarkably with the pale yellow of the more recently formed layers of sapwood. It is valued and in request' by cabinet-makers, and for articles of turnery, such as cups and ladles, and for flutes and other musical instruments. It is frequently used for pegs, wedges, pulleys and blocks, and also for those instruments in which strength and elasticity are required. A cubical foot of timber in a dry state weighs from fifty-two to fifty-three pounds. It is seldom produced beyond a foot in diameter, and its value has considerably decreased since the art of staining wood has become so perfect. The soil and climate of Scotland are very congenial to the growth of the tree. At Sauchie, in Stirling-shire, a tree stands forty feet high, with a trunk six feet in circumference; but the largest laburnum of British growth lately stood at Castle Leod, in Ross-shire, of the English variety, the trunk of which measured nine feet in circumference—a size very remarkable for a tree of the species.

C. Laburnum purpureum (Hort.), the purple-flowering Laburnum, or *C. Adami* (Poiret), originated in France. It forms a hybrid between the common laburnum (*Alpina*) and the *Cytisus purpureus*, a dwarf, spreading shrub. The purple blossoms of the hybrid are tinged with buff, and of the size and shape of those of the common laburnum, and its habit of growth is vigorous and erect. The tree is in the habit of producing blossoms of both the parents, and of the variety, all three at the same time; and in cool situations, at a great altitude, it has been known to throw off the small twigs of *C. purpureus*, and the shoots of the hybrid variety in the form of dead-wood, and to appear entire a tree of *C. Laburnum Alpina*. This and several other kinds of laburnum are propagated by being engrafted on the common kinds.

XLVIII

THE LOCUST TREE.

Locust Tree—*Robinia* or *Pseud-Acacia* (L.)—The common Robinia, or False Acacia, is a leguminous tree, and belongs to *Diadelphia Decandria* in the Linnæan system. It is a native of North America, where, in rich, dry, well-sheltered soil, it grows rapidly, and becomes a tree of considerable height. According to Loudon, the first plant of the species that was brought to Europe was planted in the Jardin des Plantes, Paris, in 1635, and in 1835 it still existed, and stood seventy-eight feet high. In the climate of Britain the tree requires the earliest and best-sheltered situations, and in such its chief use is ornament. Like many trees of small stature, such as laburnum, elder, cherry, plum, and sallow, it produces shoots of great vigour during the first few years of its growth ; but being far more tender than the kinds referred to, it is only in the best situations, and during the most favourable seasons, that its luxuriant growths become sufficiently matured to resist the influence of frost ; and from a third to a half of the extremities of the branches of young trees are commonly cut off, which reduces the progress of the tree, in many situations, to that of very ordinary growth. Notwithstanding this circumstance, the plant, though rendered branchy, has a natural tendency to grow erect. In after life the tree assumes a more spreading habit, with growths less vigorous and better fitted to endure the winter, but it is seldom found to girth in proportion to its height.

It is after the tree has attained the age of ten or twelve years that it appears in its most captivating form, and produces white and yellowish racemes of great beauty and fragrance. As an ornament, however, it is very precarious; it is much influenced by the seasons, and seldom blooms abundantly for a few years in succession. It is generally late in coming into leaf, and unless the opening up of the summer is mild and genial, it retains a bare and uninteresting appearance after trees in general are arrayed in their summer dress. As it continues to grow to a late period in the year, to ripen the more vigorous' shoots requires a summer possessed of a temperature warm and more protracted than that enjoyed by most situations in this country. This is particularly the case in soils moist and rich, and where shoots are produced from trees after being broken or lopped.

The tree is sometimes raised from cuttings of the roots, but it is more frequently produced from seeds, which generally ripen by the end of October. The best seeds are imported from America. They should be sown early in spring, after being soaked in water. The soil should be well drained, light, and friable, and where the plants will have full advantage of sunshine. It is only where the climate is of the best description that the soil should be made rich for seedling plants.

The seeds should be sown about two inches apart, and covered half an inch. The plants will appear early in summer; and during the first season they commonly grow from one and a half to two feet in height, and seldom ripen their tops sufficiently when of a much greater size. The plants should be transplanted, when one year old, into nursery lines two feet apart, and the plants ten or twelve inches asunder, where they may remain one or two years, when they are commonly from five to eight feet high, and suitable for being removed to their final destination. One-year seedling plants are sometimes thinned out, and a proportion of them left, six or eight inches apart, in the seed-beds, when they become two years old. When treated in this way, they are often from five to six feet high, and fit for being permanently planted out. Thus in early life the locust is apt to excite the belief that as a timber tree it would outgrow every sort.

Perhaps there is no other foreign tree that has had its properties and uses so enthusiastically recommended to the notice of planters. It is one of the first introduced American trees into Britain, and it has been eulogized in various countries and at different periods; in this country, under the different names of "Robinia," "Acacia," "the Locust Tree." But by far the greatest impulse was given to its cultivation by Cobbett, who, between the years 1820 and 1825, imported large quantities of seeds, and grew immense numbers of plants, which he sold under the name of locust-tree, the popular name in its native districts in America, but then hardly known in this country to be the Robinia, or pseud-Acacia. Many therefore believed the locust to be a newly-introduced tree, while Cobbett, in his *Woodlands*, and other publications, extolled it as far superior to any other, both for rapidity of growth and durability of timber. Of its timber he says, "it is absolutely indestructible by the power of earth, air, and water." He also states that "the time will come, and it will not be very distant, when the locust-tree will be more common in England than the oak; when a man would be thought mad if he used anything but locust in the making of sills, posts, gates, joists, feet for rick-stands, stocks and axle-trees for wheels, hop-poles, pails, and for anything where there is liability to rot. This time will not be distant, seeing that the locust grows so fast. The next race of children but one, that is to say, those who will be born sixty years hence, will think the locust trees have always been the most numerous trees in England, and some curious writer of a century or two hence will tell his readers that, wonderful as it may seem, the locust was hardly known in England until about the year 1823, when the nation was introduced to a knowledge of it by William Cobbett." These and similar statements of this writer created a sensation throughout the country, and for a few years the tree was planted to an unprecedented extent; but at the present time we do not know any district in which the locust continues to be planted for the sake of its timber. Yet in favourable situations it very quickly produces small timber, suitable for props for flowers, stakes for peas, and such purposes, preferable to almost any other tree. As a coppice it becomes feeble after being frequently lopped, and it does not thrive as an underwood ; neither is it adapted for a situation of great exposure, as its branches are remarkably brittle, and more apt to be broken by wind than any other tree. Like many other timber trees which spread their roots near the surface of the ground, the trunk has a tendency to become hollow in the centre before it attains a great age. One of the finest recorded specimens of the tree in South Britain has been grown at Claremont. It stood about seventy feet high, with a trunk four feet in diameter, and fifty feet diameter of top. In North Britain the best specimen stands at Beaufort Castle, about forty feet high, with a trunk upwards of two feet in diameter at three feet from the surface.

As a post its timber is remarkably durable, even when young, and in some districts in France it is cultivated in coppice and in pollards, and cut every four years for vine-props ; and the leaves and young shoots are sometimes used in feeding cattle.

In the native districts of the tree, even where it grows best, only a small proportion of timber is produced of sufficient size for shipbuilding, the trees generally not exceeding one foot in diameter. The timber has a high reputation for strength and durability, and its lateral strength in resisting fracture is greater than that of the best oak. In America it is more extensively manufactured into tree-nails, used in ship-building, than any other timber. It is also esteemed by the cabinet-maker, and in articles of turnery it is often substituted for box-wood.

XLIX.

THE HOLLY.

THE HOLLY.—*Ilex* belongs to the natural order *Aqui-foliaceæ* and to *Triandria tetragynia* in the Linnæan system.

I. aquifolium (L.)—The prickly leaved, or common Holly, is an evergreen tree, usually attaining a height of from twenty to thirty feet in a wild state, but a much greater height when cultivated. It is found indigenous throughout Britain, on dry ground of various qualities, and most frequently associated with other trees. Its services in a living state are particularly valuable. When partially shaded, in a congenial soil it presents a dark glossy polish on its foliage, of the richest description ; and as an ornamental tree it has no rival. The closeness of its evergreen leaves yields a shelter and seclusion at all seasons of the year, very great in proportion to the space occupied by the tree. It is quite hardy in ground moderately elevated, although it does not luxuriate at a great altitude.

As a hedge plant it is much esteemed, particularly when grown in the vicinity. of a residence, as a boundary to the flower-garden, lawn, or pleasure-ground ; it is also esteemed as a hedge plant for the protection and division of fields ; but in some parts it is unsuitable, from the circumstance of its under branches being subject to injury from sheep and vermin, which eat them during protracted snow-storms, and unless the hedge is young and luxuriant, the plants do not readily protrude young shoots at this point to fill up the deficiencies, which are commonly rendered the more unsightly and conspicuous from the closeness and polish which generally prevail throughout the other parts of the fences. Apart from casualties, the plant forms a strong and durable hedge, almost, if not altogether, exempt from disease.

We have found the method of mixing holly with hawthorn to suit remarkably well in forming hedges on dry soil of various descriptions. Some writers on this subject have doubted the eligibility of the practice, and have stated that they could not recommend it, believing that the hollies, without relief, would become overpowered by the hawthorn. Experience, however, shows that this is not the case, and that the contrary is the more likely occurrence. Hollies inserted of the same size as the hawthorn, and placed a few paces apart, although they almost disappeared for a few years after being planted, have come up, and spread so extensively as to give the hedges during the months of winter the appearance of being almost entirely evergreen, which adds greatly to its shelter and beauty. The progress of the holly thus situated is just what might be expected from the nature of the tree. It advances fastest when partially shaded, while the opposite is the case with the hawthorn, which, at the opening up of the season, has the disadvantage of being to some extent confined by the evergreen.

Being possessed of the rare qualification of growing vigorously under the shade or drip of other trees, the holly affords a rich verdure and seclusion when interspersed between the bare stems of lofty timber trees. For this purpose, with the exception of the yew, it has no equal among evergreen trees. It generally rises in a handsome conical form, and during autumn and winter the brilliancy of its scarlet berries, contrasted with its deep green foliage, and these associated with the fading leaves of deciduous trees, render it a most desirable object.

The holly is commonly propagated by seed. The berries are ripe, and should be collected during winter; each contains several seeds; they should be mixed up with about double their bulk of sand or dry earth, and turned over once every month or six weeks. Frequent turnings serve to decompose the berries more equally than is the case when they are allowed to remain undisturbed. In open weather during the following winter, the seeds should be sown into rich, dry soil, of an open quality, partially shaded. Such a situation as admits either the morning or the evening sunshine is suitable. The beds should be four feet wide, and the cover about half an inch deep. In sowing it is not necessary to separate the seeds from the mixture of sand. One bushel, half seed half sand, is usually allowed to eight or ten lineal yards of a bed. A proportion of the plants will break through the ground in the month of May ensuing, but it frequently happens that a portion which has not become sufficiently decomposed will remain in the ground dormant until the end of the second spring after sowing, which frequently forms the principal crop, the success of which depends greatly on the openness of the ground. In this case the springing of the seeds is sixteen or eighteen months after the time of sowing; and as the ground is not disturbed during that period, where it is of a clayey nature it is apt to become so consolidated that the young plants perish from being unable to penetrate it with their roots, hence the necessity of the soil being naturally free and soft. Another mode of growing the holly, which I have found very simple and successful, is to sow the seeds in the usual way as soon as they are gathered from the tree, cropping the ground at same time, or in spring, with any annual crop of plants or vegetables, till the holly begins to vegetate; but even then the holly seeds do not vegetate equally,—some will remain in the ground dormant for two, and often, to some extent, for three summers. After two summers' growth the young plants should be transplanted into nursery lines.

The best time for transplanting hollies is September, and moist weather should be preferred, particularly for their being removed for the first time; the lines should be about one foot apart, and the plants a few inches asunder in the lines. They should afterwards be transplanted every second or third year, which serves to keep their roots in a proper state, and to afford the plants sufficient space as they advance in size. It is common to recommend their removal every other year, but it frequently happens that the plants lose their leaves, and are so severely interrupted in their progress, by being transplanted, that they require more than two years to recover. This casualty never occurs when the roots of the plants are in a proper or fibrous state. The frequency of their removal should be regulated by the health or growth of the plants. It seldom happens, however, that the plants sustain much injury by being transplanted, provided they have not grown many years in one place, that the proper season for transplanting them has been adopted, and that care has been taken in preserving their roots from exposure to drought. The plants in the nursery lines should be kept clear of weeds, and the space between the lines should be dug over every September. The opening up of the soil, and disturbing the roots at this season, has a wonderful effect in causing the roots to form numerous spongioles, which adapt the plant for removal, even when of a large size.

In Scotland, the holly abounds in the natural woods on the banks of the Dee, the Spey, and the Findhorn. At Gordon Castle, and Orton on the Spey, and at Darnaway Castle, on the Findhorn, the best specimens of the tree measure about fifty feet high, with boles from two to two feet six inches in diameter.

The bark of the tree appears to be an indispensable article in the manufacture of birdlime. The timber is white and hard, and susceptible of a fine polish. It is therefore much esteemed in cabinet-making, turnery, and wood-engraving. When dyed black it is a substitute for ebony.

The number of the cultivated varieties of the tree is very great, particularly of the gold and silver variegated sorts. These differ in their shade of colouring, and in the breadth of the margin and structure of their leaves and prickles. They are cultivated in nurseries by budding and grafting on the common holly, and are known under the popular name of gold-edged, silver-edged, gold-blotched, silver-blotched, hedgehog, laurel-leaved, etc. The last named, and all its varieties yield smooth, massive foliage of a dark glossy green, and are exceedingly beautiful.

L

THE YEW TREE.

YEW TREE.—The genus *Taxus* belongs to *Diœcia Monadelphia* of Linnæus, and it forms the type of the natural order *Taxaceæ*. The blossoms of the tree are unisexual, and generally the sexes are placed on distinct plants; instances, however, sometimes occur of a tree yielding both male and female flowers. The genus may be divided into two distinct species, the common or English yew, and the upright or Irish yew, each of which has given rise to a few varieties.

Taxus baccata (L.), the English or common Yew, is found indigenous throughout Britain and in most parts of Europe. It is also a native of North America. In a wild state it is found in solitary trees, and very seldom in close woods of the genus. It is partial to a north or shaded locality, where it can only enjoy a soil cool and moist; and it luxuriates, though considerably confined, among tall deciduous trees. It is therefore valuable as an underwood, in forming a shelter and seclusion, where trees in general are apt to become feeble, and yield only a scanty foliage. This species becomes a timber tree, and generally attains to the height of forty or fifty feet.

It is propagated from seed. The berries become ripe in autumn, when they should be collected and cleared from the pulp by washing. The stones or seeds should then be mixed up with twice their bulk of sand, and placed into a pit for twelve or fourteen months; they may be sown in open weather, either in winter or early in spring. The soil should be rich and loamy, but such as is not apt to become stiff or hard on the surface. The seeds should be sown into beds, as regularly as possible, with the view of the seedling plants standing about two inches apart, and the cover of soil on them should be about half an inch deep. The seeds, however, under any treatment, are apt to spring irregularly, and frequently only a part of the crop comes up the first season after sowing, and the remainder the following spring. As the young seedlings, on the appearance above ground, are apt to suffer from frost, a light cover of straw, fir, twigs, or some other herbage, should be placed as a protection on the surface of the beds.

At the age of two or three years the seedling plants should be removed into nursery lines, and transplanted every third year, affording the plants sufficient space to keep them in full foliage till they are finally transplanted. September is the best month for planting, and moist weather should be preferred; nursery ground partially shaded with standard trees, thus rendered unsuitable for plants in general, is adapted for the yew; in such, its progress is fastest, and the colour of its foliage of the deepest green. The plant is remarkable for its slow growth. Plants five years old from seed, once transplanted, do not average more than one foot in height, and at the age of ten years, with nursery treatment, they are seldom more than three feet high.

The yew is sometimes affected by frost in the early part of the season, during the formation of its growth; but in winter it is quite hardy, resisting the severest weather without injury to a leaf. As it is in situations cool and moist, and to some extent darkened by the presence of taller trees, that it grows best, it is frequently employed in forming a cover for game among the stems of tall trees—for as a shelter and thicket in confined situations, it yields a foliage and growth more rich and compact than any other tree, not excepting the holly. From its naturally fibrous roots the yew may be transplanted of a large size. In nurseries, plants a yard in height are often placed at that distance asunder, and clipped into the form of a hedge, which, by occasional removals in September, form a mass of fibrous roots, adapting the plants for removal at any time. In the formation of pleasure ground, boundary, or screen-fences, such plants furnish an immediate and ornamental effect. When dressed for a few years the yew forms a hedge of close growth, so compact that small birds cannot readily pass through it.

As an ornamental tree for the lawn or pleasure ground, in a cold climate, the yew possesses some good qualities. In early life it forms a conical bush, and never receives injury from the influence of wind, nor the intensity of frost in winter. The richness of its deep green foliage forms a good contrast with deciduous trees, particularly during the snows of winter. Its berries are the choice food of singing birds, and the density of its foliage forms their best shelter.

The practice of planting the yew in churchyards is of very great antiquity, and commenced at a time when the choice of trees was comparatively limited. It bears a resemblance to the cypress, the chief cemetery tree of many countries, and its being indigenous, quite hardy, yielding a persistent, dark, and sombre foliage, and being possessed of greater longevity than perhaps any other tree, render it a fit substitute for the cypress, the emblem of immortality. With roots small and fibrous, calculated to fix the soil, and easily cut through in excavating graves, the yew, even now, with the increased choice of plants in the present age, appears more in harmony with the scenes with which it is associated than any other tree.

In growing the yew for timber it should be closely planted, and interspersed with faster-growing trees. By this means its upward growth is more rapid than when allowed to branch in its ordinary form, and through confinement in a suitable soil its timber is clean and free of knots. Every agriculturist should be aware of the poisonous nature of the tree. No doubt numerous instances occur of horses and cattle eating the spray of the yew with impunity. This may arise from habit, after small beginnings, or from its being associated with their food, or from some other cause ; but cases are too numerous where the fresh twigs, or half-dried twigs, have proved fatal, and it is more frequently so when the animal receives them to a considerable extent into an empty stomach.

Many yew trees throughout the country are celebrated for extreme old age. The lapse of a century generally makes no great change on a yew of considerable size,—much less indeed than on any other tree. Of all the old trees we have seen of this species, perhaps that at Fountain Abbey, in Yorkshire, is the most remarkable. Tradition reports it to have been an old spreading tree, under the shadow of which the monks encamped before the foundation of the abbey in 1132, the widespread ruins of which now occupy many acres. Many of the boughs of the tree are still covered with luxuriant foliage ; the trunk presents a gnarled and knotty appearance, grooved and wrinkled by time, overcome and bent down by the weight of years ; it ranges in circumference from twenty to twenty-eight feet, indicating an existence at a very remote period.

In Scotland, the Fortingal yew stands in a churchyard in the vicinity of a small Roman camp, in a romantic district at the entrance to Glen-Lyon, in Perthshire. Various writers of the last century have described the trunk of this tree at different dimensions, all recording it upwards of fifty feet. The *Edinburgh Philosophical Journal* for 1833 states that "the side of the trunk now existing gives a diameter of more than fifteen feet, so that it is easy to conceive that the circumference of the trunk, when entire, should have exceeded fifty feet. Happily, further depredations have been prevented by means of an iron rail, which now surrounds the sacred spot ; and the venerable tree, which in all probability was a flourishing tree at the commencement of the Christian era, may yet survive for centuries to come."

The timber of the yew is celebrated for its strength and elasticity ; it is more enduring than any other wood indigenous to Europe. When employed as a post in the open ground, it has been known to last for ages, with very little decay, even at that most trying and corruptible point, the surface of the ground. The heart-wood is of a rich brown colour, and contrasts with the sap-wood, which is white. It is finely grained, capable of a high polish, retains dye, and forms a good representation of ebony. It is esteemed by the cabinet-maker and turner in the manufacture of ornamental wares.   In ancient times, previously to the introduction of gunpowder, the tree was carefully cultivated for the manufacture of bows, then the principal implement of war—an instrument fatal to some of the ancient British kings, among whom were Harold and William Rufus.   It was to the skill of the English with the long-bow that the conquest of Ireland by Henry II., in 1172, was attributed, and many subsequent victories; and some of the ancient statutes of British sovereigns prohibited the exportation of yew-tree timber. . In course of time, however, the reliance of our country descended from the English yew to the British oak.

T. fastigiata (Lindley), the Upright or Irish Yew, is a well-known plant, of a shrubby habit of growth, and does not attain to the size of a timber tree.   It is propagated by seed, or more frequently by cuttings, inserted in sand or sharp soil, in August or September, and covered with a hand-glass, and shaded.   From a single stem at the surface of the ground, it sends out numerous tapering branches, richly foliated, of the deepest green ; forming an object always narrow at the surface of the ground, and broad at top, like the inverted figure of the common species.   The peculiarity in the shape of the plant renders it conspicuous, and it is generally esteemed as one of the handsomest and hardiest of evergreens.   Several varieties of variegated yews are cultivated as ornamental trees.

LI.

THE USUAL PRICES OF NURSERY PLANTS.

COMPARING the prices of nursery plants one year with another, they are frequently found to vary to some extent. This often arises from the casualties of seasons, and most frequently on account of late spring or early summer frosts, which destroy the blossom, and consequently the seed crop. The following are the usual prices for the leading kinds in ordinary demand noticed in this work :—

Pine, Scotch, native Highland, 1 year seedling, 1s. to 1s. 3d. per 1000.

„ „ „ 2 year seedling, 2s. per 1000.

„ „ „ 1 year seedling, 1 year transplanted, 3s. 6d. per 1000.

„ „ „ 1 year seedling, 2 years transplanted, 6s. to 8s. per 1000.

„ „ „ 2 year seedling, 1 year transplanted, 4s. 6d. to 5s. 6d. per 1000.

„ „ „ 2 year seedling, 2 years transplanted, twelve to eighteen inches high, 10s. to 15s. per 1000.

Older plants should be at least twice transplanted.

Pine, Corsican, or *Laricio*, 1 year seedling, 6s. per 1000.

„ „ „ 2 year seedling, 8s. to 10s. per 1000.

„ „ „ 1 year seedling, 1 year transplanted, 10s. to 12s. per 1000.

„ „ „ 2 year seedling, 1 year transplanted, 12s. to 14s. per 1000.

„ „ „ 2 year transplanted, 15s. to 25s. per 1000.

Pine, Austrian, 1 year seedling, 2s. 6d. per 1000.

„ „ 2 year seedling, 4s. per 1000.

„ „ 2 year seedling, 1 year transplanted, 8s. to 10s, per 1000.

„ „ 2 year seedling, 2 years transplanted, 12s. to 15s. per 1000.

Pine, pineaster sorts, 1 year seedling, 2s. 6d. per 1000.

„ „ 1 year seedling, 1 year transplanted, 8s. per 1000.

„ „ 1 year seedling, twice transplanted, 12s. to 15s. per 1000.

Spruce Fir, Norway, 2 and 3 years' seedling, 2s. to 3s. per 1000.

   „   „   „   1 year transplanted, 4s. to 6s. per 1000.

   „   „   „   2 years transplanted, 8s. to 10s. per 1000.

   „   „   „   3 years transplanted, twelve to eighteen inches, 12s. to 15s. per 1000.

   „   „   „   Twice transplanted, fifteen to eighteen inches, 20s. per 1000.

Spruce Fir, Douglas's.—This useful tree, about a foot high, sells at about 30s. per 100, but it gets cheaper and more plentiful every year.

Silver Fir, 2 year seedling, 5s. to 6s. per 1000.

   „   „   2 year seedling, 2 years transplanted, 15s. to 18s. per 1000.

   „   „   3 years transplanted, 18s. to 20s. per 1000.

   „   „   Twice transplanted, eighteen to twenty-four inches, 40s. per 1000.

Larch from Scotch seed, 1 year seedling, 1s. 6d. to 2s. per 1000.

   „   „   2 year seedling, 3s. to 4s. per 1000.

   „   „   1 year, 1 year transplanted, 5s. to 6s. per 1000.

   „   „   2 year, 1 year transplanted, 8s. to 9s. per 1000.

   „   „   2 years transplanted, one and a half to three feet high, 10s. to 12s. per 1000.

Cedar of Lebanon, often transplanted, one to two feet, 80s. to 100s. per 100.

   „   Deodar, 2 year seedling, 40s. per 100.

   „   „   2 years transplanted, 60s. per 100.

   „   „   Often transplanted, one to three feet, 100s. to 200s. per 100.

Wellingtonia gigantea, 1 year seedling, 40s. per 100.

   „   „   1 year, 1 year transplanted, 60s. per 100.

   „   „   Transplanted, one to three feet, 1s. to 3s. each.

Oak, British, 1 and 2 year seedling, 3s. to 6s. per 1000.

   „   „   Transplanted 1 year, 10s. to 12s. per 1000.

   „   „   Transplanted 2 years, one and a half to two feet, 18s. to 20s. per 1000.

   „   „   Transplanted twice, two to three feet, 25s. to 30s. per 1000.

Beech, 1 and 2 years seedling, 3s. to 5s. per 1000.

   „   2 years transplanted, one to two feet, 15s. to 20s. per 1000.

   „   Twice transplanted, two feet, 25s. per 1000.

Chestnut, Spanish, 2 years transplanted, two feet, 35s. per 1000.

   „   „   Twice transplanted, two, three, and four feet, 40s. to 50s. per 1000.

Alder, 2 year seedling, 4s. per 1000.

   „   1 year transplanted, 8s. per 1000.

   „   2 years transplanted, 10s. to 15s. per 1000.

Birch, common and weeping, 1 and 2 years seedling, 3s. to 4s. per 1000.

„   „   „   1 and 2 years transplanted, 10s. to 20s. per 1000.

Willow sorts, 1 year from cuttings, 3s. per 100.

„   „   2 year from cuttings, 5s. per 100.

Poplar, sorts raised by cuttings, 1 year, 3s. per 100.

„   „   „   2 years, 3s. to 5s. per 100.

„:   „   From layers 2 years transplanted, three to five feet, 10s. to 15s. per 100.

Elm, Scotch or Wych, 2 year seedling, 4s. to 5s. per 1000.

„   „   1 year transplanted, 8s. to 10s. per 1000.

„   „   2 years transplanted, three feet, 20s. per 1000.

„   „   Twice transplanted, three to five feet, 25s. per 1000.

Maple, Norway, 1 and 2 year seedling, 6s. to 10s. per 1000.

„   „   1 and 2 years transplanted, two to four feet, 25s. to 40s. per 1000.

„   Sycamore, 1 and 2 years' seedling, 3s. to 5s. per 1000.

„   „   1 year transplanted, 10s. per 1000.

„   „   2 years transplanted, three feet, 15s. to 20s. per 1000.

„   „   Twice transplanted, four to six feet, 25s. to 35s. per 1000.

Lime, common, 2 year transplanted, three to five feet, 20s. per 100.

„   „   Twice transplanted, five to seven feet, 30s. per 100.

Ash, 2 years transplanted, two to three feet, 20s. per 1000.

„   Twice transplanted, three to five feet, 25s. to 30s. per 1000.

Horse-chestnut, 2 years transplanted, three to four feet, 5s. per 100.

„   Twice transplanted, four to eight feet, 10s. to 30s. per 100.

Mountain Ash, 2 years transplanted, three to four feet, 30s. per 1000.

„   „   Twice transplanted, four to eight feet, 4s. to 8s. per 100.

Service Tree, transplanted, four to six feet, 16s. per 100.

Thorn or Quick, 2 year seedling, 4s. to 5s. per 1000.

„   „   2 years transplanted, 8s. to 10s. per 1000.

„   „   Strong, twice and thrice transplanted, 12s. to 20s. per 1000.

Holly, 2 year seedling, 2 years transplanted, 30s. per 1000.

„   Twice transplanted, twelve inches, 60s. per 1000.

Yew Tree, English, transplanted, twelve inches, 20s. per 100.

„   Often transplanted, two to four feet, 50s. to 100s. per 100.

INDEX.

414

EDINBURGH : T. CONSTABLE,
PRINTER TO THE QUEEN, AND TO THE UNIVERSITY.

88 PRINCES STREET,
Edinburgh.

EDMONSTON & DOUGLAS'
LIST OF WORKS

——*oOo*——

Wanderings of a Naturalist in India,
The Western Himalayas, and Cashmere. By DR. A. L. ADAMS of the 22d Regiment. 1 vol. 8vo, with illustrations, price 10s. 6d.

Mr. Lowe's Educational Theories, examined from a practical
point of view. By H. H. ALMOND, M.A. Price 1s.

Dr. Rainy's Position Indefensible:
Or the Real Question at issue in the Union Question. By REV. WILLIAM BALFOUR. Price 6d.

Essays and Tracts:
The Culture and Discipline of the Mind, and other Essays. By JOHN ABERCROMBIE, M.D., Late First Physician to the Queen for Scotland. New Edition. Fcap. 8vo, cloth, 3s. 6d.

The Malformations, Diseases, and Injuries of the Fingers
and Toes, and their Surgical Treatment. By THOMAS ANNANDALE, F.R.C.S., Assistant Surgeon, Royal Infirmary, Edinburgh. The Jacksonian Prize for the Year 1864. 1 vol. 8vo, with Illustrations, price 10s. 6d.

Odal Rights and Feudal Wrongs.
A Memorial for Orkney. By DAVID BALFOUR of Balfour and Trenaby. 8vo, price 6s.

Basil St. John.
An Autumn Tale. 1 vol. 8vo, price 12s.

Aunt Ailie.
Second Edition. By CATHARINE D. BELL, Author of 'Cousin Kate's Story,' 'Margaret Cecil,' etc. Fcap. 8vo, cloth, 3s. 6d.

By the Loch and River Side.
Forty Graphic Illustrations by a New Hand. Oblong folio, handsomely bound, 21s.

1.5.68.

Charlie and Ernest; or, Play and Work.

A Story of Hazlehurst School, with Four Illustrations by J. D. By M. BETHAM EDWARDS. Royal 16mo, 3s. 6d.

Homer and the Iliad.

In three Parts. By JOHN STUART BLACKIE, Professor of Greek in the University of Edinburgh. In 4 vols. demy 8vo, price 42s.

> PART I.—HOMERIC DISSERTATIONS.
>
> II.—THE ILIAD IN ENGLISH VERSE.
>
> III.—COMMENTARY, PHILOLOGICAL AND ARCHÆOLOGICAL.

Blindpits.

A Novel. 3 vols. crown 8vo.

By the same Author.

On Democracy.

Sixth Edition, price 1s.

On Greek Pronunciation.

Demy 8vo, 3s. 6d.

Political Tracts.

No. 1. GOVERNMENT. No. 2. EDUCATION. Price 1s. each.

On Beauty.   Lyrical Poems.

Crown 8vo, cloth, 8s. 6d.   Crown 8vo, cloth, 7s. 6d.

Works by Margaret Maria Gordon (nee Brewster).

LADY ELINOR MORDAUNT; or, Sunbeams in the Castle. Crown 8vo, cloth, 9s.

LETTERS FROM CANNES AND NICE. Illustrated by a Lady. 8vo, cloth, 12s.

WORK ; or, Plenty to do and How to do it. Thirty-fourth thousand. Fcap. 8vo, cloth, 2s. 6d.

LITTLE MILLIE AND HER FOUR PLACES. Cheap Edition. Fiftieth thousand. Limp cloth, 1s.

SUNBEAMS IN THE COTTAGE; or, What Women may do. A narrative chiefly addressed to the Working Classes. Cheap Edition. Forty-first thousand. Limp cloth, 1s.

PREVENTION ; or, An Appeal to Economy and Common-Sense. 8vo, 6d.

THE WORD AND THE WORLD. Price 2d.

LEAVES OF HEALING FOR THE SICK AND SORROWFUL. Fcap. 4to, cloth, 3s. 6d. Cheap Edition, limp cloth, 2s.

THE MOTHERLESS BOY; with an Illustration by J. NOEL PATON, R.S.A. Cheap Edition, limp cloth, 1s.

France under Richelieu and Colbert.

By J. H. BRIDGES, M.B., late Fellow of Oriel College, Oxford. In 1 vol. small 8vo, price 8s. 6d.

Memoirs of John Brown, D.D.
By the Rev. J. CAIRNS, D.D., Berwick, with Supplementary Chapter by his Son, JOHN BROWN, M.D. Fcap. 8vo, cloth, 9s. 6d.

Works by John Brown, M.D., F.R.S.E.
LOCKE AND SYDENHAM, with other Professional Papers. By JOHN BROWN, M.D. A New Edition in 1 vol. extra fcap. 8vo, price 7s. 6d.

HORÆ SUBSECIVÆ. Sixth Edition, in 1 vol. extra fcap. 8vo, price 7s. 6d.

LETTER TO THE REV. JOHN CAIRNS, D.D. Second Edition, crown 8vo, sewed, 2s.

ARTHUR H. HALLAM ; Extracted from 'Horæ Subsecivæ.' Fcap. sewed, 2s. ; cloth, 2s. 6d.

RAB AND HIS FRIENDS ; Extracted from 'Horæ Subsecivæ.' Thirty-fifth thousand. Fcap. sewed, 6d.

MARJORIE FLEMING : A Sketch. Fifteenth thousand. Fcap. sewed, 6d.

OUR DOGS ; Extracted from 'Horæ Subsecivæ.' Nineteenth thousand. Fcap. sewed, 6d.

RAB AND HIS FRIENDS. With Illustrations by George Harvey, R.S.A., J. Noel Paton, R.S.A., and J. B. New Edition, small quarto, cloth, price 3s. 6d.

"WITH BRAINS, SIR ;" Extracted from 'Horæ Subsecivæ.' Fcap. sewed, 6d.

MINCHMOOR. Fcap. sewed, 6d.

JEEMS THE DOORKEEPER : A Lay Sermon. Price 6d.

THE ENTERKIN. Price 6d.

Tragic Dramas from History.
With Legendary and Other Poems. By ROBERT BUCHANAN, M.A., late Professor of Logic and Rhetoric in the University of Glasgow. 2 vols. fcap. 8vo, price 12s.

The Biography of Samson
Illustrated and Applied. By the REV. JOHN BRUCE, D.D., Minister of Free St. Andrew's Church, Edinburgh. Second Edition. 18mo, cloth, 2s.

The Post Office and its Money-Order System :
With Proposal for a cheap System of conducting Money-Order business by private enterprise. By HENRY CALLENDER. Price 6d.

My Indian Journal,
Containing descriptions of the principal Field Sports of India, with Notes on the Natural History and Habits of the Wild Animals of the Country—a visit to the Neilgherry Hills, and the Andaman and Nicobar Islands. By COLONEL WALTER CAMPBELL, author of 'The Old Forest Ranger.' 8vo, with Illustrations, price 16s.

Popular Tales of the West Highlands,
Orally Collected, with a translation by J. F. CAMPBELL. 4 vols., extra fcap., cloth, 32s.

Inaugural Address at Edinburgh,

April 2, 1866, by THOMAS CARLYLE, on being Installed as Rector of the University there. Price 1s.

Book-keeping,

Adapted to Commercial and Judicial Accounting, giving Systems of Book-keeping for Lawyers, Factors and Curators, Wholesale and Retail Traders, Newspapers, Insurance Offices, and Private Housekeeping, etc. By F. H. CARTER, C.A. 8vo, cloth, price 10s.

On the Constitution of Papal Conclaves.

By W. C. CARTWRIGHT. 1 vol. fcap. 8vo, price 6s. 6d.

Characteristics of Old Church Architecture, etc.,

In the Mainland and Western Islands of Scotland. 4to, with Illustrations, price 25s.

Ballads from Scottish History.

By NORVAL CLYNE. Fcap. 8vo, price 6s.

Life and Works of Rev. Thomas Chalmers, D.D., LL.D.

MEMOIRS OF THE REV. THOMAS CHALMERS. By REV. W. HANNA, D.D., LL.D. 4 vols., 8vo, cloth, £2 : 2s.

—— Cheap Edition, 2 vols., crown 8vo, cloth, 12s.

POSTHUMOUS WORKS, 9 vols., 8vo—

Daily Scripture Readings, 3 vols., £1 : 11 : 6. Sabbath Scripture Readings, 2 vols., £1 : 1s. Sermons, 1 vol., 10s. 6d. Institutes of Theology, 2 vols., £1 : 1s. Prelections on Butler's Analogy, etc., 1 vol., 10s. 6d.

Sabbath Scripture Readings. Cheap Edition, 2 vols., crown 8vo, 10s.

Daily Scripture Readings. Cheap Edition, 2 vols., crown 8vo, 10s.

ASTRONOMICAL DISCOURSES, 1s. COMMERCIAL DISCOURSES, 1s.

SELECT WORKS, in 12 vols., crown 8vo, cloth, per vol., 6s.

Lectures on the Romans, 2 vols. Sermons, 2 vols. Natural Theology, Lectures on Butler's Analogy, etc., 1 vol. Christian Evidences, Lectures on Paley's Evidences, etc., 1 vol. Institutes of Theology, 2 vols. Political Economy ; with Cognate Essays, 1 vol. Polity of a Nation, 1 vol. Church and College Establishments, 1 vol. Moral Philosophy, Introductory Essays, Index, etc., 1 vol.

' Christopher North ;'

A Memoir of John Wilson, late Professor of Moral Philosophy in the University of Edinburgh. Compiled from Family Papers and other sources, by his daughter, MRS. GORDON. Third Thousand. In 2 vols. crown 8vo, price 24s., with Portrait, and graphic Illustrations.

Chronicle of Gudrun ;

A Story of the North Sea. From the mediæval German. By EMMA LETHERBROW. With frontispiece by J. NOEL PATON, R.S.A. New Edition for Young People, price 5s.

Creeds and Establishments,
Price 1s.

Dainty Dishes.
Receipts collected by LADY HARRIET ST. CLAIR. Sixth edition, with many new Receipts. 1 vol. crown 8vo. Price 7s. 6d.

"Well worth buying, especially by that class of persons who, though their incomes are small, enjoy out-of-the-way and recherché delicacies."—*Times.*

Notes on the History, Methods, and Technological importance of Descriptive Geometry. By ALEX. W. CUNNINGHAM. Price 1s.

The Annals of the University of Edinburgh.
By ANDREW DALZEL, formerly Professor of Greek in the University of Edinburgh; with a Memoir of the Compiler, and Portrait after Raeburn. In 2 vols. demy 8vo, price 21s.

Gisli the Outlaw.
From the Icelandic. By G. W. DASENT, D.C.L. 1 vol. small 4to, with Illustrations, price 7s. 6d.

The Story of Burnt Njal;
Or, Life in Iceland at the end of the Tenth Century. From the Icelandic of the Njals Saga. By GEORGE WEBBE DASENT, D.C.L. In 2 vols. 8vo, with Map and Plans, price 28s.

Popular Tales from the Norse,
With an Introductory Essay on the origin and diffusion of Popular Tales. Second Edition, enlarged. By GEORGE WEBBE DASENT, D.C.L. Crown 8vo, 10s. 6d.

Select Popular Tales from the Norse.
For the use of Young People. By G. W. DASENT, D.C.L. New Edition, with Illustrations. Crown 8vo, 6s.

On the Application of Sulphurous Acid Gas
to the Prevention, Limitation, and Cure of Contagious Diseases. By JAMES DEWAR, M.D. Eleventh edition, price 1s.

Memoir of Thomas Drummond, R.A., F.R.A.S.,
Under-Secretary to the Lord-Lieutenant of Ireland, 1835 to 1840. By JOHN F. M'LENNAN, M.A. 1 vol. demy 8vo, price 15s.

Studies in European Politics.
By M. E. GRANT DUFF, Member for the Elgin District of Burghs. 1 vol. 8vo. Price 10s. 6d.

"We have no hesitation in saying that there is no work in the English Language which has anything like the same value to persons who wish to understand the recent history and present position of the countries described."—*Saturday Review.*

A Glance over Europe.
By M. E. GRANT DUFF, M.P. Price 1s.

Inaugural Address to the University of Aberdeen, on his
Installation as Rector, by M. E. GRANT DUFF, M.P. Price 1s.

Notes on Scotch Bankruptcy Law and Practice.
By GEORGE AULDJO ESSON, Accountant in Bankruptcy in Scotland. Second edition, price 2s. 6d.

Karl's Legacy.
By the REV. J. W. EBSWORTH. 2 vols. ex. fcap. 8vo. Price 6s. 6d.

Social Life in Former Days;
Chiefly in the Province of Moray. Illustrated by letters and family papers. By E. DUNBAR DUNBAR, late Captain 21st Fusiliers. 2 vols. demy 8vo., price 19s. 6d.

Veterinary Medicines; their Actions and Uses.
By FINLAY DUN. Third Edition, revised and enlarged. 8vo, price 12s.

The Secret of Happiness.
A Novel. By ERNEST FEYDEAU. 2 vols. fcap. 8vo, price 7s.

Forest Sketches.
Deer-stalking and other Sports in the Highlands fifty years ago. 8vo, with Illustrations by Gourlay Steell, price 15s.

L'Histoire d'Angleterre. Par M. LAMÉ FLEURY. 18mo, cloth, 2s. 6d.

L'Histoire de France. Par M. LAMÉ FLEURY. 18mo, cloth, 2s. 6d.

Christianity viewed in some of its Leading Aspects.
By REV. A. L. R. FOOTE, Author of 'Incidents in the Life of our Saviour.' Fcap., cloth, 3s.

Frost and Fire;
Natural Engines, Tool-Marks, and Chips, with Sketches drawn at Home and Abroad by a Traveller. Re-issue, containing an additional Chapter. In 2 vols. 8vo, with Maps and numerous Illustrations on Wood, price 21s.

"A very Turner among books, in the originality and delicious freshness of its style, and the truth and delicacy of the descriptive portions. For some four-and-twenty years he has traversed half our northern hemisphere by the least frequented paths; and everywhere, with artistic and philosophic eye, has found something to describe—here in tiny trout-stream or fleecy cloud, there in lava-flow or ocean current, or in the works of nature's giant sculptor—ice."—*Reader*.

A Girl's Romance.
1 vol. ex. fcap. cloth, price 6s.

Camille.
By MADAME DE GASPARIN, Author of 'The Near and Heavenly Horizons.'
1 vol. fcap. 8vo, price 3s. 6d.

By the Sea-shore.
By MADAME DE GASPARIN, Author of 'The Near and Heavenly Horizons.'
1 vol. fcap. 8vo, price 3s. 6d.

An Ecclesiastical History of Scotland,
From the Introduction of Christianity to the Present Time.   By GEORGE GRUB,
A.M.   In 4 vols. 8vo, 42s.   Fine Paper Copies, 52s. 6d.

The Earlier Years of our Lord's Life on Earth.
By the Rev. WILLIAM HANNA, D.D., LL.D.   Extra fcap. 8vo, price 5s.

The Last Day of our Lord's Passion.
By the Rev. WILLIAM HANNA, D.D., LL.D.   46th thousand, extra fcap. 8vo,
price 5s.

The Forty Days after our Lord's Resurrection.
By the Rev. WILLIAM HANNA, D.D., LL.D.   Extra fcap. 8vo, price 5s.

The Passion Week.
By the Rev. WILLIAM HANNA, D.D., LL.D.   Extra fcap. 8vo, price 5s.

The Ministry in Galilee.
By the Rev. WILLIAM HANNA, D.D., LL.D.   1 vol. ex. fcap. 8vo.

Homely Hints from the Fireside.
By the author of 'Little Things.'   Cheap Edition, limp cloth, 1s.

Herminius.
A Romance.   By I. E. S.   In 1 vol. fcap. 8vo, price 6s.

Sketches of Early Scotch History.
By COSMO INNES, F.S.A., Professor of History in the University of Edinburgh.
1. The Church ; its Old Organisation, Parochial and Monastic.   2. Universities.
3. Family History.   8vo, price 16s.

Concerning some Scotch Surnames.
By COSMO INNES, F.S.A., Professor of History in the University of Edinburgh.
1 vol. small 4to, cloth antique, 5s.

The New Picture Book.
Pictorial Lessons on Form, Comparison, and Number, for Children under Seven
Years of Age.   With Explanations by NICHOLAS BOHNY.   36 oblong folio
coloured Illustrations.   Price 7s. 6d.

Instructive Picture Books.
Folio, 7s. 6d. each.

"These Volumes are among the most instructive Picture-books we have seen, and we know of none better calculated to excite and gratify the appetite of the young for the knowledge of nature."—*Times.*

I.

The Instructive Picture Book. A few Attractive Lessons from the Natural History of Animals. By ADAM WHITE, late Assistant, Zoological Department, British Museum. With 58 folio coloured Plates. Seventh Edition, containing many new Illustrations by Mrs. BLACKBURN, J. STEWART, GOURLAY STEELL, and others.

II.

The Instructive Picture Book. Lessons from the Vegetable World. By the Author of 'The Heir of Redclyffe,' 'The Herb of the Field,' etc. Arranged by ROBERT M. STARK, Edinburgh. New Edition, with many New Plates.

III.

Instructive Picture Book. The Geographical Distribution of Animals, in a Series of Pictures for the use of Schools and Families. By the late Dr. GREVILLE. With descriptive letterpress by ADAM WHITE, late Assistant, Zoological Department, British Museum.

IV.

Animals and Plants in their Homes. 60 Illustrations.

The History of Scottish Poetry,
From the Middle Ages to the Close of the Seventeenth Century. By the late DAVID IRVING, LL.D. Edited by JOHN AITKEN CARLYLE, M.D. With a Memoir and Glossary. Demy 8vo, 16s.

The Circle of Christian Doctrine;
A Handbook of Faith, framed out of a Layman's experience. By LORD KINLOCH, one of the Judges of the Supreme Court of Scotland. Third and Cheaper Edition. Fcap. 8vo, 2s. 6d.

Time's Treasure;
Or, Devout Thoughts for every Day of the Year. Expressed in verse. By LORD KINLOCH. Third and Cheaper Edition. Fcap. 8vo, price 3s. 6d.

Devout Moments.
By LORD KINLOCH. Price 6d.

Studies for Sunday Evening.
By LORD KINLOCH. Second Edition. Fcap. 8vo, price 4s. 6d.

The Philosophy of Ethics:
An Analytical Essay. By SIMON S. LAURIE, A.M., Author of 'The Fundamental Doctrine of Latin Syntax: being an Application of Psychology to Language.' 1 vol. demy 8vo, price 6s.

Notes, Expository and Critical, on certain British Theories
of Morals. By SIMON S. LAURIE. 1 vol. 8vo, price 6s.

Supplemental Descriptive Catalogue of Ancient Scottish Seals.
By HENRY LAING. 1 vol. 4to, profusely illustrated, price £3 : 3s.

Life of Father Lacordaire.
By DORA GREENWELL. 1 vol. fcap. 8vo. Price 6s.

A Memoir of Lady Anna Mackenzie,
Countess of Balcarres, and afterwards of Argyle, 1621-1706. By ALEXANDER
LORD LINDSAY. Fcap. 8vo, price 3s. 6d.

The Reform of the Church of Scotland
In Worship, Government, and Doctrine. By ROBERT LEE, D.D., Professor of
Biblical Criticism in the University of Edinburgh, and Minister of Greyfriars.
Part I. Worship. Second Edition, fcap. 8vo, price 3s.

The Clerical Profession,
Some of its Difficulties and Hindrances. By ROBERT LEE, D.D. Price 6d.

The Early Races of Scotland and their Monuments.
By LIEUT.-COL. FORBES LESLIE. 2 vols. demy 8vo, profusely Illustrated, price
32s.

"This learned and elaborate book presents the closest and most satisfactory
investigation of the character of the primitive races who inhabited the British
Islands yet given to the public. Whether the readers agree with Colonel Leslie or
not, they must of necessity allow that he has produced the most complete book on
this subject that has ever been published."—Daily News.

Life in Normandy;
Sketches of French Fishing, Farming, Cooking, Natural History, and Politics,
drawn from Nature. By an ENGLISH RESIDENT. Third Edition, 1 vol. crown
8vo, price 6s.

Specimens of Ancient Gaelic Poetry.
Collected between the years 1512 and 1529 by the REV. JAMES M'GREGOR, Dean
of Lismore—illustrative of the Language and Literature of the Scottish Highlands
prior to the Sixteenth Century. Edited, with a Translation and Notes, by the Rev.
THOMAS MACLAUCHLAN. The Introduction and additional Notes by WILLIAM F.
SKENE. 8vo, price 12s.

The Development of Science among Nations.
By BARON JUSTUS LIEBIG, F.R.S., President of the Royal Academy of Science,
Member of the French Institute, etc. etc. Price 1s.

Reasons for the Study of Jurisprudence as a Science.
By JAMES LORIMER, Professor of Public Law in the University of Edinburgh.
Price 1s.

Primary and Classical Education.
By the Right Hon. ROBERT LOWE, M.P. Price 1s.

Little Ella and the Fire-King,
And other Fairy Tales. By M. W., with Illustrations by HENRY WARREN. Second
Edition. 16mo, cloth, 3s. 6d. Cloth extra, gilt edges, 4s.

10 EDMONSTON AND DOUGLAS,

Macvicar's (J. G., D.D.)
THE PHILOSOPHY OF THE BEAUTIFUL; price 6s. 6d. FIRST LINES OF SCIENCE SIMPLIFIED; price 5s. INQUIRY INTO HUMAN NATURE; price 7s. 6d.

Max Havalaar;
Or, The Coffee Auctions of the Dutch Trading Company. By MULTATULI; translated from the original MS. by ALPHONSE JOHAN BERNARD KORSTMAR, Baron Nahuys. 1 vol. 8vo, with Maps, price 14s.

Heroes of Discovery.
By SAMUEL MOSSMAN, Author of 'Our Australian Colonies,' 'China : its Inhabitants,' etc. 1 vol. crown 8vo, price 5s.

Medical Officers of the Navy.
Everthing about them. For the information of Medical Students, and of the Parents of Young Gentlemen intended for the Medical Profession. Price 1s.

The Correct Form of Shoes.
Why the Shoe Pinches. A contribution to Applied Anatomy. By HERMANN MEYER, M.D., Professor of Anatomy in the University of Zurich. Translated from the German by JOHN STIRLING CRAIG, L.R.C.P.E., L.R.C.S.E. Fcap., sewed, 6d.

The Herring :
Its Natural History and National Importance. By JOHN M. MITCHELL, F.R.SS.A., F.S.A.S., F.R.P.S., etc. Author of 'The Natural History of the Herring, considered in Connection with its Visits to the Scottish Coasts,' 'British Commercial Legislation,' 'Modern Athens and the Piræus,' etc. With Six Illustrations, 8vo, price 12s.

Speech on the Union Question in the Free Church Presbytery of Edinburgh. By REV. THOMAS MAIN. Price 3d.

Political Sketches of the State of Europe—from 1814-1867 ;
Containing Ernest, Count Münster's Despatches to the Prince Regent from the Congress of Vienna and of Paris. By GEORGE HERBERT, Count Münster. 1 vol. demy 8vo, price 9s.

The Insane in Private Dwellings.
By ARTHUR MITCHELL, A.M., M.D., Deputy Commissioner in Lunacy for Scotland, etc. 8vo, price 4s. 6d.

Ancient Pillar-Stones of Scotland :
Their Significance and Bearing on Ethnology. By GEORGE MOORE, M.D. 1 vol. 8vo, price 6s. 6d.

North British Review.
Published Quarterly. Price 6s.

Reflections on the Relation of Recent Scientific Inquiries
to the Received Teaching of Scripture. By JAMES MONCREIFF, Esq., M.P., LL.D., Dean of the Faculty of Advocates. Price 1s.

The Extension of the Suffrage.
By JAMES MONCREIFF, Esq., M.P., LL.D., Dean of the Faculty of Advocates. Price 1s.

The Education of a Lawyer.
By JAMES MONCREIFF, Esq., M.P., LL.D., Dean of the Faculty of Advocates. Price 1s.

National Education and the Church of Scotland.
Price 1s.

Biographical Annals of the Parish of Colinton.
By THOMAS MURRAY, LL.D., Author of 'The Literary History of Galloway, etc. etc. Crown 8vo, price 3s. 6d.

A New-Year's Gift to Children.
By the author of 'John Halifax, Gentleman.' With Illustrations, price 1s.

Man: Where, Whence, and Whither?
Being a glance at Man in his Natural-History Relations. By DAVID PAGE, LL.D. 1 vol. fcap. 8vo, price 3s. 6d.

Practical Water-Farming.
By WM. PEARD, M.D., LL.D. 1 vol. fcap. 8vo, price 5s.

The Great Sulphur Cure.
By ROBERT PAIRMAN, Surgeon. Thirteenth Edition, price 1s.

The Bishop's Walk and The Bishop's Times.
By ORWELL. Fcap. 8vo, price 5s.

Suggestions on Academical Organisation,
With Special Reference to Oxford. By MARK PATTISON, B.D., Rector of Lincoln College, Oxford. 1 vol. crown 8vo, price 7s. 6d.

Popular Genealogists;
Or, The Art of Pedigree-making. 1 vol. crown 8vo, price 4s.

Reminiscences of Scottish Life and Character.
By E. B. RAMSAY, M.A., LL.D., F.R.S.E., Dean of Edinburgh. Fifteenth Edition, price 1s. 6d.

"The Dean of Edinburgh has here produced a book for railway reading of the very first class. The persons (and they are many) who can only under such circumstances devote ten minutes of attention to any page, without the certainty of a dizzy or stupid headache, in every page of this volume will find some poignant anecdote or trait which will last them a good half-hour for after-laughter : one of the pleasantest of human sensations."—*Athenæum.*

*** The original Edition in 2 vols. with Introductions, price 12s., and the Sixteenth Edition in 1 vol. cloth antique, price 5s., may be had.

Memoirs of Frederick Perthes ;
Or, Literary, Religious, and Political Life in Germany from 1789 to 1843.　By C. T. PERTHES, Professor of Law at Bonn.　Crown 8vo, cloth, 6s.

Report on the Condition of the Poorer Classes of Edinburgh,
and of their Dwellings, Neighbourhoods, and Families.　Price 1s.

Scotland under her Early Kings.
A History of the Kingdom to the close of the 13th century.　By E. WILLIAM ROBERTSON, in 2 vols. 8vo, cloth, 36s.

Doctor Antonio
A Tale.　By JOHN RUFFINI.　Cheap Edition, crown 8vo, boards, 2s. 6d.

Lorenzo Benoni ;
Or, Passages in the Life of an Italian.　By JOHN RUFFINI.　With Illustrations. Crown 8vo, cloth gilt, 5s.　Cheap Edition, crown 8vo, boards, 2s. 6d.

A Quiet Nook in the Jura.
By JOHN RUFFINI, Author of 'Doctor Antonio,' etc.　1 vol. extra fcap. 8vo, price 7s. 6d.

The Salmon ;
Its History, Position, and Prospects.　By ALEX. RUSSEL.　8vo, price 7s. 6d.

Twelve Years in China :
The People, the Rebels, and the Mandarins, by a British Resident.　With coloured Illustrations.　Second Edition.　With an Appendix.　Crown 8vo, cloth, price 10s. 6d.

A Handbook of the History of Philosophy.
By Dr. ALBERT SCHWEGLER.　Second Edition.　Translated and Annotated by J. HUTCHISON STIRLING, LL.D., Author of the 'Secret of Hegel.'　Crown 8vo, price 6s.

Supplementary Notes to the First Edition of Schwegler's
History of Philosophy.　Price 1s.

John Keble :
An Essay on the Author of the 'Christian Year.'　By J. C. SHAIRP, Professor of Humanity, St. Andrews.　1 vol. fcap. 8vo, price 3s.

Studies in Poetry and Philosophy.
By J. C. SHAIRP, Professor of Humanity, St. Andrews.　1 vol. fcap. 8vo, Price 6s.

The Sermon on the Mount.
By the Rev. WALTER C. SMITH, Author of 'The Bishop's Walk, and other Poems, by Orwell,' and 'Hymns of Christ and Christian Life.'　1 vol. crown 8vo, price 6s.

On Archaic Sculpturings of Cups and Circles upon Stones
and Rocks in Scotland, England, etc. By Sir J. Y. SIMPSON, Bart., M.D., D.C.L.,
Vice-President of the Society of Antiquaries of Scotland, etc. etc. 1 vol. small 4to,
with Illustrations, price 21s.

Proposal to Stamp out Small-pox and other Contagious
Diseases. By Sir J. Y. SIMPSON, Bart., M.D., D.C.L. Price 1s. ↖

The Law and Practice of Heraldry in Scotland.
By GEORGE SETON, Advocate, M.A., Oxon, F.S.A., Scot. 8vo, with numerous
Illustrations, 25s.

₊ A few copies on large paper, half-bound, 42s.

' Cakes, Leeks, Puddings, and Potatoes.'
A Lecture on the Nationalities of the United Kingdom. By GEORGE SETON,
Advocate, M.A., Oxon, etc. Second Edition. Fcap. 8vo, sewed, price 6d.

The Roman Poets of the Republic.
By W. Y. SELLAR, M.A., Professor of Humanity in the University of Edinburgh,
and formerly Fellow of Oriel College, Oxford. 8vo, price 12s.

Theories of Classical Teaching.
By W. Y. SELLAR, M.A., Professor of Humanity in the University of Edinburgh.
Price 1s.

The Four Ancient Books of Wales,
Containing the Kymric Poems attributed to the Bards of the Sixth century. By
WILLIAM F. SKENE. With Maps and Facsimiles. 2 vols. 8vo, price 36s.

Life and Work at the Great Pyramid
During the Months of January, February, March, and April A.D. 1865 ; with a
Discussion of the Facts Ascertained (Illustrated with 36 Plates and several
Woodcuts). By C. PIAZZI SMYTH, F.R.SS.L. and E., F.R.G.S., F.R.SS.A.,
Hon. M.I.E. Scot., P.S. Ed., and R.A.A.S. Munich and Palermo, Professor of
Practical Astronomy in the University of Edinburgh, and Astronomer-Royal for
Scotland. 3 vols. demy 8vo, price 56s.

On the Antiquity of Intellectual Man from a Practical and
Astronomical Point of View. By C. PIAZZI SMYTH, F.R.SS.L. and E., Astronomer-Royal for Scotland. 1 vol. crown 8vo.

Dugald Stewart's Collected Works.
Edited by Sir WILLIAM HAMILTON, Bart. Vols. I. to X. 8vo, cloth, each 12s.
        Vol. I.—Dissertation. Vols. II. III. and IV.—Elements of the Philosophy
        of the Human Mind. Vol. V.—Philosophical Essays. Vols. VI. and VII.—
        Philosophy of the Active and Moral Powers of Man. Vols. VIII. and IX.—
        Lectures on Political Economy. Vol. X.—Biographical Memoirs of Adam
        Smith, LL.D., William Robertson, D.D., and Thomas Reid, D.D. ; to which
        is prefixed a Memoir of Dugald Stewart, with Selections from his Corre-
        spondence, by John Veitch, M.A. Supplementary Vol.—Translations of the
        Passages in Foreign Languages contained in the Collected Works ; with
        General Index.

History Vindicated in the Case of the Wigtown Martyrs.
By the Rev. ARCHIBALD STEWART.   Price 1s.

Jerrold, Tennyson, Macaulay, and other Critical Essays.
By JAMES HUTCHISON STIRLING, LL.D., Author of "The Secret of Hegel."
1 vol. fcap. 8vo, price 5s.

"The author of 'The Secret of Hegel' here gives us his opinions of the lives
and works of those three great representative Englishmen whose names appear on
the title-page of the work before us.   Dr. Stirling's opinions are entitled to be heard,
and carry great weight with them.   He is a lucid and agreeable writer, a profound
metaphysician, and by his able translations from the German has proved his grasp
of mind and wide acquaintance with philosophical speculation."—*Examiner.*

Natural History and Sport in Moray.
Collected from the Journals and Letters of the late CHARLES ST. JOHN, Author
of 'Wild Sports of the Highlands.'   With a short Memoir of the Author.   Crown
8vo, price 8s. 6d.

Christ the Consoler;
Or Scriptures, Hymns, and Prayers for Times of Trouble and Sorrow.   Selected and
arranged by the Rev. ROBERT HERBERT STORY, Minister of Roseneath.   1 vol. fcap.
8vo, price 3s. 6d.

Shakespeare.
Some Notes on his Character and Writings.   By a Student.   8vo, price 4s. 6d.

Works by Professor James Syme.
OBSERVATIONS IN CLINICAL SURGERY.   Second Edition.   1 vol. 8vo, price 8s. 6d.
STRICTURE OF THE URETHRA, AND FISTULA IN PERINEO.   8vo, 4s. 6d.
TREATISE ON THE EXCISION OF DISEASED JOINTS.   8vo, 5s.
ON DISEASES OF THE RECTUM.   8vo, 4s. 6d.
EXCISION OF THE SCAPULA.   8vo, price 2s. 6d.

Lessons for School Life;
Being Selections from Sermons preached in the Chapel of Rugby School during his
Head Mastership.   By THE RIGHT REVEREND THE LORD BISHOP OF LONDON.   Fcap.,
cloth, 5s.

What is Sabbath-Breaking?
8vo, price 2s.

The Dynamical Theory of Heat.
By P. G. TAIT, Professor of Natural Philosophy in the University of Edinburgh.
1 vol. fcap. 8vo.

Day-Dreams of a Schoolmaster.
By D'ARCY W. THOMPSON. Second Edition. Fcap. 8vo, price 5s.

Ancient Leaves;
Or Metrical Renderings of Poets, Greek and Roman. By D'ARCY W. THOMPSON. Fcap. 8vo, 6s.

Sales Attici:
Or, The Maxims, Witty and Wise, of Athenian Tragic Drama. By D'ARCY WENT-WORTH THOMPSON, Professor of Greek in Queen's College, Galway. Fcap. 8vo, price 9s.

Antiquities of Cambodia.
By J. THOMSON, F.R.G.S., F.E.S.L. Sixteen Photographs, with Explanatory Text. Imperial 4to, handsomely bound, half-morocco. Price Four Guineas.

An Angler's Rambles among the Rivers and Lochs of Scotland.
By THOMAS TOD STODDART, Author of "The Angler's Companion." 1 vol. crown 8vo, price 9s.

Travels by Umbra.
8vo., price 10s. 6d.

Hotch-Pot.
By UMBRA. An Old Dish with New Materials. Fcap. 8vo, price 3s. 6d.

Life of Dr. John Reid,
Late Chandos Professor of Anatomy and Medicine in the University of St. Andrews. By the late GEORGE WILSON, M.D. Fcap. 8vo, cloth, price 3s.

Researches on Colour-Blindness.
With a Supplement on the danger attending the present system of Railway and Marine Coloured Signals. By the late GEORGE WILSON, M.D. 8vo, 5s.

Dante's—The Inferno.
Translated line for line by W. P. WILKIE, Advocate. Fcap. 8vo, price 5s.

Westfield.
A View of Home Life during the American War. 1 vol. crown 8vo, price 8s. 6d.

History of Anæsthetics, from an American point of view.
Extracted from "Surgical Observations, with Cases and Opinions." By J. MASON WARREN, M.D., Surgeon to the General Hospital, etc. etc. With a Preface by JAMES SYME, Professor of Clinical Surgery in the University of Edinburgh.

Railway Management and Accounts.
By WILLIAM WOOD. Price 6d.

16   EDMONSTON AND DOUGLAS, 88 PRINCES STREET.

ODDS AND ENDS—*Price 6d. Each.*

Now Ready, Vol. I., in Cloth, price 4s. 6d., containing Nos. 1-10,

1. **Sketches of Highland Character—**
SHEEP FARMERS AND DROVERS.

2. **Convicts.**
By a PRACTICAL HAND.

3. **Wayside Thoughts of an Asophophilosopher.**
By D'ARCY W. THOMPSON. No. 1. RAINY WEATHER; or, the Philosophy of Sorrow. GOOSESKIN; or, the Philosophy of Horror. TE DEUM LAUDAMUS; or, the Philosophy of Joy.

4. **The Enterkin.**
By JOHN BROWN, M.D.

5. **Wayside Thoughts of an Asophophilosopher.**
By D'ARCY W. THOMPSON. No. 2. ASSES—HISTORY—PLAGUES.

6. **Penitentiaries and Reformatories.**

7. **Notes from Paris; or, Why are Frenchmen and English-**
men different?

8. **Essays by an Old Man.**
No. 1. IN MEMORIAM—VANITAS VANITATUM—FRIENDS.

9. **Wayside Thoughts of an Asophophilosopher.**
By D'ARCY W. THOMPSON. No. 3. NOT GODLESS, BUT GODLY; A TRIANGULAR TREATISE ON EDUCATION.

10. **The Influence of the Reformation on the Scottish Character.**
By J. A. FROUDE, Author of the 'History of England.'

Now Ready, Vol. II., in Cloth, price 4s. 6d., containing Nos. 11-19.

11. **The Cattle Plague.**
By LYON PLAYFAIR, C.B., LL.D., F.R.S., etc.

12. **Rough Nights' Quarters.**
By ONE OF THE PEOPLE WHO HAVE ROUGHED IT.

13. **Letters on the Education of Young Children.**
By S. G. O.

14. **The Stormontfield Piscicultural Experiments. 1853-1866.**
By ROBERT BUIST.

15. **A Tract for the Times.**

16. **Spain in 1866.**

17. **The Highland Shepherd.**
By the Author of 'The Two Queys.'

18. **The Doctrine of the Correlation of Forces: its Develop-**
ment and Evidence. By the Rev. JAMES CRANBROOK, Edinburgh.

19. **'Bibliomania.'**

20. **A Tract on Twigs, and on the best way to Bend them.**

www.ingramcontent.com/pod-product-compliance
Lightning Source LLC
Chambersburg PA
CBHW031827270326
41932CB00008B/583